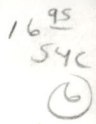

NORTH CAROLINA
STATE BOARD OF COMMUNITY COLLEGES
LIBRARIES
ASHEVILLE-BUNCOMBE TECHNICAL COLLEGE

DISCARDED

JUN 2 3 2025

THE COMPLETE HANDBOOK OF DRAFTING

Other TAB books by the author:

No.	788	*Handbook of Practical Boat Repairs*
No.	860	*The Woodworker's Bible*
No.	894	*Do-It-Yourselfer's Guide to Furniture Repair & Refinishing*
No.	910	*How to Make Your Own Built-In Furniture*
No.	937	*Modern Sailmaking*
No.	1004	*The Upholsterer's Bible*
No.	1008	*Woodworking with Scraps*
No.	1044	*The Woodturner's Bible*
No.	1114	*How to Make Early American & Colonial Furniture*
No.	1130	*How to Make Your Own Knives . . . etc.*
No.	1179	*The Practical Handbook of Blacksmithing & Metalworking*
No.	1188	*66 Children's Furniture Projects*
No.	1237	*Practical Knots & Ropework*
No.	1257	*The Master Handbook of Sheetmetalwork . . . with projects*
No.	1312	*The GIANT Book of Wooden Toys*

No. 1365
$16.95

THE COMPLETE HANDBOOK OF DRAFTING
BY PERCY W. BLANDFORD

FIRST EDITION

FIRST PRINTING

Copyright © 1982 by TAB BOOKS Inc.

Printed in the United States of America

Reproduction or publication of the content in any manner, without express permission of the publisher, is prohibited. No liability is assumed with respect to the use of the information herein.

Library of Congress Cataloging in Publication Data

Blandford, Percy W.
    The complete handbook of drafting.

    Includes index.
    1. Mechanical drawing. I. Title.
T353.B58   604.2'4   81-18245
ISBN 0-8306-0049-3   AACR2
ISBN 0-8306-1365-X (pbk.)

Contents

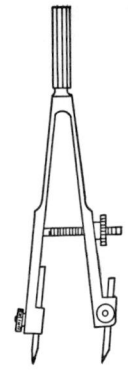

Introduction   vii

1 What Is Drafting?   1
Orthographic Projection—History—Communications

2 Basic Drawing Instruments   11
Drawing Boards—T Squares and Drafting Machines—Pencils—Erasers—Triangles—Scales and Straightedges—Compasses

3 Other Drawing Instruments   25
Compasses—Protractors—Curves—Ink Instruments—Pens

4 Scales   44
Proportions—Decimal Scales—Oversize Scales—Scale Rules—Graphic Scales—Fine Scale—Verniers

5 Lines   54
Hidden Lines—Center Lines—Dimension Lines—Break Lines—Section Lines

6 Handling Instruments   63
Main Lines—Borders—Using Compasses—Using Angular Instruments—Using Curves and Templates

7 Drafting Geometry   77
Right Angles—Bisecting—Triangles—Polygons—Tangents—Ellipses

8 Lettering   101
Style—Upright Lettering—Proportions—Guidelines—Fractions

9 Dimensions   113
Dimension Lines—Locating Dimensions—Small Dimensions—Circles—Angles and Center Lines—Tolerances

10 Sections 126
Cutting Plane—Cross-Hatching—Sectioned Materials—Multiple Parts—Revolved Sections—Thin Sections

11 Auxiliary Views 139
Examples—Curved Parts—Conic Projections—Revolved Views

12 Developments 156
Tubes—Bend Allowances—Curved Developments—Conic Developments—Round Cones—Other Truncated Cones—Other Shapes—Internal Cuts and Junctions

13 Fasteners 178
Screw Threads—Representing Threads—Nuts and Bolts—Rivets—Other Fasteners—Keys—Welding

14 Inking 197
Tracing Materials—Indian Ink—Ruling Pens—Technical Fountain Pens—Drawing with Ink—Sequence of Inking—Erasing—Reproduction

15 Three-Dimensional Drawings 211
Pictorial Views—Inset Views—Oblique Drawing—Circles Drawn Obliquely—Isometric Views—Isometric Axes—Isometric Curves—Choice of Views

16 Laying Out Drawings 231
Number of Drawings—Material Schedules and Lists—Proportions—Basic Designing—Chair Drawing—Engineering Drawing

17 Engineering Drawings 247
Dimensions—Cams—Gears—Finish Marks—Electrical Engineering—Airplane Drafting

18 Architecture and Civil Engineering 265
House Plan—Dimensions—Site Plans—Ancillary Services—Structural Drafting—Civil Engineering

19 Lofting 280
Lines—Right Angles—Curves—Angles

20 Technical Illustrating 287
Drawing Sizes—Exploded Views—Pictorial Views—Graphs—Charts—Style

Glossary 296

Index 312

Introduction

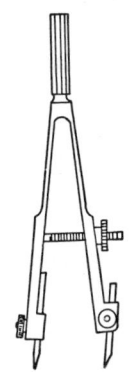

If another person is to make something for you or to work under your direction, it is very difficult to convey to him in words only what you want. A picture is more easily understood. If you want to describe a piece of land or a building exactly, you will need drawings. The layout of electric wiring, plumbing pipes, and many other things can only be described by one person to another with drawings.

The person who makes drawings is a draftsman. He might also be an artist, but he does not have to be. Most draftsmen are not artists. Drafting is a means of expressing graphically from one person to another a message containing the sort of information that could not be passed in words only. Quite often the drawing and words are complementary.

Practical drawings have been used ever since primitive man needed to show others his constructional ideas, even if all he did was scratch lines in mud or dust. Drafting has developed into a method of conveying information on practical subjects that come anywhere between a simple place mat and an oil refinery. Techniques are basically the same, even if the more complex subjects call for hundreds of separate drawings.

Modern drafting involves the use of standard techniques and accepted conventions that allow drawings to be produced with instruments, which give another person all the information on the subject he needs to know. The draftsman may be fully occupied doing nothing else, or he may be a designer or craftsman who wants to put his ideas on paper.

This book provides an introduction to drafting methods, covering both those things which are general to all draftsmen and some things which are peculiar to the more specialized worker. Drafting is a subject that requires practice. It cannot be learned solely from a book, so I hope that you acquire the few simple instruments needed to start drafting and begin to practice.

Drafting skill is valuable to anyone concerned with practical subjects as an extension of reading, writing, and calculations. I hope that this book may lead many readers to become full-time draftsmen, but I also hope that other readers will learn enough about drafting to use the skill as a valuable addition to various activities in other directions.

Chapter 1
What Is Drafting?

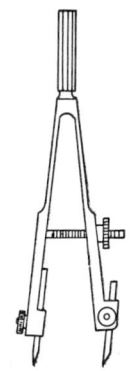

Drafting has been described as a graphical language, and that is as concise a description as any. The person who prepares the graphical language is a *draftsman*. Drafting is sometimes described as *draftsmanship*.

So what is this graphical language? It is mainly a means of making drawings that convey all the information about a three-dimensional object on the two-dimensional paper. If it is a drawing of something to be made, it gives the craftsman all the information he needs to work from.

A pictorial view may give a good idea of the general appearance of an article (Fig. 1-1), but it does not show true views in various directions and cannot be to size at scale. A pictorial view may have advantages in showing what the finished thing will look like. Architects may produce pictorial views of buildings for the benefit of the customers, but the drawings prepared for the builder to work to are very different.

For such pictures it is advantageous for the draftsman to also be an artist. For mechanical drawing, though, it is possible for anyone who is not artistic, but understands the graphical language, to prepare satisfactory drawings that anyone else who understands the language can work to.

Draftsmen are concerned with all branches of engineering: all of the crafts and trades using wood, metals, plastics, and other materials; all of the work involved in building, architecture, and civil engineering; and everywhere that a designer has to convey his

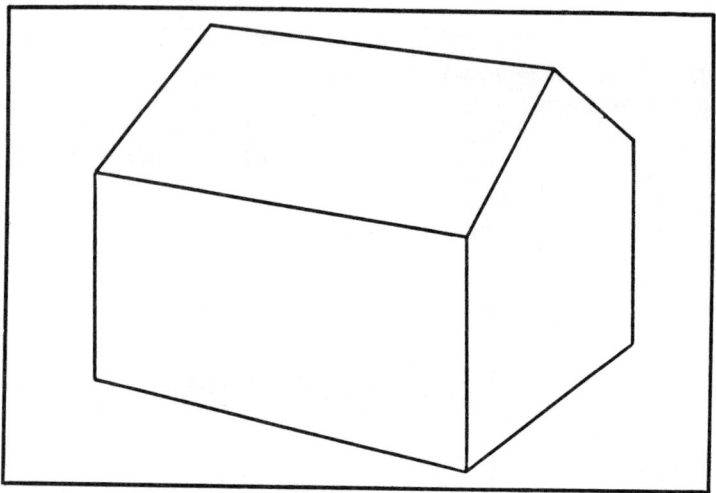

Fig. 1-1. A pictorial view shows appearance, but it cannot be measured.

ideas to other people. A draftsman may work to other people's ideas and prepare drawings from them, but in many cases it helps if the originator can make at least the preliminary drawings. Drafting is an activity which should be understood by a vast number of people besides those calling themselves draftsmen.

ORTHOGRAPHIC PROJECTION

The basic method of making a working drawing is by *orthographic projection*. Most things can be shown by three views. There may have to be others for some things, and it may be necessary to give details in other drawings, but general outlines can be shown by three views for the majority of subjects. These can be a *front view*, a *side view*, and a view from above (Fig. 1-2A). It would be possible to draw these views anywhere on the paper, but in orthographic projection they have a particular relationship to each other. The front view, usually called *front elevation*, is drawn first. Below that and located by projection lines from it is the view from above called a *plan*. To one side there is a projected side view or *side elevation* (Fig. 1-2B). The side view has the same thickness as the plan and could be found by projecting around (Fig. 1-2C). These lines show the method of projecting, but it is unusual to actually draw the lines.

Note that in this projection the plan is as viewed from above the front elevation (Fig. 1-2D), and the side elevation is as viewed from the side remote from where it is drawn (Fig. 1-2E) and could be to either side. Sometimes two side elevations are needed because of different things to show on the other surface (Fig. 1-2F). This is the relation given to the three views in nearly all modern drafting techniques. Older draftsmen may describe it as *third-angle projection* to distinguish it from other methods where views have different relations to each other, but these other ways are best forgotten.

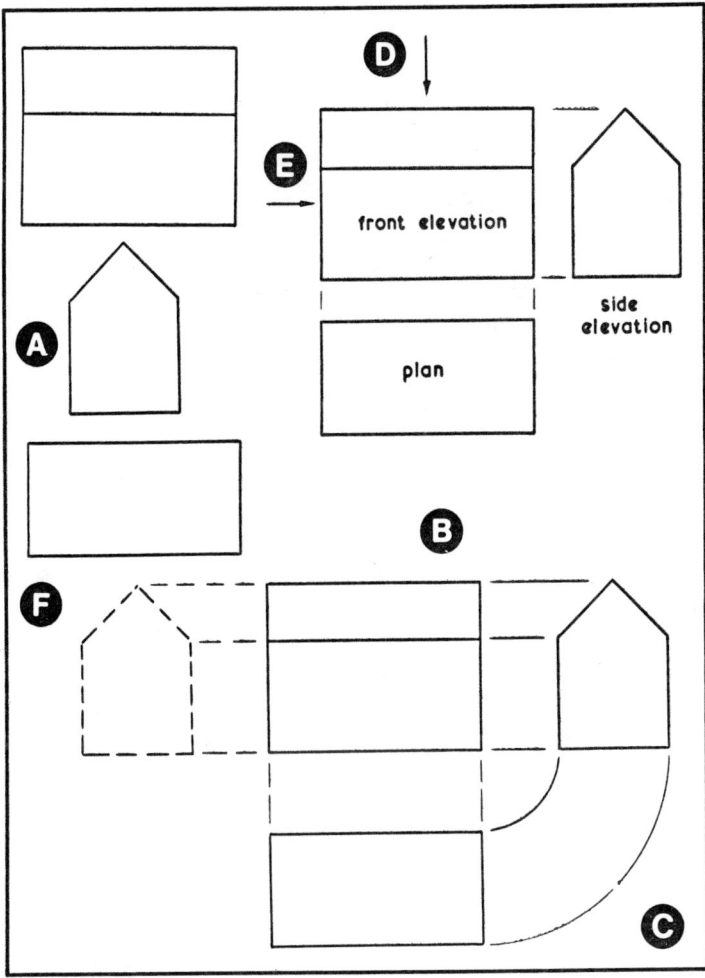

Fig. 1-2. A view of three sides can show sizes and may be drawn so each part is drawn as viewed from the far side of another one in orthographic projection.

Other methods of laying out drawings should be understood as they provide semipictorial views that may be easier for some people to understand, although they cannot give all sizes for anything even slightly complicated. In *pictorial projection* (Fig. 1-3A) there is a normal front view. Then the article is given an illusion of thickness with lines at 45°, but the lines are drawn half-length so they do not give an appearance of being too thick. A similar drawing with the projected lines full-length is termed an *oblique projection* (Fig. 1-3B).

A way of drawing a more conventional pictorial view with instruments is *isometric projection* (Fig. 1-3C). Diagonal lines are at 30° to horizontal, and dimensions in all directions are full-size. The effect is to make the drawing larger than actual size, but it does allow measuring from it.

An isometric view does not look correct pictorially, as the diagonal lines should be getting closer if projected to a vanishing point (Fig. 1-3D). That brings drawing into the realm of an artist's work and is not a requirement for normal drafting.

A draftsman should know about these other views. The views have uses when it is necessary to show the appearance of an item particularly to viewers unused to more conventional mechanical drawing. In some assemblies it may help to supplement a plan and two elevations with one of these more pictorial representations to give a better idea of how certain things look when completed. The drawings made by orthographic projection will still be the ones worked to.

In complicated engineering or building projects, the number of drawings needed to inform the workmen may run into hundreds. Preparation of these drawings calls for skill, particularly if several draftsmen are involved and their work has to be related. Where the thing is drawn three-dimensional, whether a whole thing or some part of it, the method of laying out the views needed is the same as in the simple example.

Not all drawings represent three-dimensional objects. An electrical wiring diagram may be a conventional layout to show connections. One view is all that is needed. Similarly, a plan or map of a tract of land can be imagined to be viewed from an airplane, and that is the only view required.

A great many details and complications are described later in the book, but if you remember the accepted way to lay out two elevations and a plan, you will be able to see how intricate and complex items can be represented by following these basic princi-

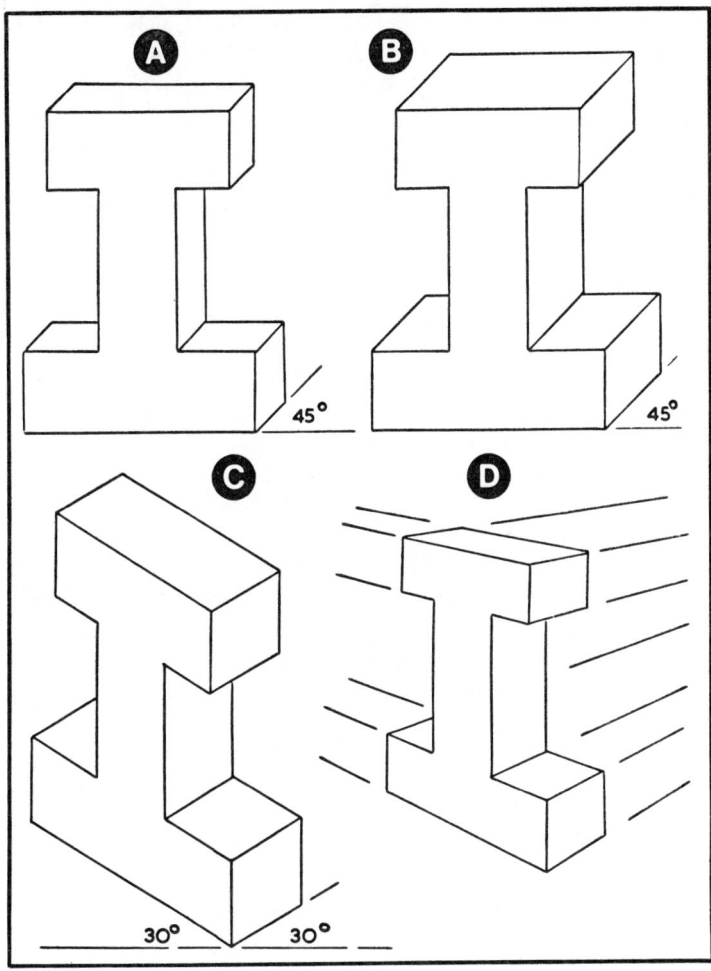

Fig. 1-3. Pictorial views can be in oblique projection (A and B), isometric projection (C), or as actually appearing, with lines leading away to vanishing points (D).

ples. There are certain accepted ways of drawing lines and symbols that make clear the meaning of a drawing, and practice allows the reader of this graphic language to interpret the example of drafting and visualize the object drawn.

HISTORY

Drafting as practiced today, with its accepted ways of showing three-dimensional objects on flat paper, may not go back many centuries, but there must have been technical drawings for the great things built thousands of years ago. The pyramids could not have been built just by starting to put one stone on top of another. The palaces of the ancient Persians and other early civilizations must have been designed by people who were able to transfer their ideas in some way to the workers who actually did the building.

Early drawings may have been scratched in clay or soft stone, and they must have been to scale. A few of these drawings have survived. Early paperlike materials, such as papyrus, were not as durable, so few of these records are available. One way of showing what is wanted is by making a scale model. That has the advantage of being three-dimensional. It not only provides something to measure in all directions, but it shows what the finished thing should look like. Models must have been made of many large things before they were built, and the process is not yet dead.

Scale would have been something understood locally. Measurements based on the span of a man's fingers, his stride, or the spread of his arms could not be very precise. What these measurements meant in one place might not have stood a very close comparison with what was allegedly the same thing elsewhere. Vagueness of measurement continued for a long time. It did not matter when all the work carried out to a set of measurements was in one place and done by one man or a group of workers. After all, measurement in feet and inches or any other system is only a way of comparing one thing with another. Even things like screw threads were not to any standard. Providing a nut fitted a bolt, it did not matter if the nut would not fit any other bolt if the two were made by one man for one purpose.

The need for standardization came with increased travel and communication. With the lead-up to the *Industrial Revolution,* things began to be made in factories, usually in towns, and these were distributed to other craftsmen who might be working in places remote from them. There was a need for standardization of measurements; otherwise, things might not match. For instance, a ½-inch bolt made in a factory might be found to be too big for a ½-inch hole drilled to the local man's standard.

Even today standards vary between countries, but the differences are so minute that they can be ignored for practical purposes. Recognized measuring standards are not much more than two centuries old. They accompanied the development of precise methods of manufacture, which not only meant things could be made with interchangeable parts, but rules and scales, as well as other measuring instruments, could be made in quantity so they all matched.

Architectural models are often made either in wood or by folding paper or card. They are particularly useful in showing buildings in relation to each other. Models can be moved around when laying out a new housing development. Model furniture or equipment can be moved about in a room. If other nontechnical people have to be shown such a housing complex, models are better than flat drawings for conveying the ideas.

In ship and boat building, half models of hulls are sometimes made from wood, often built up with layers. The outside is carved, and the hull surface is smoothed until the designer sees a satisfactory shape. The layers are then taken apart, and each gives the shape of the hull outline at different levels. From these curves the hull lines are *lofted* full-size on the floor, and templates are made for the parts of the boat. This is still the method used by some boat builders/designers, and in medieval days it was the normal way of deciding what shape a hull was to be.

The *Greeks* and *Romans* were great builders of houses, places, and other large buildings. The Romans in particular used building methods in *hydraulic engineering* to bring water long distances and even to provide central heating. The surviving Roman baths at Bath, England, show how hot springs were used for heating and medical treatment.

Plans showing views from above seem to have been made from quite early times, and there were side views, but very often the designer resorted to pictorial views to show what was wanted. In that case detail sizes may have been left to the man carrying out the work, and he depended on experience—either his own or that of the man in charge. Some information on intended detail sizes may have been included in an accompanying written description.

A medieval designer with a considerable reputation as an inventor and designer was *Leonardo da Vinci*. Thousands of his drawings still survive. Fortunately, he had artistic ability. His drawings are mostly pictorial, but they show technically correct details. Where enlarged details are needed or parts are better shown separated, these things are drawn so they lead off relevant parts. His drawings are not laid out as you would make a mechanical drawing today, but they are showing the way.

Many drawings were freehand sketches, possibly with dimensions written on. There must have been straightedges, and compasses were certainly around, but until the coming of the Industrial

Revolution, more exact drawings made with instruments were unusual. That great engineering revolution, with the use of steam and other power making large and more complex machines possible, called for detailed accurate drawings to which craftsmen could work, using the new means of greater precision. No longer could the blacksmith sketch in chalk on sheet metal what he intended to make and leave the rest to the way he worked the iron by eye under the hammer. Man could work to close tolerances, and he had to if a machine was to function. He needed more accurate drawings.

A Frenchman named *Gaspard Monge* (1746-1818) is credited with discovering the methods of graphic projection which are used today. His methods were considered a French military secret for some time, but they became known elsewhere and spread to become the accepted and recognized methods of technical drawing throughout the Western world. A drawing made in one country could be understood in another, even when the language was different.

Obviously, much detail development has occurred. New symbols and conventions have been added to Monge's original work. Improvements in instruments have made better drawing possible. Information on instruments is given in later chapters, but you should get the best you can afford. A few good instruments covering immediate needs are better than an elaborate collection of inferior instruments, some of which may never be needed.

COMMUNICATIONS

The object of a technical drawing is to communicate information. The information usually concerns the appearance of some object, which has to be made, although it might show the reader how electrical equipment is wired, what the layout of a piece of land is, or even a graph showing how sales have shrunk in the past year. The draftsman must insure that what he puts on paper tells the whole story. The user of the drawing should not have to come back to him and ask what something means or what a missing measurement should be.

The value of this graphical language can be appreciated if an attempt is made to describe how to make quite a simple thing in words only. The text will have to become quite long and involved in every aspect is covered, and the possibility of error is to be avoided. Yet quite a simple drawing will show what is needed much more clearly and with far less risk of error. A classic example is supposed to be the tying of a knot in rope. Some knots are impossible to describe in words without the aid of a diagram.

In modern industry it is often necessary to convey what is wanted to workers who are unskilled in reading standard drawings. A skilled man may prefer the standard drawings, but a worker on an assembly line or doing other semiskilled work may need the information on a standard drawing elaborated so he can understand what is required. This means that there may have to be pictorial view alongside the normal views. *Exploded drawings* may have to be used. They show all the items that make up an assembly drawn in line, away from each other, but in the sequence they have to assemble. This could be a simple side view of the things to go on to a rod (Fig. 1-4A), or it might be the parts of a box, drawn apart, but in the correct relative positions (Fig. 1-4B). A draftsman doing this sort of work can use isometric projection or pictorial views, but it helps if he can visualize the exploded parts from his conventional drawing of them. A freehand sketch often helps in deciding how to deal with the problem.

Most draftsmen are concerned with the basic task of communicating technical information. They may originate an idea or design, in which case they are designers as well, but quite often their task is to interpret the ideas of a designer. A professional designer working in the same sphere as the draftsman may have no difficulty in passing on to him information on what is wanted on the

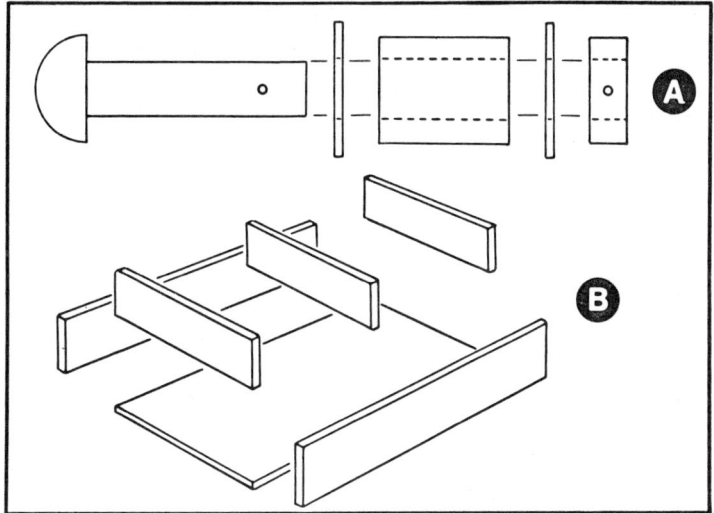

Fig. 1-4. Exploded drawings show parts of an assembly separated.

drawing, but sometimes it is necessary to take the ideas of a lay person who is unable to express himself graphically. Getting what he wants on to a drawing may make a tricky but interesting task.

A draftsman may have to make a drawing of something that already exists. A piece of furniture by one of the famous cabinetmakers of the last century may have to be measured and drawn for the sake of record, or because it is to be reproduced. A marine draftsman may have to "take off the lines" of an existing boat hull to make a record of it for historical reasons, or to discover what it is about a racer that puts it ahead of others apparently identical.

Draftsmen also produce charts and graphs, which may be for scientific reasons or for display to salesmen. Besides conventional graphs, there are many ways used to indicate progress from one thing to another, and a draftsman may have to originate or follow instructions to get such trend indicators correct. Basically this is drafting, but an artist may add human figures or other things to make the technical information more attractive.

Another draftsman may be mainly concerned with technical book illustrating. His final work will be in ink, and he has to judge sizes in relation to pages and make his lines of suitable thickness to give the right result when reduced. If the lettering is not typeset, he has to do his lettering to a size and thickness that will be legible when reduced and printed.

A map maker or *cartographer* needs to be a draftsman. His work is very different from that of an engineering draftsman, but it bears some similarity to part of the work of an architectural draftsman. Maps, with their diagrammatic way of showing details, their contour lines, and compact lettering, call for great care and accuracy.

Some electrical draftsmen may do little besides wiring diagrams, with symbols representing almost everything concerned. Comparable drawings may be done by draftsmen concerned with hydraulic and *pneumatic* work.

No draftsman can cope with moving from one sphere of activity to another—at least not without retraining. Most draftsmen specialize and stay with their specialization. The basic methods of drafting are common to all. It is the engineering and building draftsman who makes most use of the basic methods and their ramifications, so much of this book is concerned directly with the needs of those professions. Other specializations have the same roots, though and such draftsmen will benefit by a grounding in the same work.

Chapter 2
Basic Drawing Instruments

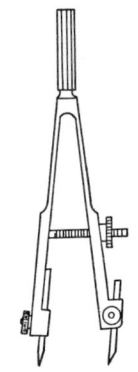

There are many drawing instruments and aids to mechanical drawing. Some add to convenience, accuracy, speed, or all three, while others are only of use to draftsmen concerned with particular types of work. Much depends on how much drafting is to be done and its purpose. A professional engaged on large work constantly may stand at a drafting machine that has many functions and costs a considerable amount. At the other extreme is the student doing drafting work only occasionally and managing with the minimum of equipment.

Even when a person is used to fairly sophisticated equipment, there may be occasions when he is glad to make do with the basics. For instance, I have prepared some drawings for other books while traveling in a motor home, using only a board and instruments that could be fitted into a small case. If you are new to drafting, provide yourself with the minimum pieces of equipment, particularly avoiding large prepared sets of instruments until you know what you need. The things bought should be of good quality and not just the types sold for school geometry and similar work. The kit can be added to as needs arise.

The absolute minimum is something to form a base for the paper, a pencil and eraser, something to act as a guide for drawing straight lines, something to measure with, and a means of drawing circles. Something for checking right angles is also essential.

DRAWING BOARDS

The simplest drawing board is a piece of thick plywood, with its edges planed squarely and preferably mounted on battens,

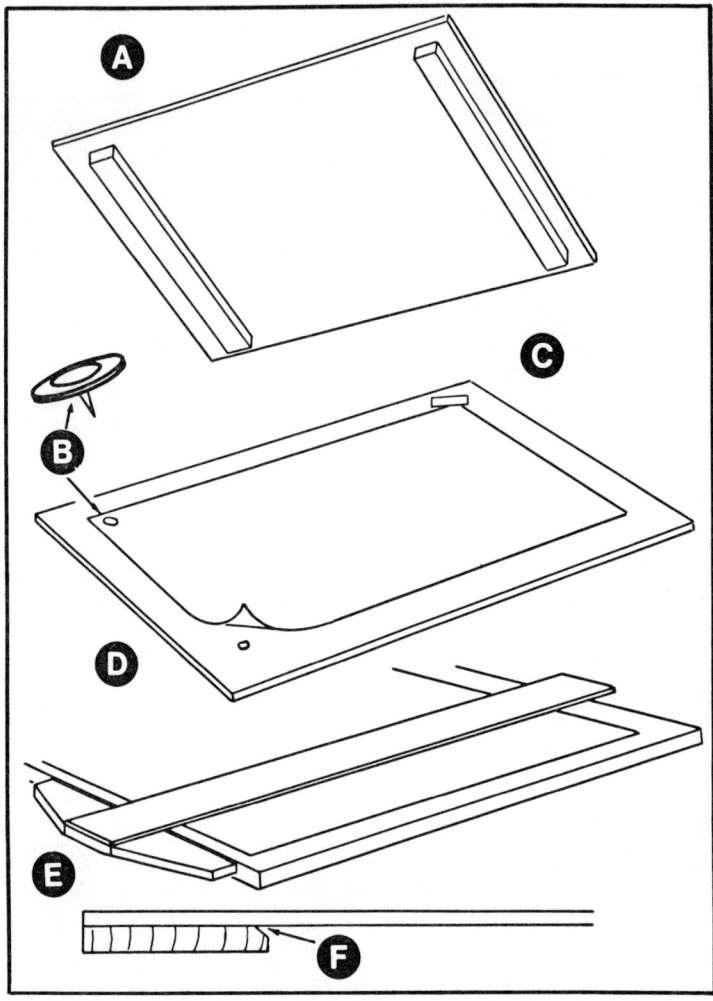

Fig. 2-1. A simple drawing board is made of plywood stiffened with battens.

tape is similar to masking tape, which sticks without leaving a residue when peeled off (Fig. 2-1C). Stik-tacks are little discs with a similar adhesive on both sides which can be put under a corner of a drawing (Fig. 2-1D). They may be peeled off and reused.

T SQUARES AND DRAFTING MACHINES

For use with the drawing board there should be a *T square*, of which there are several styles. You may pay a high price for one, but basically what is needed is a good straightedge to go across the

12

which hold it clear of the table (Fig. 2-1A). Its size should be a few inches bigger than the paper you expect to use.

Practice drawings can be made on any available paper, but paper is an unstable material that expands and contracts according to the moisture it absorbs from the air. If a drawing is made, from which measurements have to be taken, it is advisable to extract what dimensions you need soon after making the drawing, in case the paper size varies slightly. Drawing paper is stouter and less likely to change its size. It also has a better surface for taking pencil, and it will stand up to considerable erasing without suffering.

Much professional drawing is done on paper or other material with varying degrees of transparency. It is convenient to be able to put one drawing over another for comparison by looking through. The main reason for transparency, though, is to make the drawing printable. Reproduction is still often called *blue printing,* but that method of copying with white lines on a blue background has been largely superseded by other methods of dyeline printing. Tracing paper is convenient for original drawing or for making copies. A cheap grade of tracing paper is called *detail paper.* At one time tracings that had to be strong and durable were made on tracing cloth or linen. Although this had a cloth base, it was treated to have a smooth surface, usually pale blue, which would take ink or pencil. Today there are plastic films that serve the same purpose.

Perspiration from your hand may be enough to spoil a surface. For most work in pencil there is usually no problem, but for ink work you may use *talc,* usually filtered through a cloth, to wipe over a surface. Do not leave any powder on the surface.

The paper to be drawn on must be held to the drawing board so it will not move while you are drawing; yet it should be easy to

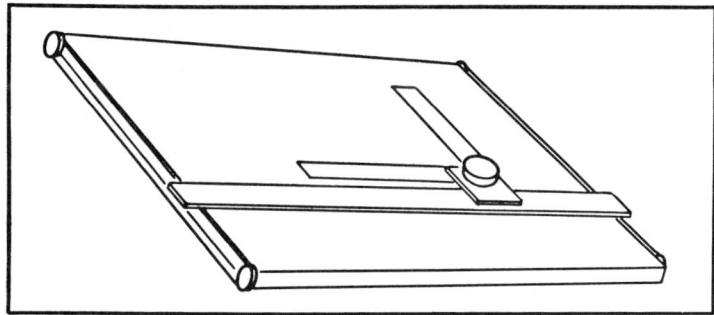

Fig. 2-2. One form of drafting machine has a straightedge kept parallel as it moves up and down the board. It may carry adjustable straightedges.

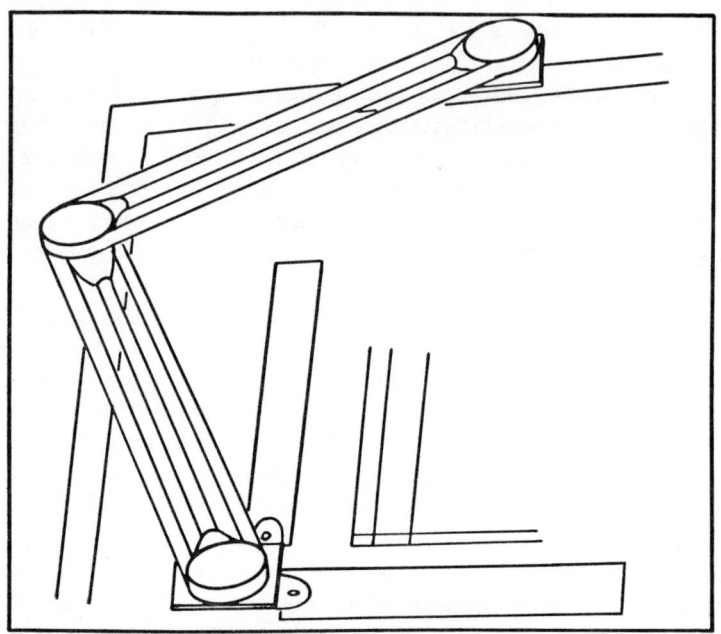

Fig. 2-3. Another type of drafting machine carries an adjustable head on arms that keep it set accurately, whatever the position.

remove. At one time all draftsmen used specially made thumbtacks with large tapered heads (Fig. 2-1B). There are occasional uses for them, but more often drafting tape or *stik-tacks* are used. Drafting board and a head to slide against the side of the board (Fig. 2-1E). The head should be at right angles to the blade. If it is not perfect, that is less of a fault than lack of straightness in the blade, as any error in the angle will be the same at any position on the board. The blade may be parallel, so both sides can be used. A left-handed person may be glad to use it with the head on the other side. The head may be beveled so it runs closely and smoothly against the board (Fig. 2-1F). It is important that the head and blade be securely fastened so that they cannot move in relation to each other. T squares are usually polished wood, although there are metal ones. The working edge may be reinforced with harder wood or plastic. If you make your own T square, and there is no reason why any amateur woodworker should not, glue and screw the parts firmly together.

There will probably always be uses for T squares, but today it is better for anyone doing much drafting to have some sort of mechanical aid instead. This may be a straightedge across the board that can be moved up and down, as it is controlled by cables around pulleys at the sides of the board. Lines drawn along it will all be parallel. It may have heads for drawing vertical and other lines mounted on it (Fig. 2-2).

Another drafting machine uses a *pantograph* action of rods from the side of the board to move a pair of straightedges at constant angles to any position on the board (Fig. 2-3). Some of these machines are large and appear complicated, but the basic movements to maintain parallel action are the same (Fig. 2-4).

For normal work on moderate sizes of drawing, the operator can sit fairly high at a table and use the board flat or slightly tilted with a packing at the far side. For larger drawings, the board that goes with the drafting machine is mounted on legs or a stand and can be set anywhere between horizontal and vertical.

In normal use the head of the T square has to be kept tight against the side of the board with the left hand. This is best done by spreading the fingers. The smaller two or three fingers are hooked around the head pressing inwards, while the first finger and thumb press the blade down on the board. Once the T square has been positioned, the left hand may come entirely on the blade to help the

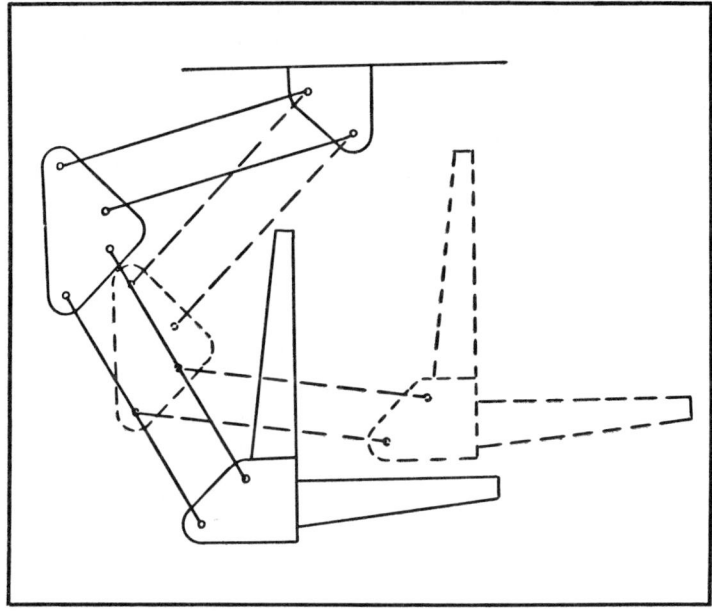

Fig. 2-4. The accuracy of the settings of the drafting machine is controlled by parallel pantograph arms.

other hand, but all the time it is maintaining sideways pressure to keep the head in place.

PENCILS

Lines are drawn with a pencil for all normal drafting. Ink may be used for special drawings or on tracings because of its *opacity*, which gives clearer lines when printed.

Pencils are often described as lead, but lead gave way to graphite a long time ago. Modern pencils for general use are graphite, with clay and resin. Lead pencil is still the accepted name. The common pencil for general purposes is graded HB or F, except that some makers use their own gradings. When choosing a pencil for drawing, check what the average grade is. Then go up or down from there. With HB or F at the center, softer pencils are B, 2B, and so on to 6B. Harder pencils go up through H and 2H to 9H. For sketching you can use an average grade or a slightly softer one, but for accurate drafting choose something harder. Grade 2H is suitable for your first drawings; then you may experiment to find what suits you best. Table 2-1 shows the relative grades. Although it is possible to use ordinary pencils on drafting film, there are special pencils that give a more opaque line for printing. They have a different system of grading, and not all makes are the same.

There are many sorts of mechanical holders for holding leads. They are convenient for drafting as a piece several inches long can be fed through. Frequent sharpening becomes necessary with large or elaborate drawings. Many draftsmen favor pencils of the usual wood-cased type. Leads of extreme hardness or softness are only available in wood.

Table 2-1. Pencil Grades.

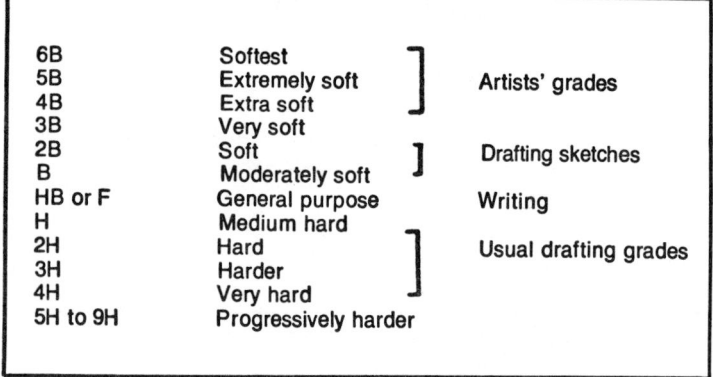

Fig. 2-5. A pencil may have a round point (A), chisel (B) point, or a bevel point as used in a compass (C). Abrasive paper is used to get a fine point (D).

A wood-cased pencil can be sharpened with any of the pocket or desk type sharpeners. It is better to have a greater extension of lead than the usual sharpener produces, so cutting away the wood with a knife is more satisfactory, although there are sharpeners arranged to cut the wood and leave the lead projecting.

For most work, the end of the pencil can be sharpened to a conical point (Fig. 2-5A). Where long lines have to be drawn, the pencil will keep its shape better if made into a chisel end (Fig. 2-5B). It is also possible to sharpen from one side only to leave a wedged elliptical shape (Fig. 2-5C). This has some advantages in a compass lead, but would be unusual in an ordinary pencil.

Sharpening of the actual point is best done on fine abrasive paper. Pads may be bought, so a worn piece can be removed to expose a new one. The pad or a single piece should be glued to wood (Fig. 2-5D) to give something firm to press against and to keep your hand away from the graphite dust, which should not be allowed anywhere near your drawing. For a conical point, rotate the pencil as you rub it along the abrasive paper.

Different people have different pressures and feel when using a pencil. If you find you are pressing so heavily that the pencil grooves the paper as you try to get a dark line, it will probably be better to change to a softer pencil and use less pressure. For most drafting, your aim should be to produce a thin line using a sharpened pencil. Pressure will change the color from light gray to near black. If you want a thicker line, the pencil tip must be slightly worn.

ERASERS

Pencil lines may have to be removed. They are not necessarily mistakes, but may be construction lines needed when laying out the drawing before putting in the final lines. For ordinary pencil lines the usual eraser is a block of rubber. Having a wedged end is useful when new, but this soon wears away. Erasers for ink have fine grit embedded in the rubber, similar in makeup to a typewriter eraser. Ordinary rubber is not very effective on plastic film, and there are special plastic erasers for pencil and ink on film. Soft plastic erasers may be described as *art gum*. They are meant for cleaning dirty marks off rather than removing pencil lines.

There are electric erasing machines which hold a length of round rubber in a chuck and rotate it at a high speed. They are light enough to hold in one hand and can be used to remove lines in the same way as a hand eraser, but quicker and, often, more cleanly. It is also possible to get holders of pencil size to hold the long round piece of rubber for hand use as an alternative to a block of rubber (Fig. 2-6). Rubber deteriorates with age and becomes less effective. If a rubber eraser is sprung to a curve and shows cracks across the curve, it has passed its prime and will soon have to be discarded, as using it will smudge rather than erase. If the surface of an eraser becomes dirty, rub it clean on scrap paper or fine abrasive before using it on your drawing.

It is sometimes difficult to erase a line without doing damage to a nearby line that has to be left. There are thin metal and plastic erasing shields that have slots and holes in them. An appropriate opening is put over the line you want to work on, so the surrounding solid part protects other lines. Similar shields are used by typists.

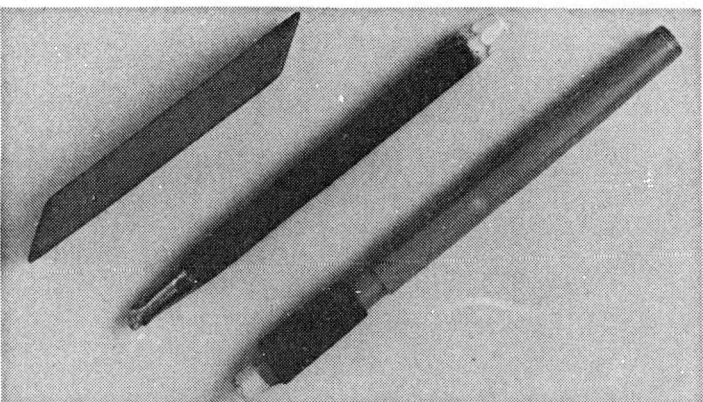

Fig. 2-6. Three types of eraser: a solid block of rubber, a fiber glass brush, and a length of round rubber in a holder.

Fig. 2-7. Transparent triangles—two 60° and one 45°.

Although it is possible to erase ink with a special eraser, that method cannot be used where only a small amount has to be removed, such as when a line overruns a meeting place with another line by a small amount. Scraping is more appropriate, using a razor blade or other sharp edged tool. This is particularly appropriate to tracing paper or film, and the method is described later. Another eraser for ink on plastic film has filaments or fiber glass in a handle like a pencil.

Whatever method of erasing is used, there will be particles of rubber, grit, graphite, or ink loose on the paper. It is unwise to wipe them off with the side of your hand as that may mark the paper. You may be able to blow them away, but a small soft brush should be kept available.

TRIANGLES

Triangles, sometimes called *set squares,* are used with a T square. At one time they were always made of wood, but now they are more usually transparent or translucent plastic, which has the advantage of allowing you to see the lines beneath them. The common patterns are described as 60° and 45°. In both cases one corner is 90°, but the 45° triangle has equal arms, while the other has corners of 60° and 30° (Fig. 2-7). They are made in many sizes from a few inches to 24 inches or more. Larger triangles have their centers cut away for convenience in handling. It is possible to get triangles with beveled edges, but these are of doubtful value. They have a distorting lens effect when you look through, and the edge is kept from the paper when you have to turn the triangle over. Wooden triangles have their uses, but metal ones are not as suitable as they may seem. They tend to make gray marks on the paper.

A triangle standing on a T square will allow you to draw a perpendicular line. Its long edge allows drawing at the appropriate angle. Two triangles held together will produce other angles. In practice, triangles get more uses than may be expected. They can be used as straightedges, being often more suitable for drawing straight lines than tilting the T square or using a scale.

SCALES AND STRAIGHTEDGES

A draftsman needs to draw straight lines and to measure. In other activities a person may have a graduated straightedge and use it for both purposes. Most craftsmen have a *rule* (sometimes incorrectly called a ruler), made of wood or metal, which is used for testing levels and straightedges as well as measuring. A draftsman tends to keep the two operations separate. His graduated straightedge is called a *scale,* whether it is marked for full-size measuring or is intended for use with drawings that are to a scale other than full-size. Straight lines are drawn with the T square or a triangle. For large drawings, there may be long independent straightedges. For precision, they may be metal because of the risk of wood warping or twisting. As metal tends to mark, the paper metal is avoided for shorter straightedges.

In a basic drafting kit there need only be one scale, with its edges marked full-size. The ordinary cheap school scale or rule is not really satisfactory. A draftsman's scale is marked with more precision, and the edge is tapered quite thinly. It is a more delicate instrument, which is one reason why it is not customarily drawn along and would certainly not be used for cutting.

What graduations to choose depends on the type of drawing to be done. Inches divided into sixteenths may be all you need, or you can have at least part of the edge divided into smaller fractions. If you expect to be dealing with metric measure, the centimeters may be divided into millimeters, although you may also have uses for part of the edge divided into half millimeters. You can buy a scale marked with the two systems on opposite edges (Fig. 1-8).

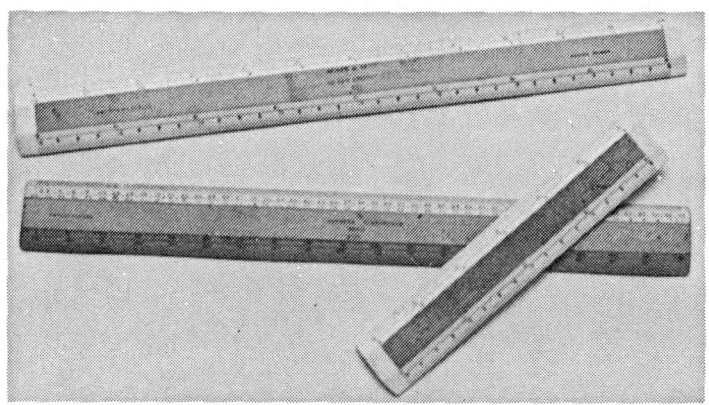

Fig. 2-8. A selection of scales—one solid wood and two with plastic edges.

At one time scales were made of hard close-grained wood. Some were made of bone and ivory. For precision, the graduations are cut into the material. Most woods do not allow very fine cutting without breaking out. Many scales were mainly made of wood, but with bone or other hard edges which carried the graduations. Today, most scales are hard plastic. There is a carry-over from the early days. Many of them are still supplied with the center parts of genuine or imitation wood.

A draftsman's scale is longer than the graduated part, unlike a carpenter's or machinist's rule, which usually measures from the end. This is so wear at the end does not affect measurement. When distances are taken off with dividers, one point can go in the end graduation and not slip over the end of the scale (Fig. 2-9).

There are many sections of scales, with some of them designed to include a variety of graduations in one instrument. These may have advantages for occasional use of a particular scale size, but they can be confusing. For general use of an instrument graduated full-size, it is better to have a clear marking with just that and no more overlaid on it. It is best arranged flat with a taper to the edges, both on the top side.

COMPASSES

A *compass* has two arms, with a point on the end of one arm and a pencil point at the end of the other arm. Its primary use is to draw circles or parts of circles. There may be an ink pen end to substitute for the pencil point or in a separate compass. If there is another steel point in place of the pencil, the instrument becomes a pair of

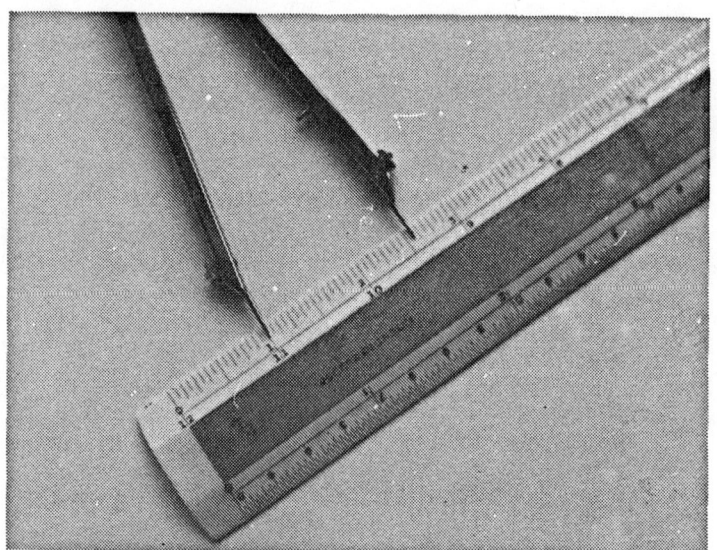

Fig. 2-9. Dividers are set using the graduations on a scale, but the measurements need not start at the end.

dividers. Dividers are used for stepping off or comparing distances.

Although the requirement of a compass is simple, there is an enormous variety available, with variations in detail or principle of operation.

For a first compass, the type which depends on friction to maintain its setting and is about 5 inches long is a good choice. In its simplest form it may be for pencil only, but sets are often offered with alternative pen and divider ends to fit in place of the pencil with a container for space leads and steel points. Whatever sort of drafting a beginner moves on to, one of these sets will have its uses (Fig. 2-10).

The sections of the compass arms varies between makers, but most have a hinge with frictional inserts and a knurled handle or grip standing above the joint and arranged to remain central at all settings of the compass. This handle extends between the two arms to hold an adjusting screw, which can be turned to vary the friction in the hinge. There are variations in detail, but that is the effect. You open the compass to the size you want, and it stays at that setting providing it is used normally.

The steel point, about which the compass turns to draw a circle, is often made reversible. There is a normal tapered point

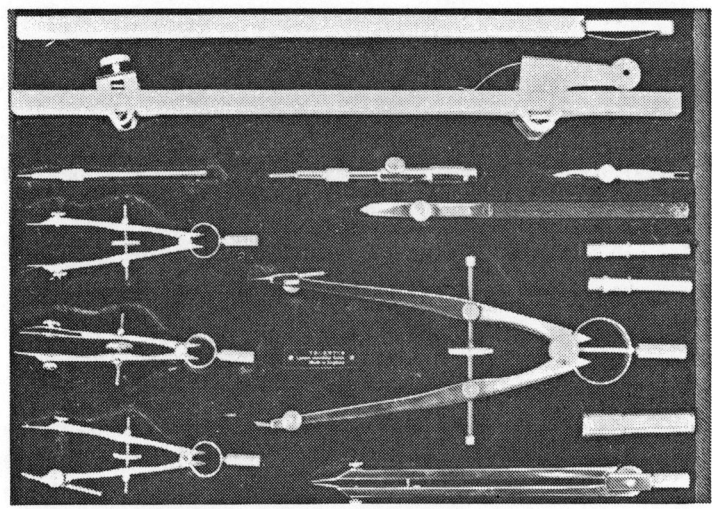

Fig. 2-10. A set of instruments in a fitted case.

(Fig. 2-11A) and one that is shouldered, with the actual point much thinner (Fig. 2-11B). The grip of the point is usually by a screw, which squeezes the split end of the compass arm, or there may be a small chuck.

It is only in simple school compasses that a wood-cased pencil can be fitted. In a draftsman's compass the lead may be offset and the split arm squeezed on it (Fig. 2-11C), or there may be a straight-line chuck (Fig. 2-11D). One or both arms can have hinges along their length, so the pencil or pen end may still be upright, even with a wide opening. For very small curves, the pencil may be angled close to the center.

For drawing circles, the pencil or pen point should be adjusted shorter than the needle point to allow for its slight penetration into

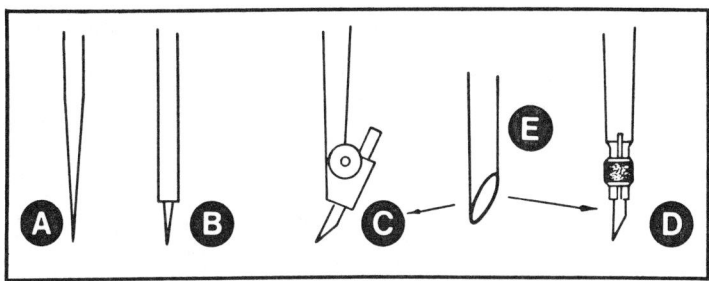

Fig. 2-11. Two steel points are used in a compass (A and B). The bevel pencil end slopes inward (C, D, and E).

the paper and drawing board. It should never be necessary to thrust the point in very deeply. Divider points should come exactly level when the instrument is closed. Sharpen the pencil point with a bevel outside only (Fig. 2-11E).

The foregoing instruments are the essential first tools of the draftsman's trade. Most drawing can be done with them, although perhaps not as expeditiously in every case as with some of the instruments described later. It is important that these basic instruments are mastered. Any competent draftsman should be able to produce good working drawings with them.

Store the instruments so edges that have to be kept straight do not run the risk of being damaged. Keep striaghtedges and triangles away from compasses, dividers, and other metal things. See that needle and pencil points cannot be damaged. This is where sets of instruments supplied in cases have an advantage. Keep instruments clean. A surprising amount of dirt comes from pencil dust. Most scales can be cleaned with water and detergent, but make sure by careful trial that markings will not be affected. If a drawing on a board has to be left between working sessions, cover it with plastic or cloth. If that is not possible, put another piece of paper over it. Keep your hands clean and roll your sleeves up.

Chapter 3
Other Drawing Instruments

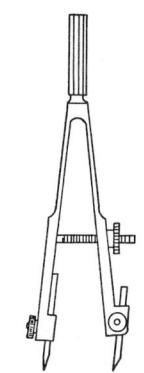

A draftsman tends to accumulate many instruments, some of which are variations on the basic ones. There are sometimes advantages in those that are bigger or smaller. There will have to be several scales, and these are dealt with in Chapter 4. The largest numbers of variations are in compasses and dividers. Apart from other considerations, it is often useful to have several pairs so one can be left set at a size that will be needed again. Then there will be less risk of errors due to slight differences in resetting.

COMPASSES

Friction joint compasses and dividers have a useful opening about the same as their length. For instance, 5-inch compasses will reach to draw a circle of 5-inch radius, but that is the maximum. Normal use is better restricted to less than that. Some of these compasses are provided with an extension piece that can be fitted between the pencil or pen end and the main part (Fig. 3-1). This allows the drawing of larger circles, but it is a rather clumsy arrangement to use. For larger circles only occasionally needed, however, it is practical.

A better arrangement for drawing large curves is a *beam compass* or *trammel*, although that is more correctly the name of a device with a beam compass for drawing ellipses. The beam can be any length, so almost any radius within the limits of a drawing can be set. There are many varieties of beam compasses. A simple

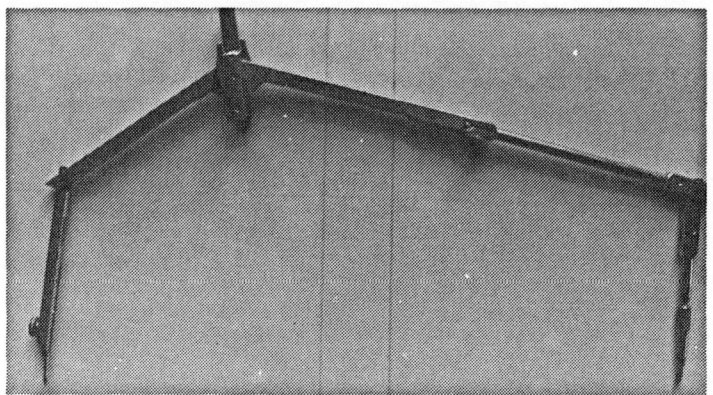

Fig. 3-1. A compass can be fitted with an extension, but the ends have to be hinged to bring the points square to the paper.

type has the tubular metal beam to take apart for storage, and the two heads slide on it (Fig. 3-2). One head carries a needle point, and the other will take pencil, pen, or another needle point. Some beam compasses have fine adjustments arranged by a secondary screw on one of the heads, either to move the point sideways or to tilt it (Fig. 3-3). For the really large curves, beam compass heads are arranged to fit on a wooden beam, which can be a lath of stock size obtained for the purpose. If a beam compass is unavailable, it is still possible to improvise a large compass with a strip of wood. A fine spike or awl is put through at the radius distance from the end, and a pencil or pen is used against the end while the strip is pulled round (Fig. 3-4).

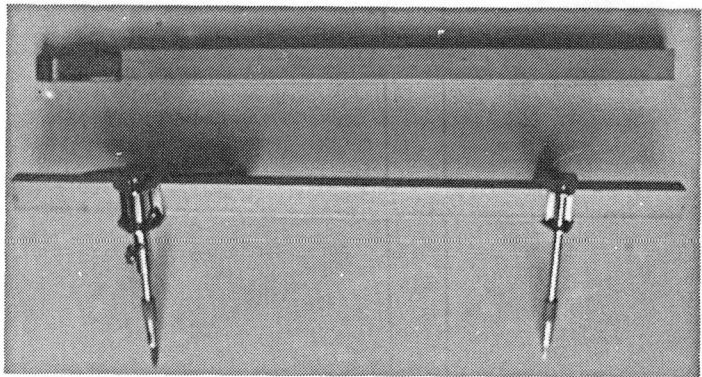

Fig. 3-2. A beam compass with its bar in two parts and two adjustable heads for drawing curves larger than would be possible with ordinary compasses.

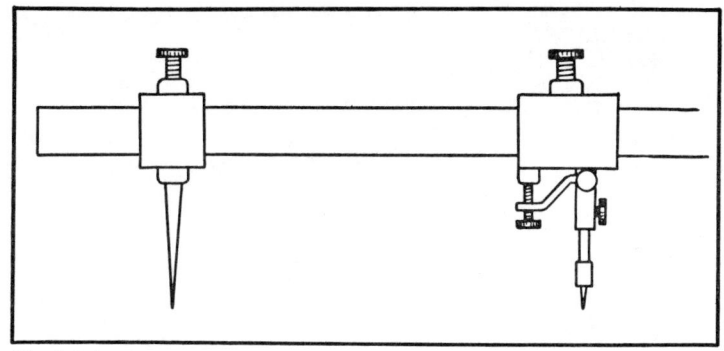

Fig. 3-3. In a beam compass, or trammel, two heads slide on a bar. One may have a fine adjustment.

At the other extreme are very small circles. If no other compass is available, it is possible to use a friction-joint compass for quite small circles. There are practical difficulties in handling for radii much less than 1-inch. The alternatives are compasses generally known as *spring bows*. They are arranged with the usual needle points, pencils, and pens, but in one construction a circular spring pinches against the top forcing the legs out. Then they are retained at the setting you want by a knurled nut on a screw. In one adjustment the nut is outside a leg (Fig. 3-5). Another type has a central knurled knob fixed to a bar with left and right-handed threads, so both legs move in or out together and the knob remains central (Fig. 3-6). There is little to choose between the two

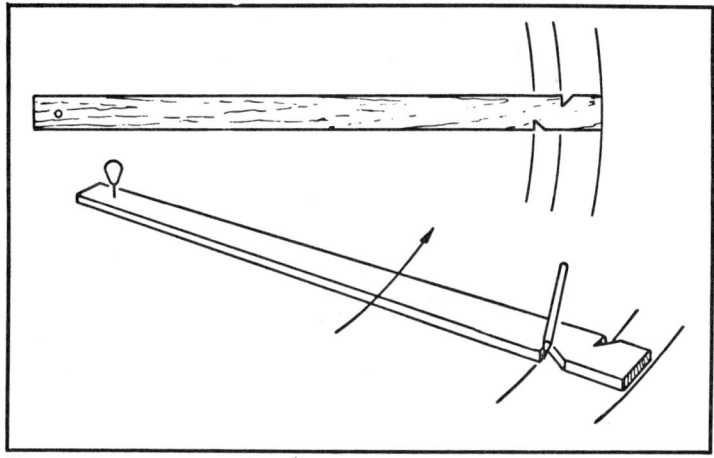

Fig. 3-4. A large compass can be improvised with a strip of wood.

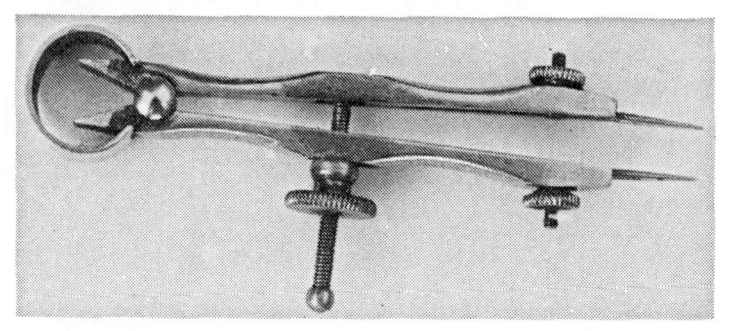

Fig. 3-5. A pair of spring bow dividers with the adjusting nut outside.

methods of adjustment, and most users find them equally satisfactory.

Spring bows are mostly about 3 inches long. These are the convenient size to have for small work, but they are also made larger in sizes comparable with friction-joint compasses. Their advantage is in the more positive locking to size that they have, with reduced risk of movement, which may sometimes be unnoticed with the other type. Such large spring bows are not very suitable for small circles, and small versions are advisable.

Another type of small compass may be called a *bow compass* to distinguish it from those just described. Bow compasses differ in having the legs sprung (Fig. 3-7A), and this limits the range of adjustment to less than that of spring bows. They are convenient for the smallest circles. They are made with the usual points. A less common variant is called a *drop bow compass* (Fig. 3-7B). Its advantages are in handling for drawing very small circles, particularly in ink. There is a stationary rod with a knob directly above the needle point. The pen or pencil is on a spring arm with screw adjustment. All this assembly is on a tube with a knurled top that

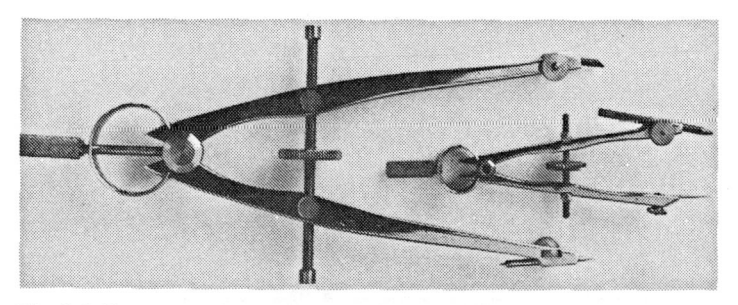

Fig. 3-6. Two pairs of spring bow compasses with central adjustments.

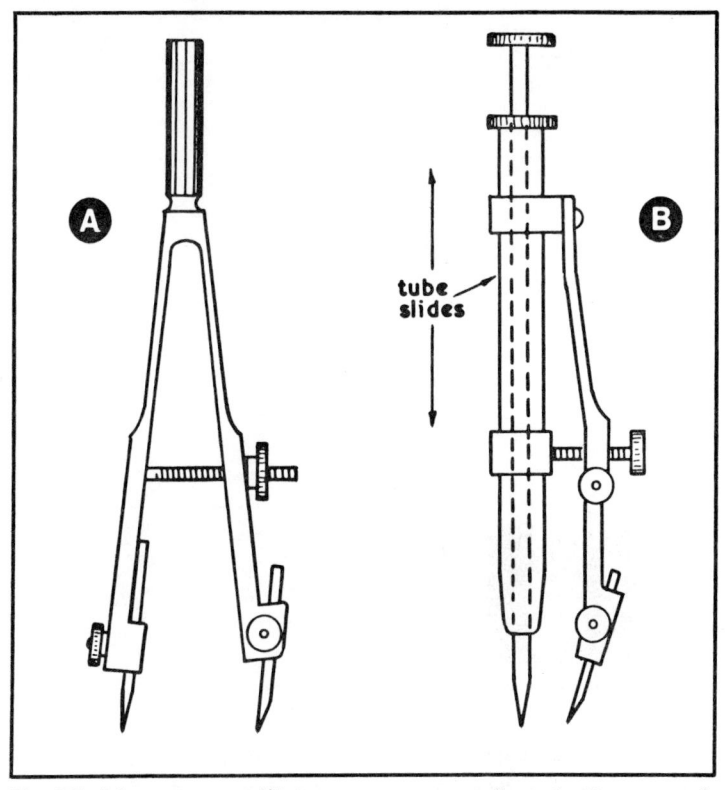

Fig. 3-7. A bow compass (A) has sprung arms, adjusted with a screw. A drop bow compass (B) has a sliding body.

will revolve around the stationary rod. In use the first finger presses down on the stationary knob, while the thumb and other fingers revolve the tube, which can be lifted on the rod so the needle point can be located before dropping the marking point.

Many compasses can have their purposes altered by exchanging points, pens, and pencils or by fitting extensions. Too many complications should be avoided as a one-purpose instrument is usually easier to use and less liable to cause trouble. While a secondary use will be infrequently needed, it may be worthwhile having it available in a combination instrument. In any compass that will be used over a large range of adjustments, the legs should have hinges so the points can be arranged approximately perpendicular to the surface. This is particularly important when using ink (Fig. 3-8). Hinging inwards reduces the effective range of the compass, but that has to be accepted and allowed for. It is possible to get

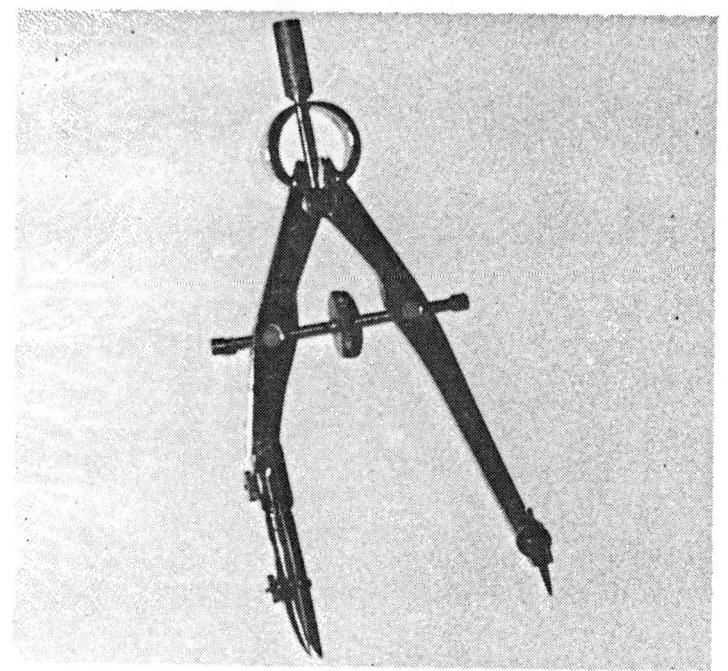

Fig. 3-8. The leg of a compass intended for ink should be hinged so the pen can be upright on the paper at any setting.

spring bows with pen and pencil on one arm that can be turned over (Fig. 3-9). This is an economy, but separate instruments are more suited to regular use.

Proportional dividers are a special type that may rarely be needed by some draftsmen, but their advantage is in the enlarging or reducing of a drawing. Proportional dividers allow the transferring of detail measurements accurately without frequent calculations or the change of measuring scale. There are points at opposite ends (Fig. 3-10). The pivot between the two arms can be adjusted along slots so a register mark comes opposite marked positions to give particular proportions. For instance, setting to "2" means that the distance between the points at one end will be twice the distance between those at the other end. Many other proportions can be set, and on the 7-inch instrument they go from ¾ to 10.

PROTRACTORS

Triangles may be used for marking angles to a number of usual sizes. There are 360° in a full circle. The common square line or

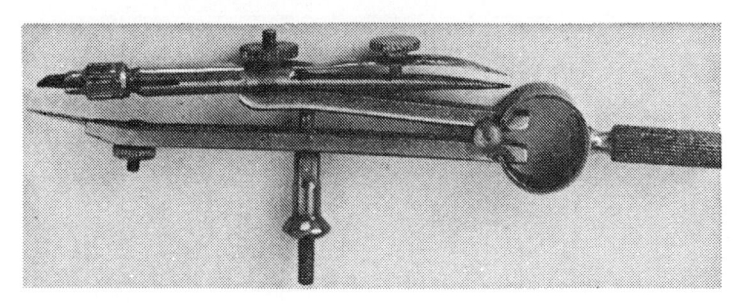

Fig. 3-9. This spring bow compass has alternative pen and pencil points on one leg.

right angle is one quarter of a circle, or 90° (Fig. 3-11A). The 45° triangle halves this (Fig. 3-11B), while the 60° triangle will mark one-third (Fig. 3-11C) or two-thirds of it (Fig. 3-11D). You can get other total angles by combining the triangles (Fig. 3-11E). Many angular markings can be arranged with these triangles, but there is often a need for other angles. Some drafting machines are arranged so their straightedges can be adjusted to any required angle. For

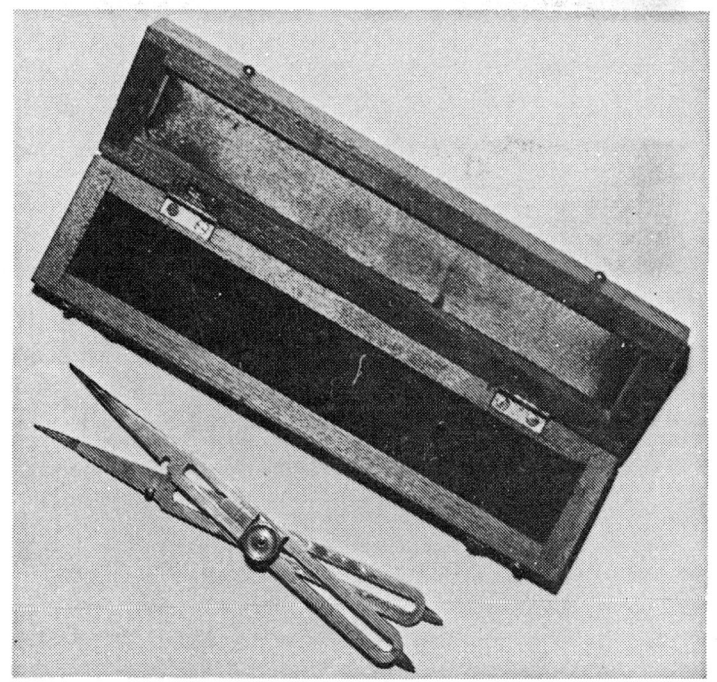

Fig. 3-10. Proportional dividers and their case.

31

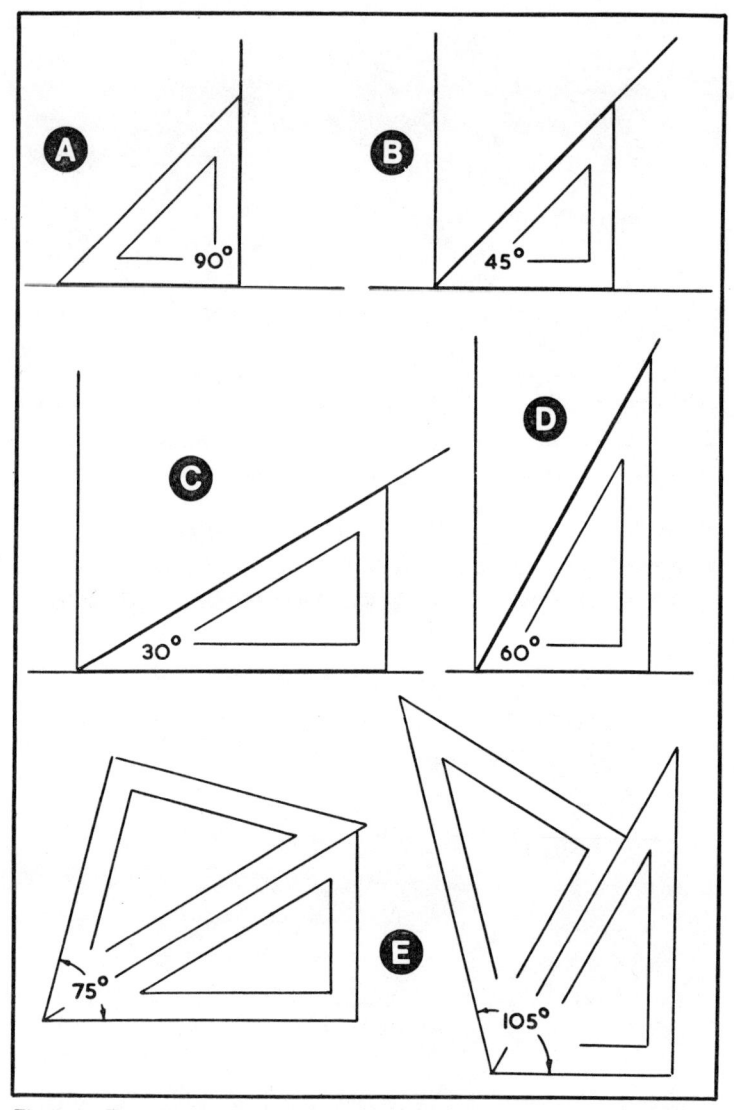

Fig. 3-11. Triangles can be used alone or together to draw many standard angles.

drawings made without their help, the tool to use for setting angles is a *protractor*.

At one time protractors were made of wood, but there is an advantage in having them made of plastic that can be seen through. One may be a full circle (Fig. 3-12) or a semicircle. The circumfer-

32

ence is marked in degrees, and it should be beveled to bring the markings very close to the paper. Usually numbering is in both directions for convenience in marking angles either way. Note where the center is, about which the degrees are arranged (Fig. 3-13A). This may be on an edge or on a line, which has to be positioned over the mark on the baseline where the angular line is to join (Fig. 3-13B). Then a mark is made at the intended angle (Fig. 3-13C) to be drawn through (Fig. 3-13D). A large protractor is of more use than a small one. There could be errors if a line has to be extended a long way from a small protractor mark, but if the protractor marking is outside the final limit of the line, any risk of inaccuracy will be minimized. A 5-inch diameter is suitable for general purposes.

Protractors are not always circular. There is a rectangular scale intended for carrying in the pocket with a variety of markings on it. It has protractor markings around three edges, based on a center on a long edge. Except for the varying distances of the marks from the center, it is used in the same way as a circular instrument.

A variation on the protractor is an adjustable triangle (Fig. 3-14). When closed, it is a 45° triangle, but the long side is pivoted and can be opened to give other angles to the base edge, as calibrated on the curved scale. In effect, it is a triangle that can be set to any angle between 45° and 90°. By combining it with triangles of other angles, any total angle can be set. As with

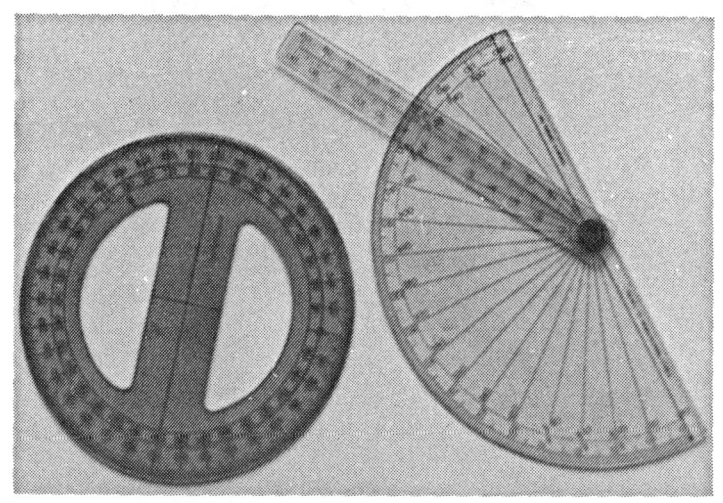

Fig. 3-12. A round protractor and a semicircular one with a swinging arm.

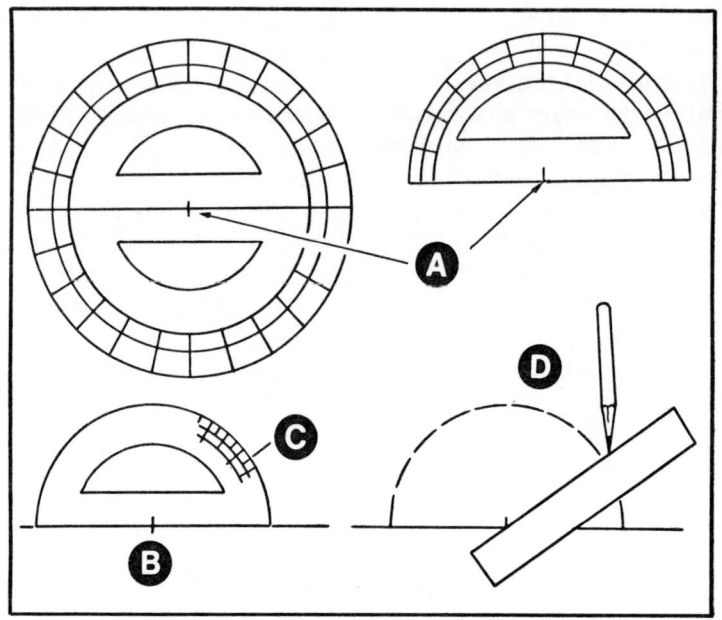

Fig. 3-13. A protractor is used to get positions of angles around its circumference; then a line is drawn through the center mark.

protractors, there is an advantage in size because extending a short marked line may lead to errors.

Other protractors are made with a straightedge pivoted about the center, so that can be set to the required angle and drawn along. The line is broken and has to be drawn again, though, so there is little advantage. Similar tools are used in shops for setting machine angles.

Triangles are made at other angles than the basic ones, and a draftsman may make one to suit his particular need for repeating certain angles. For instance, in a steel beam section the internal angle is always the same. Different angles may be used by the same draftsman in other steel sections. A three or four-sided instrument could be made out of plastic or wood for marking these angles (Fig. 3-15).

CURVES

One problem is draftsmanship is the accurate drawing of curves. A draftsman cannot do like the artist and draw in a curve that he likes, experimenting freehand until he is satisfied. The curve on a mechanical drawing usually has to be of a particular type

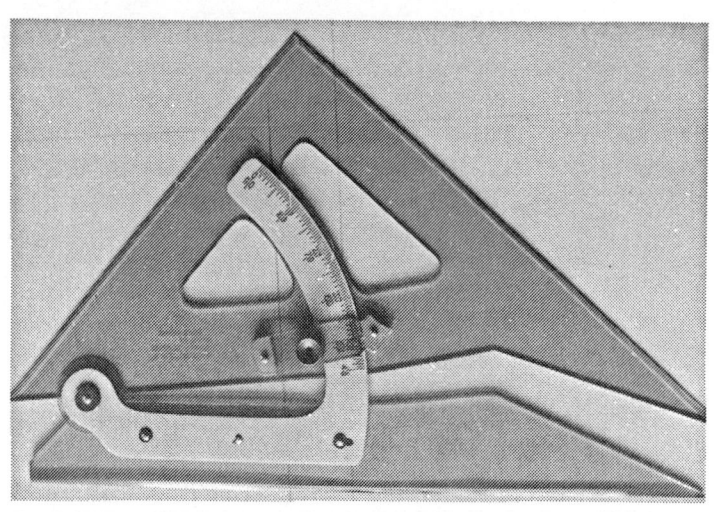

Fig. 3-14. An adjustable triangle can be set between 45° and 90°.

and pass through certain points. It is a principle of draftsmanship that no lines are drawn freehand. This is particularly so if the drawing will be finished in ink. Apart from any inaccuracies due to the freehand work, a freehand curve will be very obvious by its unevenness among the lines drawn with instruments.

A curve that is part of the circumference of a circle must be drawn with compasses if at all possible. An exception would be if

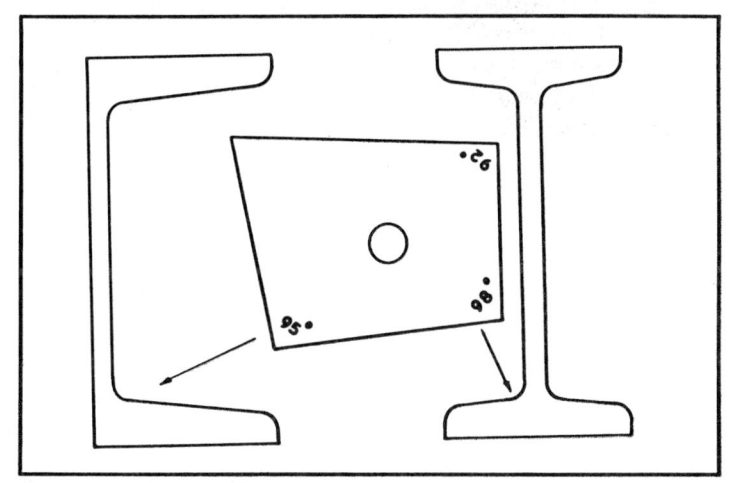

Fig. 3-15. A special triangle or template can be made for frequently used special angles.

35

the curve is so slight that its radius would take the center a long way outside the limits of the drawing board.

Ellipses

Ellipses or parts of them can have points on the curve found by geometric means (as described later); then curves can be drawn through the points. There are also trammels (Fig. 3-16) in which a beam compass with two needle points, as well as a pencil or pen point, can be adjusted to various sizes, so movement of slides in a frame permit the marking point to draw an ellipse. The trammel has its limitations of size, but it is the only thing which might be regarded as the equivalent of a compass for drawing ellipses.

To aid in drawing ellipses, there are plastic templates with various elliptical cutouts that a pencil or pen can be put through (Fig. 3-17). Unfortunately the variety of ellipses that may be required do not often match templates, which are actually of more use for pictorial drawing.

Splines

In some drafting work there is little demand for curves other than those that can be drawn with compasses. Ship and boat draftsmen are exceptions. When drawing the "lines" of a hull, nearly every line is a curve that varies in its length as it passes through marked points. The path of such a curve is taken by a *spline,* which is a flexible rod. At one time it was always wood, but it may now be plastic. To hold it in shape there are weights, often called *mice* from their appearance (Fig. 3-18). If much of this work is expected, there have to be splines of differing sizes and degrees of flexibility. Other devices have been made with stiff jointed arms

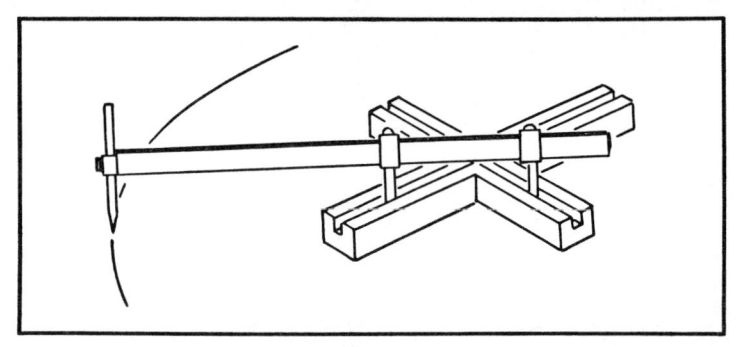

Fig. 3-16. Trammels are used for drawing ellipses. The beam compass alone may also be called a trammel.

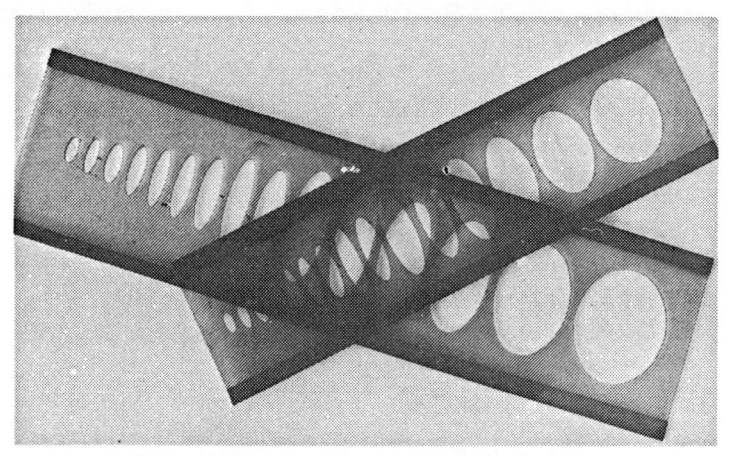

Fig. 3-17. Two templates for marking ellipses of many sizes and ratios.

to hold the curve, but most of them suffer from slight kinking so the resulting curve is not "fair." Another device for shorter curves has a lead core inside plastic, so it can be bent by hand and will stay in shape. Care is needed to avoid uneven curvature.

French Curves

Original curves should usually be drawn as just described. When it comes to lining them in with heavier pencil lines, or more especially with ink, it helps to have something with a matching curved outline to draw around. These things are collectively called

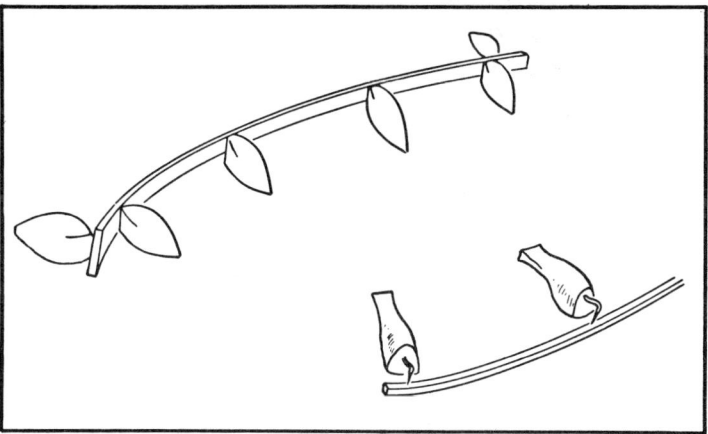

Fig. 3-18. A curve may be drawn around a spline held in place by lead weights.

French curves (Fig. 3-19), although there is no apparent connection with France.

French curves are obtainable in many sizes and shapes, so it is possible to find an edge which matches at least part of the curve you have to draw. Then you find another part to follow on. Some French curves are described as universal. These certainly have parts that will match curves of almost any sort within their outline and size. You need several other French curves as well if you draw a variety of outlines without having to deal with them, using just short parts of your French curve. Besides the range of French curves of general application, there are others for particular purposes. Ship curves are larger and mostly with flatter flowing curves to suit hull lines.

Plastic templates are made for marking many standard shapes. Some of these are meant for school use such as drawing shapes of flasks, Bunsen burners, and other things used in science lessons. Others, intended for draftsmen, contain the outlines of frequently used symbols or outlines. For instance, a draftsman mainly concerned with electrical wiring diagrams has uses for a template containing his symbols, or a draftsman dealing with maps can use a template containing the many symbols for churches, mills, etc. One problem is scale. Such standard symbols have to be in reasonable proportion to the size of drawing, and templates do not have as much value as might be expected. A keen draftsman may prefer to draw the symbols or other things each time to get the proportion right.

There are templates for lettering, and these are described later. Poor freehand lettering can spoil the appearance of a drawing, so the use of a guide is advisable. The title and key to a drawing may be arranged in one corner in a standard way. A template for the whole of the outline of this makes for standardization and neatness.

INK INSTRUMENTS

Ink work on drawings is mostly done with a dense black ink usually called *Indian ink.* Other colors are available, but black is the choice for line work because it produces the most opaque line. The usual reason for using ink is to make a drawing to print from, where the ability to obscure light is the main requirement of an ink line. Indian ink may be sold as waterproof or nonwaterproof. Use waterproof for drafting as it is a more intense black. The nonwaterproof type is used by artists for special effects.

Ink may be more economically bought in larger containers, but for use it is better in small bottles or plastic containers with a

Fig. 3-19. A selection of French curves for drawing curves that cannot be made with compasses.

stopper carrying a means of transferring ink to a pen. At one time this was a quill cut from a feather pushed into a cork, but it is more common now to have a sort of small syringe operated by the soft rubber stopper. If ink is spilled, it can be extremely messy, so it is safer to have the bottle in a stand (Fig. 3-20). One type is rubber and grips the bottle, while the base acts to prevent the whole thing from moving. Indian ink on a drawing should not be blotted, but it has to be left to dry. Fortunately, fine ink lines dry almost instantaneously. It is worthwhile to have small pieces of blotting paper available to put instruments on and deal with dropped spots of ink. The paper may also be used for cleaning pens, but a piece of cloth is better. Ink should not be allowed to harden in instruments, so they should be washed and wiped after use. Hardened ink will not easily wash out. Scraping might harm the instrument.

PENS

The basic *ruling pen* has two blades that tend to spring apart and are pulled together with a screw (Fig. 3-21A). The ink goes between the blades, and any ink getting on the outside must be cleaned off. The ends of the blades are kept very fine to the extent of not being quite sharp enough to cut the paper. If the blades are adjusted to almost touch, they will make an extremely fine line when drawn along a straightedge or around a French curve. These pens are not intended to be used freehand. The width of the line can be increased by adjusting the screw.

Fig. 3-20. A bottle of Indian ink is fitted into a stand for safety. The stopper carries a quill for filling ruling pens.

In the simplest form the two blades are similar. A better type allows one blade to swing clear for ease in cleaning (Fig. 3-21B). Compass points for ink are similar to ruling pens (Fig. 3-21C). Pen ends vary. For general use, the ends of the blades are narrow and rounded when viewed from the side. For some fine detail, it is useful to have one pen with more pointed ends. There is a limit to how much ink can be loaded into a pen. Trying to put too much ink in may cause some to spill on the drawing. When drawing a very long thick line, it is a nuisance if the pen has to be refilled before reaching the end. There are pens with extra wide blades for this purpose, as they have greater capacity. There are several special pens working on the ruling pen principle. One is a double pen, with two ends on the same handle, and they can be adjusted in relation to each other. Their special value is in drawing lines parallel to each other, as in a road on a map, but they have other uses.

A different approach to the drawing of ink lines is seen in pens which are used like fountain or ball-point pens. There is a supply of ink in the body of the pen which is fed to the point via a fine tube. Each pen will only draw a line of one width, so there has to be a separate pen for each width of line. The occasional user of ink may find it worthwhile having one of these pens for his standard lines, while the other lines are made with a ruling pen. Indian ink does not flow as easily as ordinary ink and has a tendency to clog, so makers have incorporated various features to keep these pens clear. Take care in selecting these comparatively expensive instruments so you get one that will keep clear, even if only occasionally used. Otherwise, it is better to stick to ordinary ruling pens.

Handles for ruling pens vary from slender strips to shaped grips. Choose a type you can hold without slipping. At one time bone and ivory were used for expensive instruments, and there are plastic reproductions of these, but there is a better feel to a wood handle. In some sets of instruments there is a handle to take the ruling pen point of the compass. This is not as good as a separate ruling pen since the pen for a compass is made curved inwards slightly, which is not so suitable for straight lines.

There may be some freehand work on most ink drawings, probably only lettering or the adding of arrowheads. This is done with an ordinary *nib* of the type once used for general writing. There are special fine ones intended for drafting. One type is sold as a mapping pen and may be suitable if there is difficulty in getting the others. Steel nibs do not last long. The ends thicken and they have to be discarded. Keep nibs clean by wiping off surplus ink

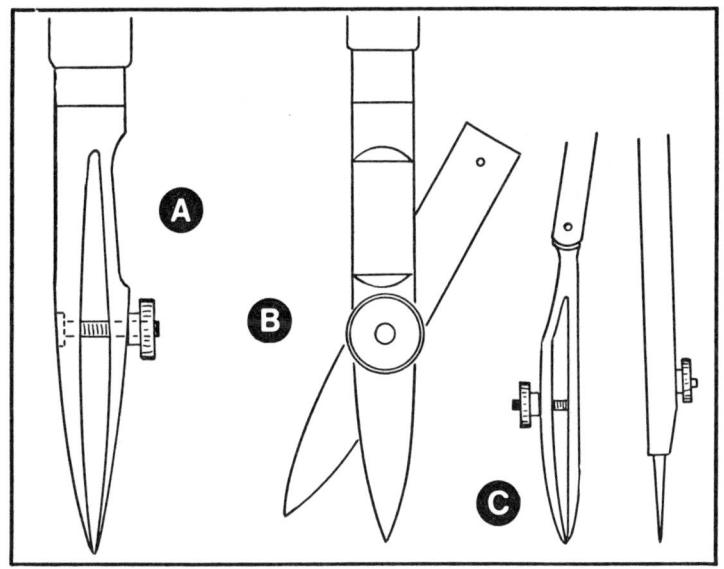

Fig. 3-21. A ruling pen has adjustable legs, which may pivot or be hinged in a compass.

before it dries, but be very careful at the tip. Damage to the split end ruins the nib. There may be a use for wider nibs for special lettering or shading, but these are artists' tools. A pen with a broad straight end will do lettering that is wide on the down stroke and narrow when the nib is used sideways. Another type has a round end and draws the same width wide lines in all directions. The larger pens have a bent metal reservoir clipped over the nib. Ordinary nibs may be dipped in the ink. Ruling pens should never be put into the ink bottle, but should be filled with the device in the stopper.

Some paper is surprisingly abrasive and will wear away the ends of pens. Smooth plastic film may be particularly bad. The effect is to thicken the ends of the blades. This means the thinnest line you can draw will not be as narrow as previously. To cure this, use an *oilstone*. It does not have to be a large one, but can be pocket size and of medium grit. Use a fine lubricating oil or kerosene on the stone. Do not use the stone dry, or it will quickly wear away and become grooved.

Examine the end of a worn ruling pen. The thickened part will shine white in a cross light. It is that which has to be rubbed away. Thin the end by rubbing with a rocking motion along the stone (Fig. 3-22A). Wipe the end and examine it. Thin both blades until they are almost knife edges. Adjust the ends close together and hold the pen upright, so you can move it along the stone with a rocking

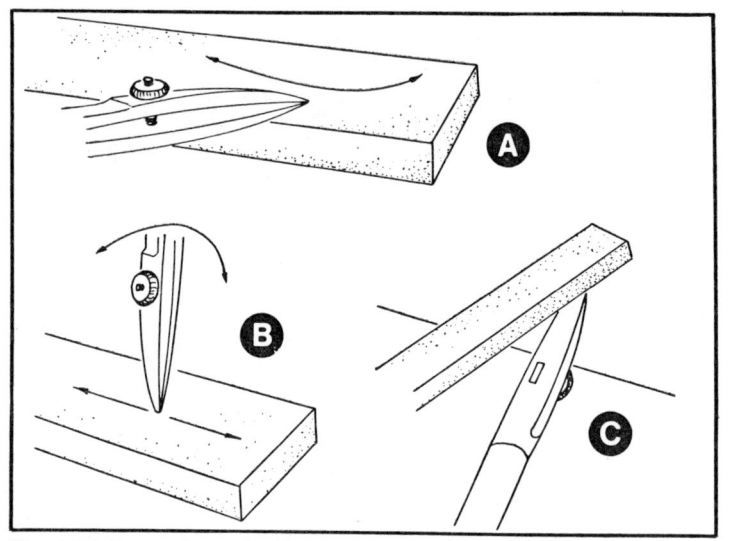

Fig. 3-22. Pen ends need occasional rubbing on an oilstone.

motion to round both blades equally (Fig. 3-22B). Do not tilt the pen sideways, or you will get uneven ends. Wipe and examine the ends. If rounding has thickened any part, you may have to do a little local rubbing flat on the stone to get an even thickness at the ends of both blades. Some draftsmen prefer to use the stone like a file, moving the stone instead of the pen. Hold the pen at an angle over the edge of the table and rub the stone on it (Fig. 3-22C). Do not rub the insides of the blades.

It should be possible to use a ruling pen along the square edge of a triangle or other guide without the ink running under and smudging. This is the best way of doing accurate work. If there is any doubt, the guide may be beveled and used with the bevel edge downwards. This has advantages when working rapidly. Smudging is more likely to occur when a straightedge is picked up from alongside a wet ink line, and using a bevel underneath reduces this risk.

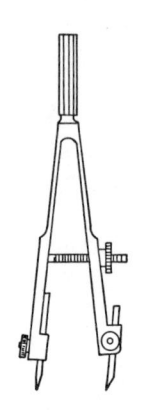

Chapter 4
Scales

The ideal mechanical drawing for practical purposes is full-size. The person working from it can then measure it directly. If something is being made from the drawing, it can be put over it for comparison. There are obviously limitations. If you are making a drawing of a house, your paper size limits the size you can draw it. Even more so would be a plan of a tract of land, where what you draw would be very small in relation to what it represents. The object being drawn may be very small. An actual size drawing would take up little space and be difficult to draw and understand, so it is drawn bigger.

The relation between the size of the drawing and the object being drawn is the *scale*. This is something which has to be decided before you start drawing. Think of the views you want to draw. You may want to include sections, part assemblies, and other things besides the usual plan, end elevation, and side elevation. It may be something that does not lend itself to the standard views. You have to visualize your needs and relate them to the size of the paper. It helps to do a rough sketch with the views in their positions and dimensions written on. On the finished drawing you have to allow for dimension lines and spaces between views. You have to arrive at a proportion between your drawing and the actual thing. That is the scale.

Suppose you want to draw three views of a building (Fig. 4-1A). Make a sketch with the outlines and their sizes written in

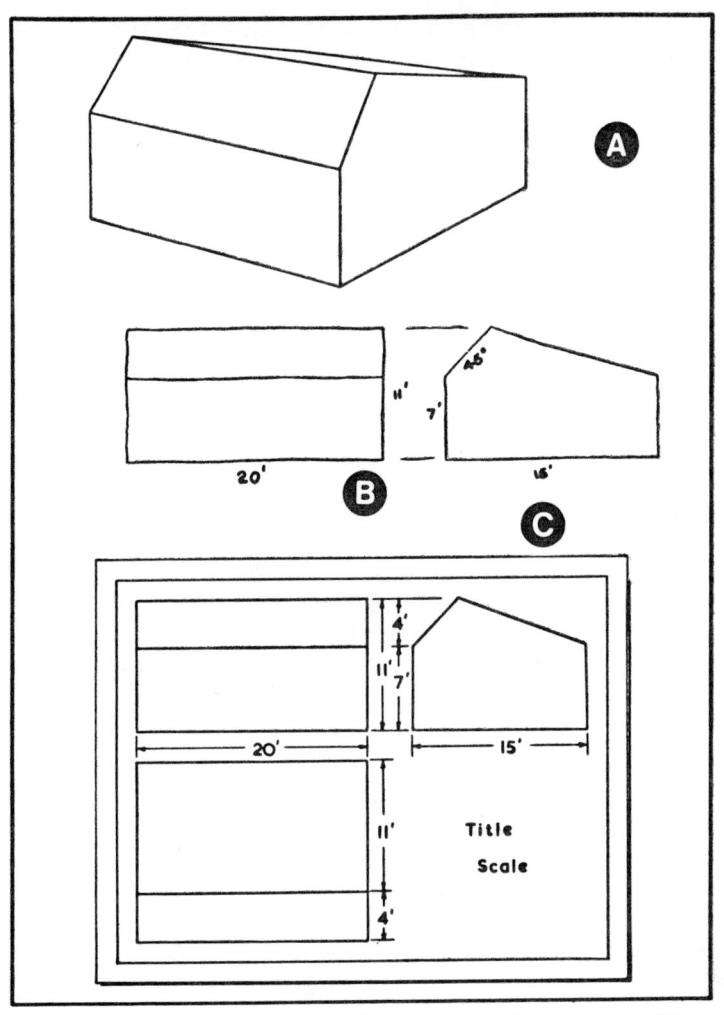

Fig. 4-1. An object to be drawn may be sketched first to note sizes. Then three views are laid out to scale on the drawing paper.

(Fig. 4-1B). Consider the size paper you want to draw on. Allow for a border. Do not pack the drawing too tightly within the border (Fig. 4-1C). The drawing would look silly and be a waste of space if it only occupied a small part of the available area. At the same time you have to relate what you are to draw to acceptable scales; otherwise, you and the user of the drawing may find yourselves with proportions for which you do not have suitable means of measuring on the drawing.

A point to remember is that it is easier to decide on a scale and make a drawing, and then put a border around and trim the paper, than it is to start with a standard size border and attempt to get the drawing within the border and well laid out. If it is not essential that the piece of paper is a standard size when finished, allow a little tolerance for trimming later. In some drawing offices, the method of filing drawings may dictate paper sizes which have to be adhered to.

PROPORTIONS

Commonly used scales are arrived at by halving. The simplest is half-size, which may also be described as 6 inches to 1 foot. Halving that again comes to quarter-size or 3 inches to 1 foot. Further steps are one-eighth, one-sixteenth, and so on. From the shop point of view, there are advantages in these proportions when dealing with feet and inches.

At one-eighth, 1½ inches on the ordinary bench rule represents 1 foot, and ⅛ inch represents 1 inch. Coming down to one-sixteenth, ¾ inch represents 1 foot, and 1/16 inch represents 1 inch. The user of the drawing can take a measurement off it without the need of a scale rule. This also applies to the draftsman, who can measure with a full-size rule if he does not have a scale rule of the right type.

Those scales, or others obtained by further halving, are the preferable ones for general work using feet and inches. There are occasions when other proportions give a drawing of more convenient size. There is the series of proportions: one-third, one-sixth, one-twelfth and further. This has the advantage of being related to the proportion of inches to feet. One-twelfth, or 1 inch to 1 foot, is a commonly used scale, but except for full inches it does not relate easily to a bench rule a scale rule is needed.

With suitable scale rules, many proportions are possible. It is unwise to choose uncommon ones if the drawing is to be used by others who may have a more limited collection of scale rules. One-third and one-sixth may be less common, but they have their uses. For instance, many of the drawings in this book started at one-sixth, or 2 inches to 1 foot, because that gave a suitable proportion for reduction to page size.

When quoting scale, make clear the proportion. On a drawing it is more useful to the user of the scales to say, for instance: "Scale: 1 inch to 1 foot," than "Scale: one-twelfth full-size." Give the scale in both ways if you wish. Do not say, "Scale 1 inch = 1 foot." This is mathematically impossible and is a wrong use of the equals sign.

DECIMAL SCALES

In architectural work and drawings connected with mapping, there is more use of decimal scales. This practice is likely to extend to other types of drawings with the coming of metrication. Scales are based on tenths. Suppose a drawing cannot be drawn half-size (0.5 inch to 1 foot). The next step down in the decimal system would be 0.2 inch to 1 foot or one-fifth, instead of one-quarter size. A further step would be to one-tenth full-size or 1/10 inch to 1 foot. If metric measurements are used, the same proportions would be 2mm to 1cm or 1mm to 1cm.

For large buildings or extensive pieces of land, the scale has to go much further down the series. It makes sense to use one-fiftieth, one-hundredth, or more. These greater proportions are often written "Scale-1:50," "Scale-1:100," or even "Scale-1:50,000," which is about 1¼ inch to one mile, and a size which gives a map of a reasonable area of land on a piece of paper easily handled. When kilometers are used for land measurement instead of miles, decimal proportions are the obvious choice.

OVERSIZE SCALES

If it is necessary to make a drawing larger than the item being detailed, choose a scale that is a simple multiple of the actual size. It would then be: "Scale-three times full-size," or whatever the proportion is. It is less common to use fractional enlargement. It would obviously be cumbersome to make the drawing 2½ times the actual size. An exception might be where the drawing is to appear in a book or magazine, and you want it to be full-size when it is printed. Printers can reduce a drawing of any size, but they prefer not to have to reproduce from the same size original. The draftsmen will prefer to draw bigger in most cases, and the reduction gives a better quality of line. The restrictions of reproduction and the quality of paper used may make too great a reduction undesirable, as fine detail may be lost or be too small for the reader to understand. A suitable proportion for a drawing that is to be produced in many publications is to a scale of 1½ times the final size.

SCALE RULES

The instruments used for measuring a drawing to scale are also commonly called scales. Where there is a risk of confusion

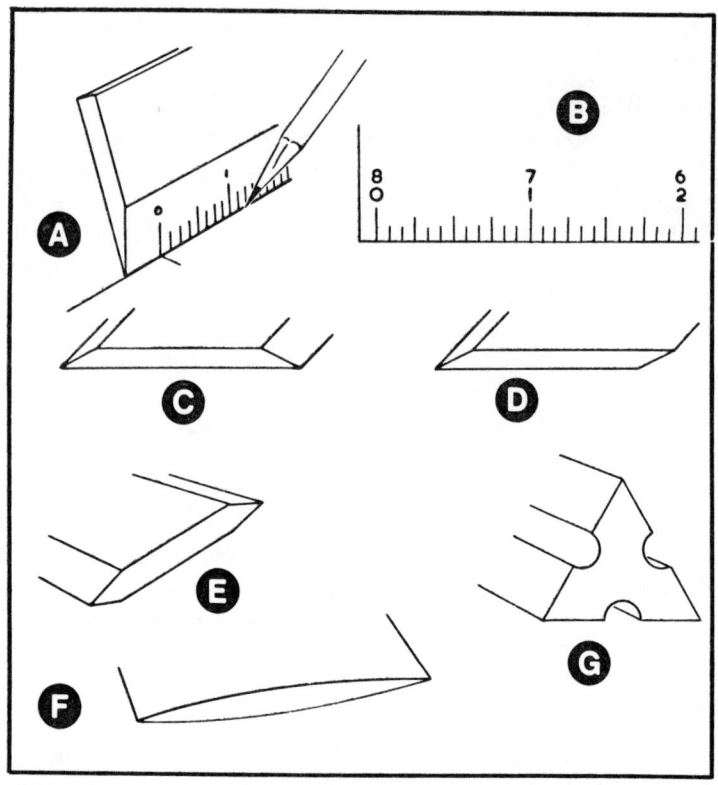

Fig. 4-2. A scale may be tilted to avoid errors (A). It may be calibrated both ways (B), and several sections are used (C to G).

between the use of the word meaning proportion and that meaning the measuring instrument, it is better to add "rule" in the latter case.

Scale rules are made in many ways. What is important is that the calibrations are clear and come close to the drawing. Most have thin edges so sighting is without difficulty, but if the edge is thick, *errors of parallax* due to sighting from one side or the other can be avoided by tilting the scale on edge (Fig. 4-2A). There are many scale rules which are marked with more than one scale on an edge. This reduces the number of scale rules that have to be obtained, but the arrangement can lead to confusion. For instance, an edge may be marked to a scale of ¾ inch to 1 foot from the left-hand end and 1½ inches to 1 foot from the right-hand end. The user needs to watch that he does not pick up a graduation the opposite way to what he intended, as the representation of 1 inch one way is 2

48

inches the other way. That sort of scale is suitable for occasional use or when there is a need to limit the number of instruments, as when traveling; otherwise, it is better to have only one set of graduations on an edge.

The scale used should be prominently indicated, often by just a figure—1½ is then known to mean a scale of 1½ inches to 1 foot. It is convenient to have some of the figures—feet, for instance—numbered both ways, so the edge can be used either way without mental arithmetic (Fig. 4-2B).

A scale may be flat with beveled edges and different scales on opposite edges (Fig. 4-2C). Another section has the two bevels on opposite sides (Fig. 4-2D). This reduces the risk of picking up a scale and starting to use the wrong edge, as the instrument would have to be turned over. Scales are also made with four bevels (Fig. 4-2E). They get four scales on the same instrument, but they have to be tilted to bring the measuring edge close to the paper. A similar effect can be obtained by curving the cross section (Fig. 4-2F), which is a little easier to handle. A scale with a triangular section (Fig. 4-2G) has six graduated edges independent of each other. This is convenient, except for the risk of changing to the wrong scale inadvertently.

GRAPHIC SCALES

Put dimension lines on a drawing, sufficient to give the user all the sizes he is likely to need. Then he does not have to use a scale rule. In some cases it is convenient to put a graphic scale on the drawing, either in place of or to supplement dimensions. This can be a simple line with measurements to scale along it (Fig. 4-3A). A common arrangement where the representation is of feet and inches is to have the zero mark scale 1 foot from the end; then whole feet extend one way. The first foot is divided into inches and fractions as well, if the scale is large enough to permit them (Fig. 4-3B).

Measurements are transferred to or from a graphic scale with dividers (Fig. 4-3C), taking in feet to the right and adding inches to the left of zero, if it is that type of scale.

Graphic scales are laid out in many ways, and on some old plans and maps they are quite decorative. The alternative sections may be filled in for emphasis (Fig. 4-3D). On a modern map there may be a scale showing miles alongside another showing kilometers (Fig. 4-3E). For comparison, 8 kilometers are almost exactly 5 miles, so 1 kilometer is approximately five-eighths of a mile.

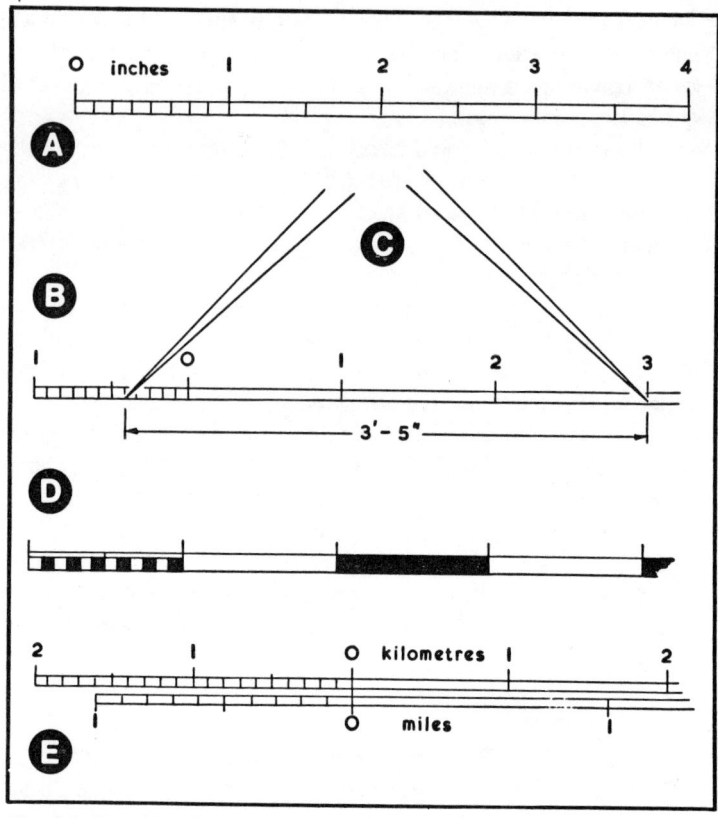

Fig. 4-3. A scale may be marked from the end (A), but it is more usual to have inches one way and feet the other (B). Parts may be filled in (D). Two scales can be related (E).

FINE SCALE

The best scale rules are "engine-divided," which is a method of producing graduations of extreme accuracy. It is possible to arrange divisions very finely, but there is a lower limit beyond which the human eye has difficulty in selecting the exact mark that is needed. There is also the problem of the material. Wood or plastic will not take very close markings without the material breaking down. Metal is more amenable. There are steel rules with graduations at 1/100-inch spacing, but anyone who has used one of these knows how difficult it is to be certain that you have the mark you want, even by using a lens.

A way to show scale graduations with less risk of confusion is to draw an open scale. It is possible to set dividers to an exact number of 1/100 inch, 1/64 inch, or other and finer differences with certainty.

As an example, suppose a distance has to be divided into four. Draw parallel lines with four equal spaces between them (Fig. 4-4A). Mark the distance, in this case 1 inch, with lines drawn squarely across them (Fig. 4-4B). Draw another line diagonally from the corners of the markings (Fig. 4-4C). Any number of whole inches can be marked further along.

It is the crossings of the diagonal line which give the measurements. At the top line it is at zero, at the next line it crosses at ¼ inch from the zero line, then at ½ inch at the next line, and ¾ inch before striking the bottom line at a complete 1 inch (Fig. 4-4D).

This can be done with any number of divisions. If there are eight spaces, the diagonal line moves on ⅛ inch as you come down the lines. If there are 10 spaces, the differences are 1/10 inch.

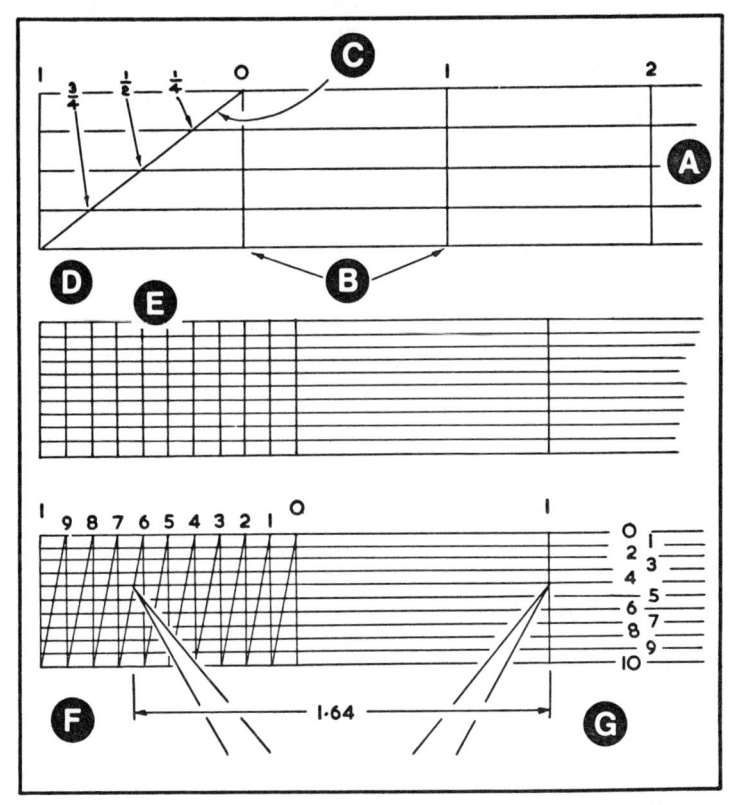

Fig. 4-4. Finer measurements are possible if the scale is divided by diagonals over several lines.

51

So far I have looked at divisions that could have been laid out directly on a straight line, without confusion, but suppose the differences are to be 1/100 inch. Draw 11 equally spaced parallel lines, and across them draw lines squarely to divide 1 inch into 10 (Fig. 4-4E). Draw diagonals in each of the 1/10-inch divisions (Fig. 4-4F). As the diagonal in each space progresses downwards, it adds 1/100 inch to the length compared with the crossing immediately above it. Distances are taken off with dividers. You have to add the number of whole inches, the number of 1/10 inches, and choose the line to measure on which gives you the number of 1/100 inches additionally that you want (Fig. 4-4G).

VERNIERS

Fine measurements in engineering may be taken with a *micrometer* or a *vernier gauge*. A draftsman does not have any instrument using the micrometer principle, but he may find verniers on some drafting machines. A vernier provides a means of adjusting to a finer measurement than can be done by direct reading from one scale. A vernier scale is a second scale moving alongside the main scale. It is not always straight. One may be found on the adjustment for setting the head carrying scales on a drafting machine (Fig. 4-5A). This allows the head to be adjusted with an accuracy of 1/10°.

Verniers commonly read to one-tenth of one-tenth, resulting in steps of one-hundredths, but there may be one-eighth of one-eighth or another combination. Engineering verniers may read to thousandths of an inch, but drafting vernier adjustments are not as fine. Comparable metric verniers may read to fiftieths of a millimeter.

As an example, a vernier arrangement to read to 1/100 of an inch has the main scale with the inches divided into tenths (Fig. 4-5B). Alongside it slides a scale with 9/10 inch divided into 10 (Fig. 4-5C), so each of these divisions is .09 inch instead of the .10 inch of the main divisions, a difference of .01 inch.

When the zero or whole inch divisions of the two scales coincide, the measurement is zero or a whole number of inches with no fractions. If the vernier scale is moved along so 1 on it coincides with a main division, the movement is .01 inch. If it is moved along so 2 on it coincides with a main division, the movement is .02 inch (Fig. 4-5D) and so on to .09 inch.

To take a reading, the number of whole inches are noted, then the number of complete tenths behind 0 on the vernier, and then

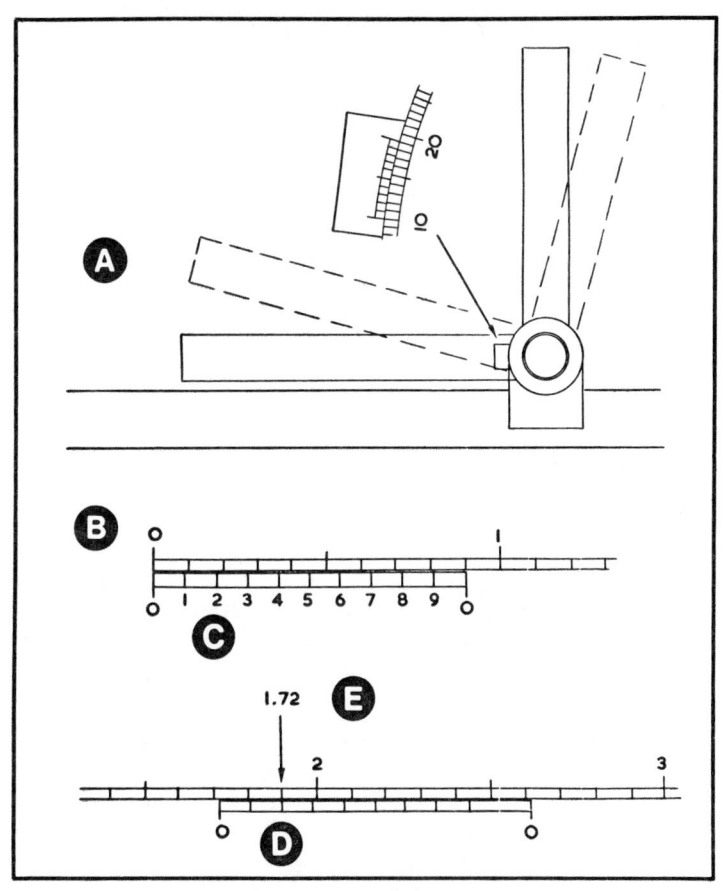

Fig. 4-5. A vernier permits finer measurements by using an auxiliary scale alongside the main one.

the position of the first line on the vernier to exactly match a line on the main scale. In the example shown (Fig. 4-5E) this gives a reading of:

One whole inch	1.00
Seven tenths on main scale	0.70
Two divisions on vernier scale	0.02
Total	1.72

On a curved scale for adjusting angles, the main protractor scale gives whole degrees. Then the vernier allows readings of 1/10°.

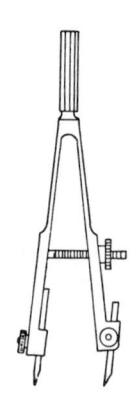

Chapter 5
Lines

Lines make a drawing—any drawing. On a practical or mechanical drawing they have a particular significance. If not the language of the drawing, they might be considered its alphabet. There are recognized types of lines with their own meanings. They are almost universal, so a mechanical drawing made anywhere should make its meaning clear. It is unwise to depart from the accepted lines, except in special circumstances which should be obvious.

Solid lines are heavy or light, which in pencil means thick or thin. Main outlines are heavy lines (Fig. 5-1A). Other solid lines which may indicate dimensions, extensions, projections, or other things that are not parts of the outline are light (Fig. 5-1B). When making a drawing, it is usually necessary to set out lines first that may be longer than finally necessary (Fig. 5-2A). Do that lightly, with just only enough pressure for you to see the lines. Then "line in" heavily where the main outlines come (Fig. 5-2B), thicken any projection or other lines that have to remain, and erase any unwanted lines.

HIDDEN LINES

When a draftsman looks at an object or visualizes one that has to be drawn, he "sees" *hidden lines* that he will have to show on his drawing. There may be a hole through or a step or other hard line on the far side, invisible from the direction the view is to be drawn. All of these are shown on dotted lines, which is the common

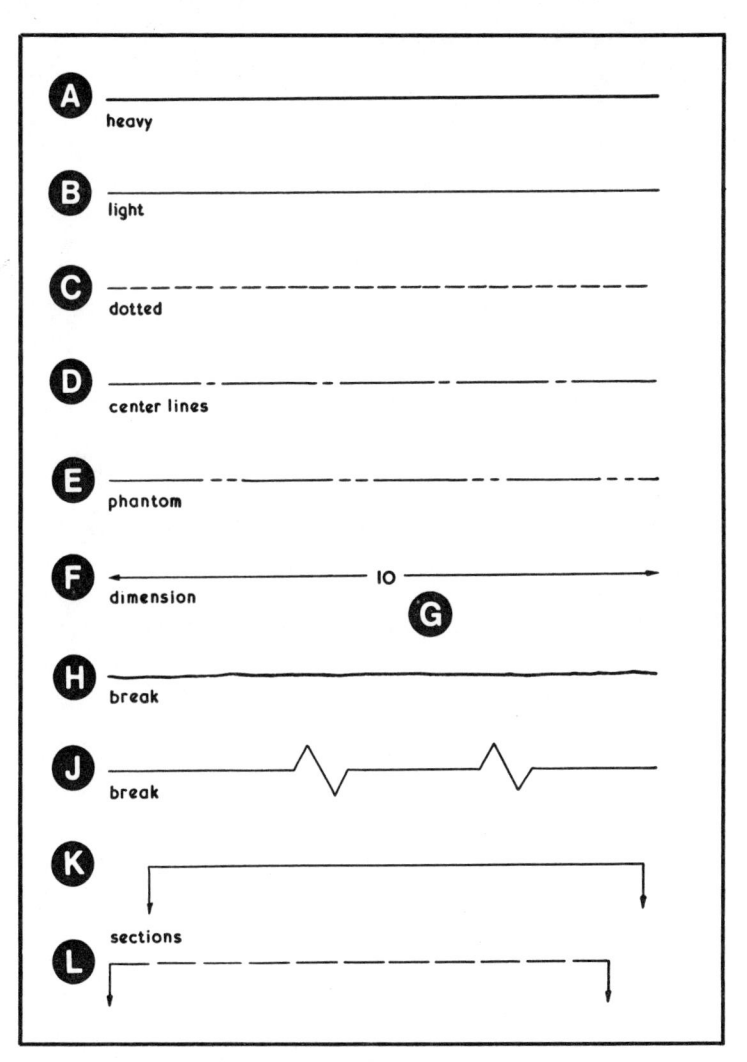

Fig. 5-1. Typical lines used on mechanical drawings.

description, but in fact the lines are made up of a series of short dashes (Fig. 5-1C). The lengths of dashes and their spacings should be consistent on a particular sheet. It is usual to regulate the lengths and spacings as you go, but aim to keep them even, with the spaces less than the lengths of the dashes, which should be short (usually less than inch) so they are not confused with other lines. When a view is drawn with hidden parts indicated (Fig. 5-2C), use light dotted lines within the main outlines (Fig 5-2D).

55

CENTER LINES

It is often necessary to indicate a *center line*. It may be just the center of a hole or other part of the whole, or it may show that the whole thing is symmetrical. If one side of the drawing is a mirror image of the other side, only one side may be drawn. Center lines are a series of long and short dashes (Fig. 5-1D). The short dashes and the spaces should be about the same as in dotted lines, but the long dashes may have to be regulated according to the size of the view they are used with. Normally they would be between ¾ inch and 1½ inches long.

Center lines are allowed to extend beyond the view to which they refer (Fig. 5-2E). This draws attention to them. If only one side of the drawing is made, the center line is at one edge, but it still extends. If it is necessary to draw attention to a center line, it can be marked with a combined C and L (Fig. 5-2F).

It is sometimes helpful to include *phantom* or *reference lines*. They may show the position of another part, that part in a different position (Fig. 5-2G), or they may be needed as a reference or guide to assembly. Such lines are like center lines, but with two short dashes in each gap between long dashes (Fig. 5-1E). If center lines or phantom lines are common to two views, they may continue through without a break.

DIMENSION LINES

Dimensions or measurements are shown with light lines having arrowheads at their ends (Fig. 5-1F). The use of an arrowhead is best, and it is the practice to be adopted unless the particular office or industry does otherwise. None of the other terminals to dimension lines are so easy to recognize. There may be closed semicircles or complete circles if lines meet (Fig. 5-3A), or just a diagonal line at the termination (Fig. 5-3B).

Arrowheads should be neatly made. The arrowheads and the lettering on a drawing can do much to make or mar the general appearance of the drawing. Arrowheads should never be open (Fig. 5-3C) and certainly not wide (Fig. 5-3D), whether open or solid. An experienced draftsman may make his arrowheads freehand in a regular standard shape. A beginner may find it worthwhile drawing a few oversize heads with instruments so as to get a satisfactory shape. He can then decide which of the variations suits his style.

A reasonable proportion has the length of the arrowhead between two and three times its width, so use an outline in these proportions for trial drawings (Fig. 5-3E). The simplest head is a

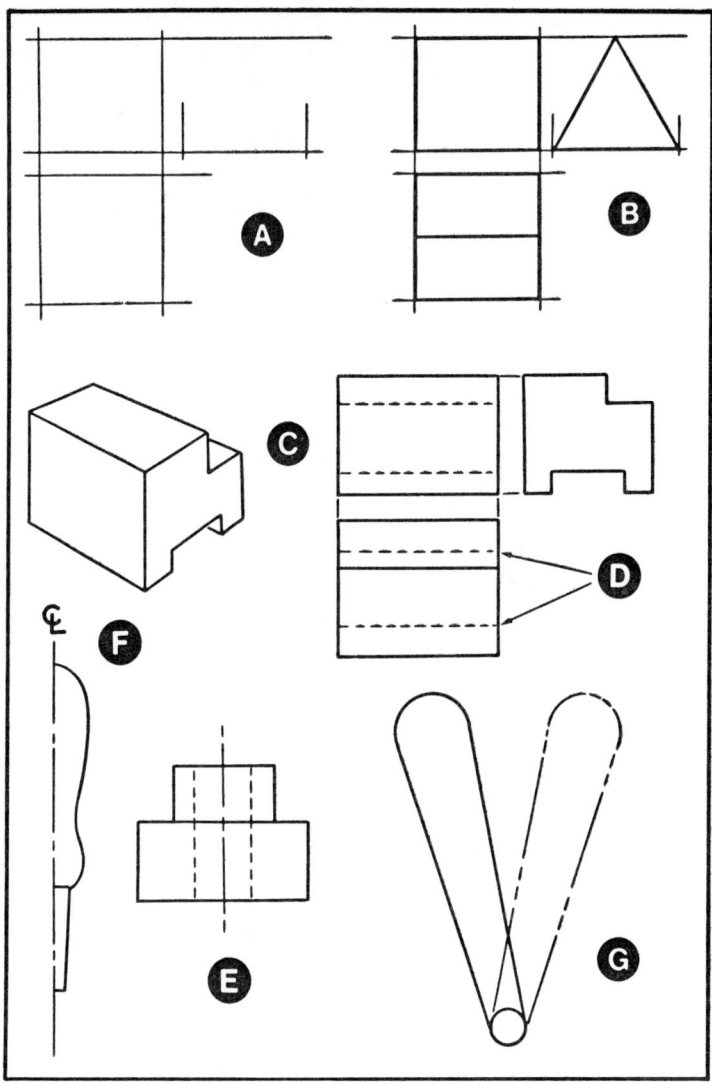

Fig. 5-2. Examples of the use of various lines.

straight-sided triangle (Fig. 5-3F). Instead of the straight end, it could be more arrowlike (Fig. 5-3G). Some draftsmen prefer hollow sides (Fig. 5-3H). Sometimes they cut in the end acutely, so there is little to fill in (Fig. 5-3J).

It is important that the arrowhead finishes with a fine point, which is easy to achieve with a pencil, but freehand inking may entail scraping the sides of heads that have finished too wide. It

57

helps to stop the dimension line fractionally short of the line, so it is the head only which completes the distance. Dimension lines should reach projection lines exactly opposite the points to which they refer (Fig. 5-3K).

If only part of a view is shown, possibly because the other part is similar or obvious from another drawing, it may be necessary to provide a dimension line that has one fixed end. The other end of it does not reach the point it refers to because that part is not drawn. In that case there can be double arrowheads (Fig. 5-3L) to indicate that the actual reference point is further.

Figures that indicate the measurements are usually put in breaks in the dimension lines (Fig. 5-1G). This is preferable, as it is clearer, to the practice in some offices of putting the figure alongside an unbroken line.

If it is necessary to add a note with a reference arrow, let this line be not parallel with the main drawing or dimension lines. Then it cannot be confused. The leader line looks neatest if it comes to the note with a parallel or horizontal line (Fig. 5-3M).

BREAK LINES

Sometimes it is necessary to show a break in a view, possibly because the object is long, but the only detail work comes at the ends. A short break is a freehand line, but if the break is longer, it is more common to use a light line with zigzag markings at intervals along it (Fig. 5-1J). How frequent these symbols are used depends on the drawing, but there should be enough of them for the line not to be confused with anything else.

The freehand short break may be used for a drawing of an object in any material, but there are other ways of indicating breaks that are more appropriate in some circumstances. For wood broken across the grain, there can be ragged fibers (Fig. 5-4A).

The freehand ragged line is usual for a break across flat or rectangular metal (Fig. 5-4B), but there is a conventional way of showing a break across round rod or tube. This does not lend itself to easy drawing with instruments, but for a large break at least part of the curves can be drawn with compasses. A rod is shown with a sort of figure eight break, and the opposite part is shown in a matching pattern (Fig. 5-4C). If the break is in a tube, matching lines are drawn for the inside (Fig. 5-4D).

SECTION LINES

There are many places where most information can be conveyed by showing a section of the object. It is common to show

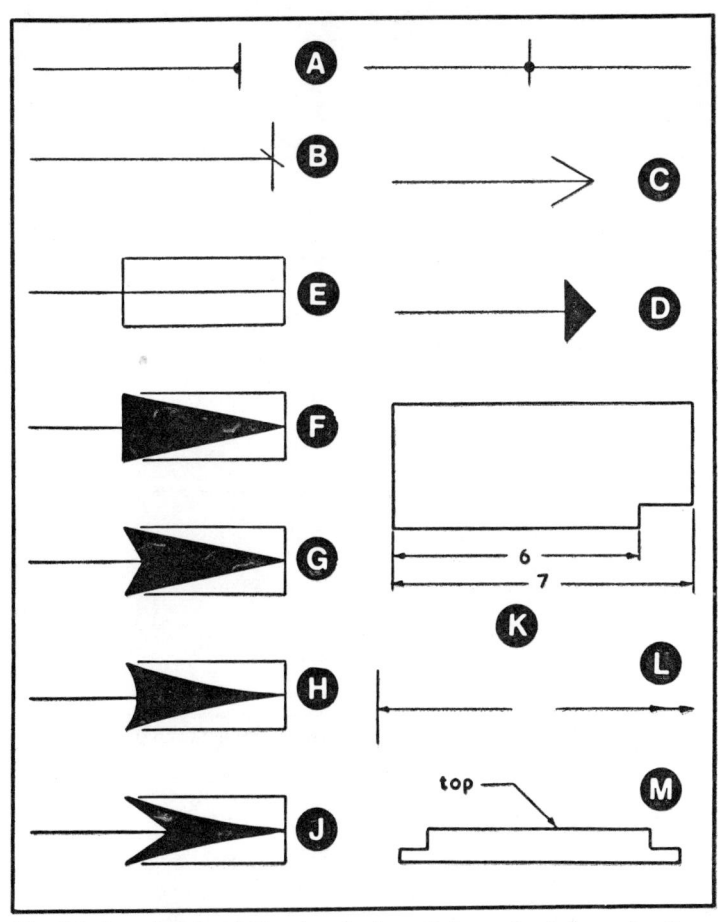

Fig. 5-3. Arrowheads or other terminals mark the ends of dimension lines and should be made neatly.

where the cut comes in one view and project the section from it. A section line may be a solid light line with arrowheads angled at the ends (Fig. 5-1K), or it can be drawn with a series of long dashes (Fig. 5-1L). Make these dashes at least three times as long as those used for hidden lines, so there can be no confusion.

Opposite the part drawn with section lines on it, draw the section. Where parts are uncut by the imagined section, they are drawn as if it is a normal solid view, but where the cut comes the area enclosed is section-lined. This is an arrangement of close light lines (Fig. 5-5A). They should be drawn usually at 45° to the main lines of the cut, or if it is round, at 45° to the border of the drawing.

59

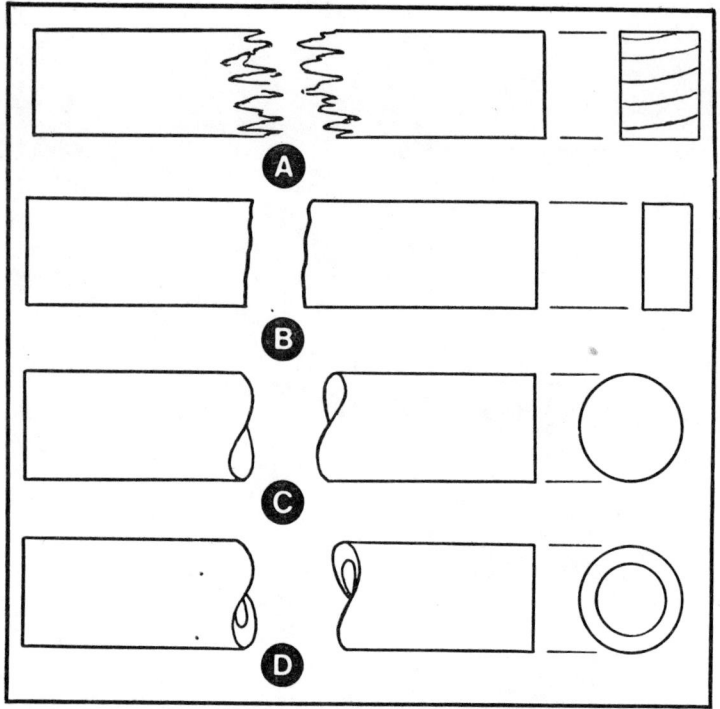

Fig. 5-4. The way breaks are shown depends on the material or its form.

It is important that the section lines are evenly spaced. There are mechanical aids for doing this which are basically straightedges that can be moved by a regular amount in steps in a holder. They are only justified where a considerable amount of sectioning has to be done regularly. Some draftsmen are able to get a regular spacing by eye, but otherwise it is helpful to draw a faint line at right angles to the direction of the section lines, and step off on this enough points with dividers (Fig. 5-5B).

If the article being sectioned is wood, it may be more appropriate to sketch in grain, either along it (Fig. 5-5C), if appropriate, or showing annual rings (Fig. 5-5D). For other materials the diagonal section lines are more usual, although there are ways of indicating concrete and other building materials which are described later.

One thing that beginners may find difficult is the exact matching of curved and straight lines, particularly if a curve has to blend into a straight line. It is worthwhile remembering that a curve drawn with a compass cannot be varied. If you try to do so, the departure from a true part of a circle will be obvious. This means that in final lining-in, if not before, draw curves before straight lines that have to couple with them. It is then possible to move a straight line marginally to make a neat joint with the curve. This is particularly important when finishing a drawing with ink.

The center on which to put the point of the compass should be found by measuring, usually with crossing lines parallel with edges, which usually means at right angles (Fig. 5-6A). If the lines are to be used for dimensions or other purposes, extend them appropriately. This applies whether you are drawing a complete circle in isolation, as when drawing holes, or when the curve has to meet straight lines (Fig. 5-6B). If all you are doing is locating the center for a rounded corner, you can measure one way and get the measurement the other way from the setting of the compass (Fig. 5-6C). You can mark the exact position where the curve meets the straight line by projecting from the center perpendicularly from the edges, which will be square in a normal corner (Fig. 5-6D), but will not be if the corner is acute or obtuse (Fig. 5-6E).

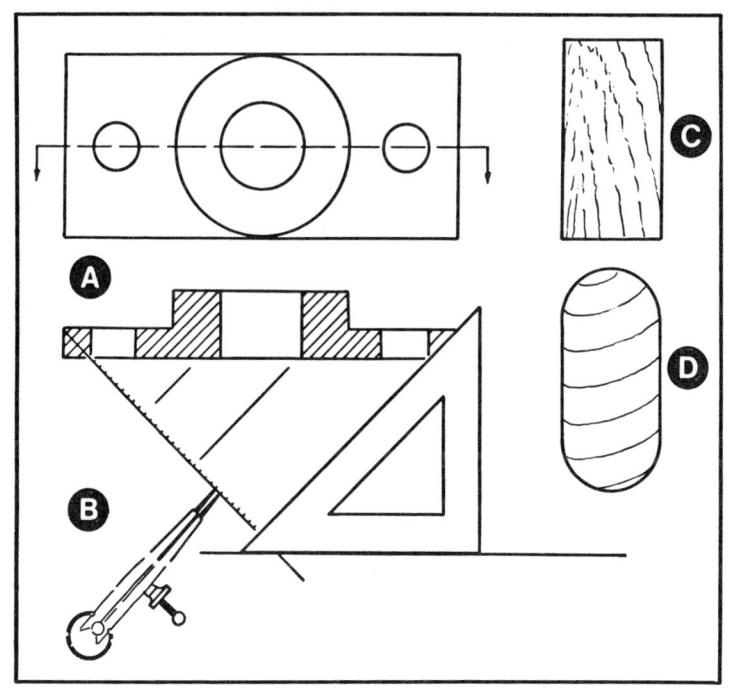

Fig. 5-5. Sectioned surfaces are shown by cross-hatching, which should be evenly spaced, or by grain marks for wood.

61

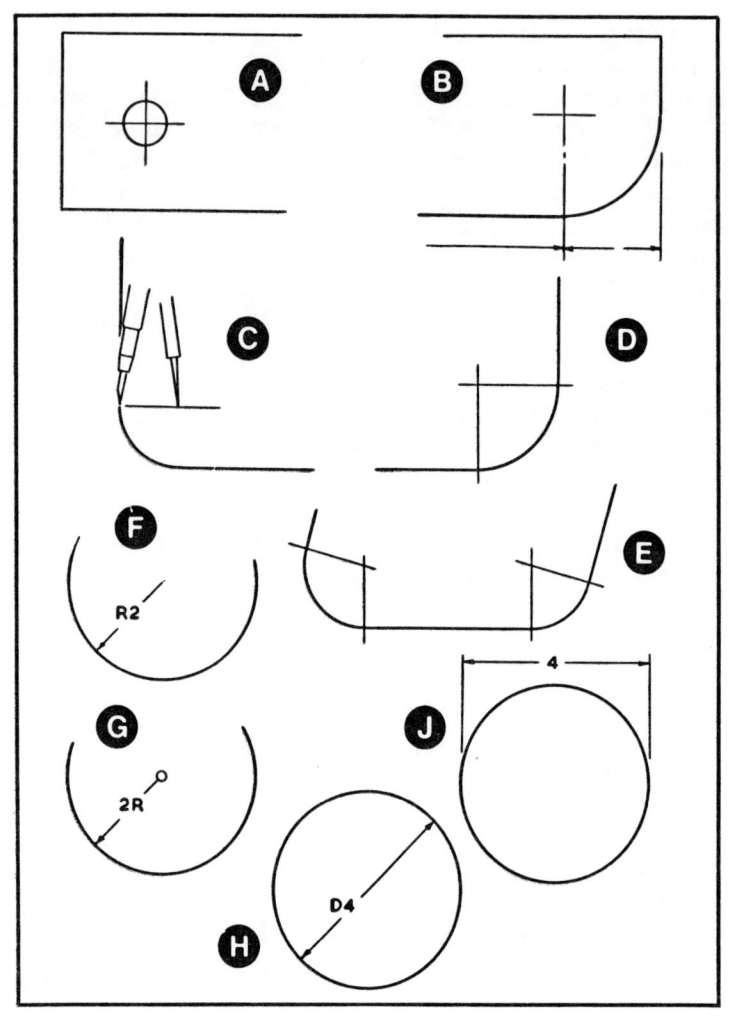

Fig. 5-6. Where curves meet straight lines, their centers and meeting points should be located. Circle dimensions are indicated inside or outside the circumferences.

When a radius has to be shown with a dimension line, put the arrowhead at the circumference, but either let the line finish unmarked at the center (Fig. 5-6F) or have a tiny circle there (Fig. 5-6G). To eliminate confusion about the meaning of the dimension, a better R may be put before or after it. If it is the diameter indicated within the circle, there may be a letter D (Fig. 5-6H), but in many cases it is better to have the dimension outside (Fig. 5-6J).

62

Chapter 6
Handling Instruments

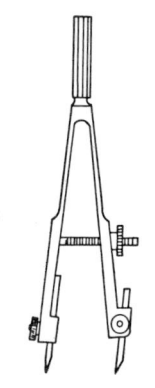

The handling of instruments and the preparation of drawings with them need not be difficult. A beginner should be able to produce results almost right away. There are ways of getting more accurate results more conveniently. Using instruments correctly will reduce the risk of errors. Setting out drawings can be arranged more expeditiously. This chapter offers hints on doing things in the expert manner.

In ordinary life many people have little use for any part of their hands except the thumb and first finger of the right hand. There is a tendency to want to bring every task to these two digits. The left hand and the other fingers of the right hand might as well not be there. In mechanical drawing it helps to be aware of all the fingers and use them frequently. A pencil or pen may be held with the first finger and thumb, but the other fingers may be spread to steady the straightedge or hold it close to the paper. If a T square is brought up to the edge of a board, the left hand holding it may also have to steady a triangle resting against it. This can only be done by spreading the fingers and using them all. Like a beginning touch typist, a beginning draftsman may find that his little fingers are quite weak, but practice will strengthen them. With a drafting machine the need to hold a T square is avoided, but triangles and straightedges have to be held against the machine edge. It is often necessary to spread pressure as much as possible to prevent slipping or lifting when dealing with a French curve or other independent shape. Use the entire hand, with fingers spread.

Having decided on a suitable scale and the relative positions of the views to be included, the first lines have to be made on the paper. It is best to get some main lines down, even if you have to make them longer than necessary at first. Then mark lengths later. It is usually bad policy to try to limit the first lines to final lengths. Do not be afraid to draw light lines to what you know will be too great lengths. These are construction lines, the actual lines of the drawing will come later.

MAIN LINES

With the usual three views, the main horizontal construction line can be the base of the front and side elevations. Locate it far enough down to get everything in (Fig. 6-1A). The first vertical line will then be the one that forms the matching edges of the front elevation and the plan (Fig. 6-1B). Lower the T square before you draw this line. Never try to erect a perpendicular directly from a straightedge on the line. With the T square below, you may be able to draw the whole vertical line. The first line will cross the baseline and can be extended (Fig. 6-1C). Use a drafting machine in a similar way.

You now have two lines crossing at 90°. Measurements can be taken along them. If you want lines to be parallel, measurements should be taken at right angles. Either measure directly or use dividers to transfer measurements. Use the T square, triangle, or drafting machine as a guide for the pencil drawing other lines (Fig. 6-1D). Even if the line finally needed will not come to one of the baselines, you can still measure along it (Fig. 6-1E). You will then know that you have measured at right angles to the line, and there cannot be errors due to a slight lack of squareness.

If you have to get lines parallel, but they are not square to either baseline, you can use the triangles if they have to be 30°, 45°, or 60°, as they would when doing isometric or oblique projections. Move the triangle along, and get all the lines to the same angle along its edge (Fig. 6-2A).

Sometimes lines have to be parallel; yet you have no datum on which to work. Two arcs with a compass may be used, as near the ends of the line as is reasonable. A straightedge touching the arcs will be parallel to the first line (Fig. 6-2B). That would be a rather cumbersome method if many lines had to be drawn parallel. Instead, you can use two triangles or one triangle and a straightedge. Position one edge of a triangle on the line to which the others have to be parallel, and bring the straightedge or other triangle up to one

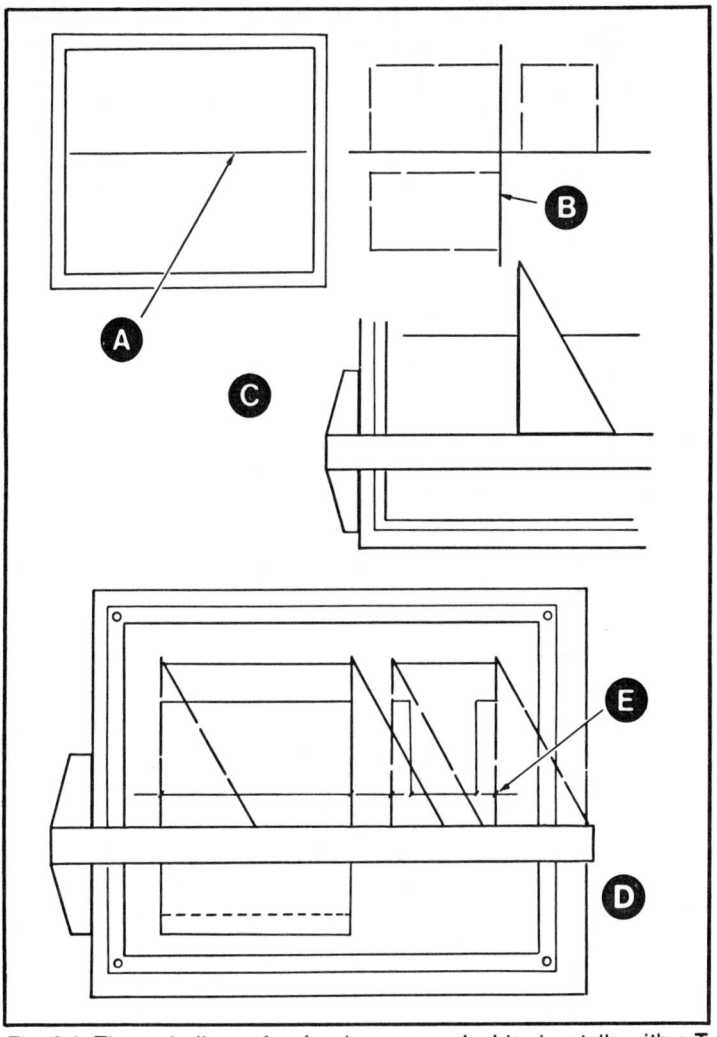

Fig. 6-1. The main lines of a drawing are marked horizontally with a T square and vertically with a triangle on it, if a drafting machine is not used.

of its other edges. Hold this firmly while sliding the ruling triangle on it. This is an occasion when you will value being able to use all fingers, with some fingers holding and some controlling the sliding. Move the triangle along so you can draw lines along it in the other marked positions (Fig. 6-2C). A little experience will show whether to use a square edge or one of the sloping ones. A preliminary experiment will show how to get the sliding triangle to all the required positions.

65

The blade of the T square can be the straightedge in this work. If lines have to be drawn perpendicular to each other but not square with your horizontal or vertical baselines, you can turn a triangle over on the straightedge without moving it. The 30° angle will then be square with the 60° angle (Fig. 6-2D), or the 45° triangle the other way will produce a perpendicular line (Fig. 6-2E). If you slide the triangle with its long edge on the straightedge, the other two edges will already be in position to draw lines perpendicular to each other (Fig. 6-2F).

BORDERS

It is normal to give the drawing a border. Sometimes the drawing is made first, the border drawn afterwards and the paper cut outside it. In most offices it is more usual to have standard sizes of paper and keep the drawings to these sizes, so borders are drawn first. Too much paper outside the border is wasteful and does not look right. For a small drawing, the margin might be ½ inch. This is increased for larger drawings, but does not usually exceed 1 inch, whatever the total size.

Horizontal border lines can be drawn with a T square against the vertical edge of the board or with a drafting machine in a similar way. The side border lines must be truly vertical; otherwise, any discrepancy will be obvious and spoil the look of the final drawing. If the drawing is being prepared with the use of the T-square, vertical lines should be drawn in the normal position, even if that means extending the lines with the T square or other straightedge. You are then certain that border corners are square. It is bad practice to move the T square to a top or bottom edge of the drawing board to draw the vertical border lines or any other vertical lines. There does not have to be much error in the squareness of the board for the lines to finish untrue.

USING COMPASSES

The distance from the pivot to the points of both legs of a pair of dividers should be the same. When stepping off large distances, any variation may not be apparent or matter. When the points are to be set fairly close to each other, the amount of their extension should be the same. Nearly all dividers are provided with adjustment on one or both legs. Set them so the points are level when close together. If the dividers are without adjustment, a longer point must be sharpened on an oilstone until it is the same length as the other.

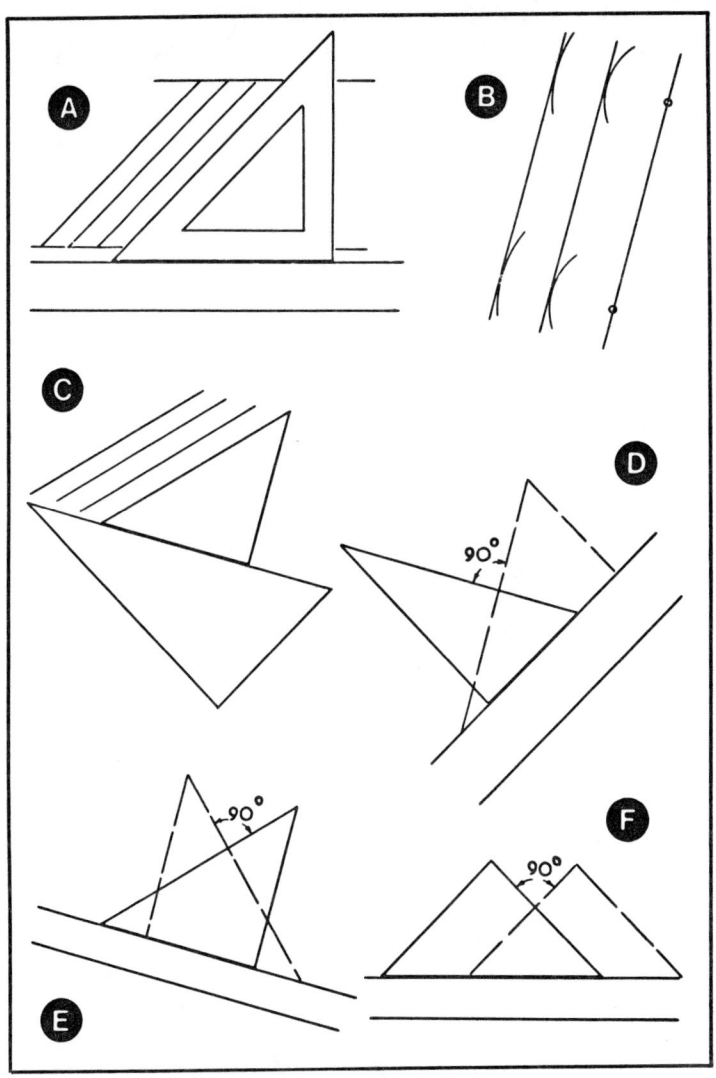

Fig. 6-2. Parallel lines can be marked with a triangle or by drawing arcs. Square lines can be drawn by choosing a second position for a triangle.

With compasses, the point on which the instrument rotates should extend slightly further to allow for penetrating the paper. This is particularly important with spring bows used for drawing small circles. With a *pencil compass*, which has to be sharpened frequently, slide the lead forward as necessary at each sharpening. With an *ink compass*, set the point slightly longer than the pen, and

67

no further adjustment should be necessary. If it is an instrument that will take pencil, ink, and divider points, set the center point to suit the pen. Adjust the pencil and divider points to match.

Most sets of instruments containing compasses have T square needle points. It is useful to be able to replace a point, but sharpening the ordinary tapered point is quite simple. Have it held by the compass leg and rub it on an oilstone, rotating it as well as moving it along the stone. Another holder is a *pricker handle*, provided with some sets to hold a point and intended for pricking through when distances have to be transferred from one drawing to another, or as a precise alternative to a pencil for marking distances on paper from the edge of a scale.

Using Dividers or Arcs

Dividers can be held and adjusted in one hand, if they are of the friction joint type. The thumb goes in front; then the first finger is outside one leg and the little finger outside the other leg, leaving the other two fingers between the legs (Fig. 6-3). For smaller settings, only one finger goes between the legs. By holding in this way, it is possible to move the legs closer or further apart by finger movement.

When dividers are to be used for stepping off a number of equal distances along a line, they are "walked" along the line with one hand on the top, using the first finger and thumb. Move the legs

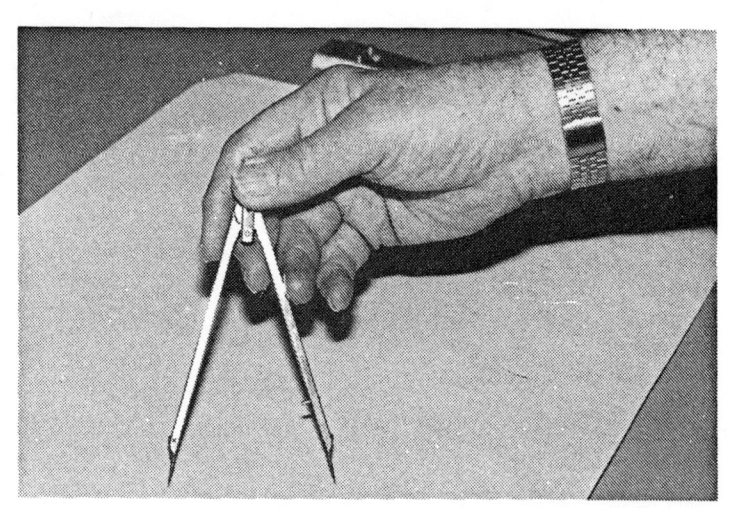

Fig. 6-3. If dividers are held with the legs between fingers, they can be opened or closed with one hand.

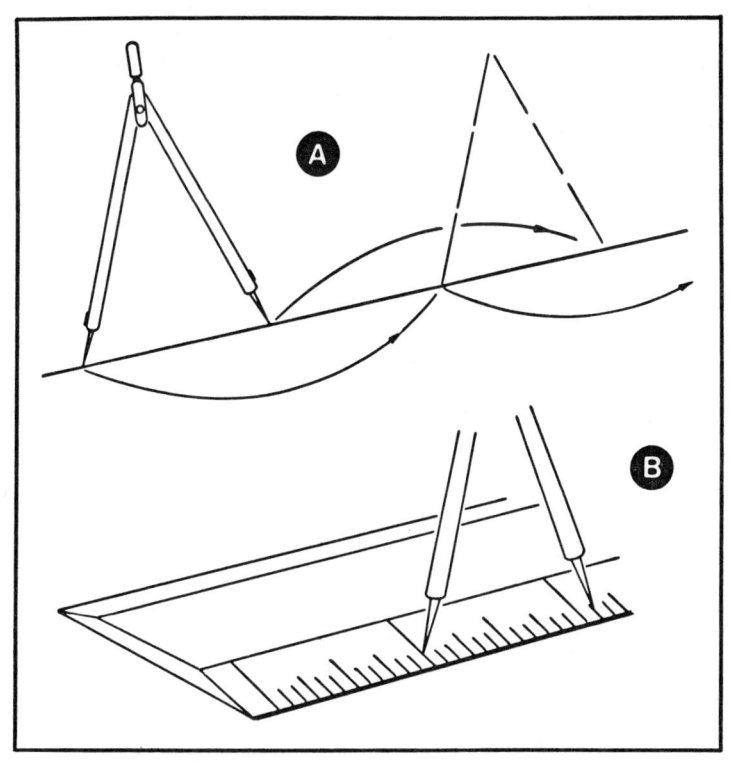

Fig. 6-4. Dividers can walk along a line (A) and be set on a scale (B).

through semicircles alternately in opposite directions (Fig. 6-4A). It is then possible to go any distance without the finger and thumb having to readjust, as they would if you turned the dividers the same way at each step.

If dividers are to be set to a distance from a scale, put one leg in a graduation indicating a whole inch or whatever the scale is. Adjust the other leg from that (Fig. 6-4B) There is no need to always put the point in the end mark, providing you allow for the distance you want. If it is a scale or rule that starts graduations from its end, instead of being set in a short distance, trying to set from the end may result in errors.

Drawing Circles or Arcs

Compasses are fairly easy to use for drawing circles or arcs. Whether friction joint or spring bows, always use a compass for normal circles with one hand only on the knurled top, so you turn the compass between finger and thumb (Fig. 6-5). Rotating through a complete circle is easy. See that the needle point penetrates the paper enough to avoid slipping, but there is no need to thrust it in excessively. While turning the compass, try to put slightly more pressure on the center than on the pencil or pen tracing the circumference. Practice drawing very small circles; those made with the compass moderately open are easy. If a compass is used with an extension, you may have to use your other hand, but the danger is putting on pressure inwards or outwards enough for the pencil or pen to vary slightly on its course. With a beam compass you must use two hands, but that is a more rigid instrument. You are unlikely to affect the curve traced.

Lubricating Bow Instruments

Bow instruments should continue to function without trouble almost indefinitely. Appearance can be improved by an occasional rub with metal polish. If any stiffness develops, the merest trace of lubricating oil can be applied with a divider point or a pointed stick. Obviously, you do not want lubricating oil finding its way on to a drawing. Another lubricant is pencil lead, which is mainly graphite. Rub the end of a pencil on a stiff screw thread. Then there will be none of the mess that comes with oil.

Friction Joint Compasses and Dividers

Friction joint compasses and dividers have the joint arranged in several ways, depending on the marker and the quality of the

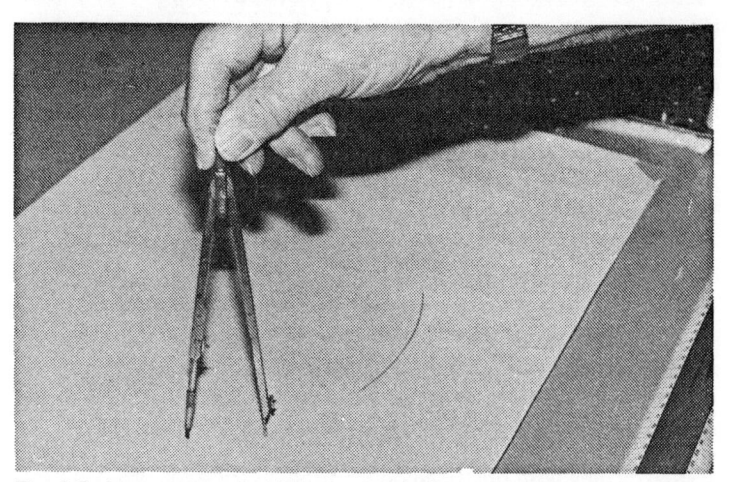

Fig. 6-5. A compass is held by the knurled handle so it can be turned between finger and thumb to draw a circle.

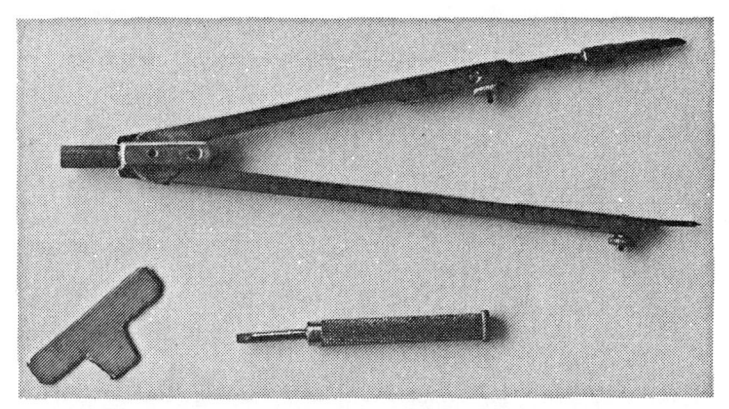

Fig. 6-6. The screws on instruments may be turned by a flat T-shaped driver, or there may be a round one with a hollow handle to accommodate spare compass points.

instrument. In some older instruments, particularly those used more for navigation than drawing, the tops of the two legs are cut with several tongues to fit into each other. A central screw binds the joint. Adjustment is minimal. Once this type of joint has become loose, it probably can not be tightened satisfactorily.

Most compasses and dividers have a handle extension above the joint, and this is on a U-shaped piece that goes over the joint between the legs. A straightening device is incorporated so the handle remains upright whatever the setting. A screw through the U-shaped piece is there to allow the adjusting of the friction between the legs. Tightening should be rarely needed and should then only be slight. Any screwdriver may be used, but there is usually one in a set of instruments, although what it is may not be obvious. It is often a flat piece of steel, shaped like a letter T—the central piece being the part to go in the screw slot (Fig. 6-6).

USING ANGULAR INSTRUMENTS

Protractors and adjustable triangles are basically simple, but a beginner can make mistakes due to unfamiliarity with the measurement of angles. Remember always that there are 360° in a full circle. If a straightedge is pivoted at one end and moved around completely, it will have traveled 360°. If it stops one quarter of the way around, it will have traveled 90°, which is the familiar angle you get when you draw one line perpendicular to another. Remember that and compare any angle you have to draw with 90°, and you are less likely to make a mistake. If the pivoted straightedge turns half a circle, it will have moved through 180°. Its two positions will make a straight line. Protractors are usually calibrated both ways. A semicircular instrument goes from 0° to 180° both ways (Fig. 6-7A). A circular protractor usually has a point marked 0 on the circumference and is calibrated to 360° in both directions (Fig. 6-7B). Check the system to avoid confusion.

Always locate a protractor with its center mark over what will be the meeting point of the lines forming the angle (Fig. 6-7C), and have the 0 mark on the line already there (Fig. 6-7D). Use the numbers that increase from that point, and you should get the correct result without difficulty (Fig. 6-7E). As a check, it is useful to note the complementary angle (the calibrations coming the other way) and see that you have got it right. For instance, with a semicircular protractor and the need to mark 37°, the complementary angle is 180° minus 37°, or 143° (Fig. 6-7F).

The mark against the protractor can be just a pencil dot, but if there are already many marks on the drawing, that might be

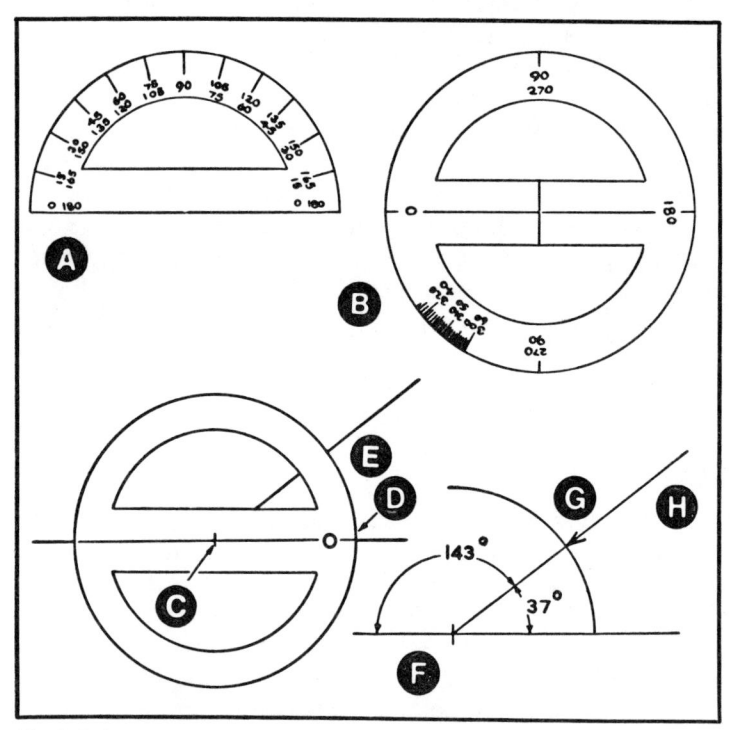

Fig. 6-7. A protractor is marked both ways, and care is needed to locate it correctly to get the intended angular marking.

72

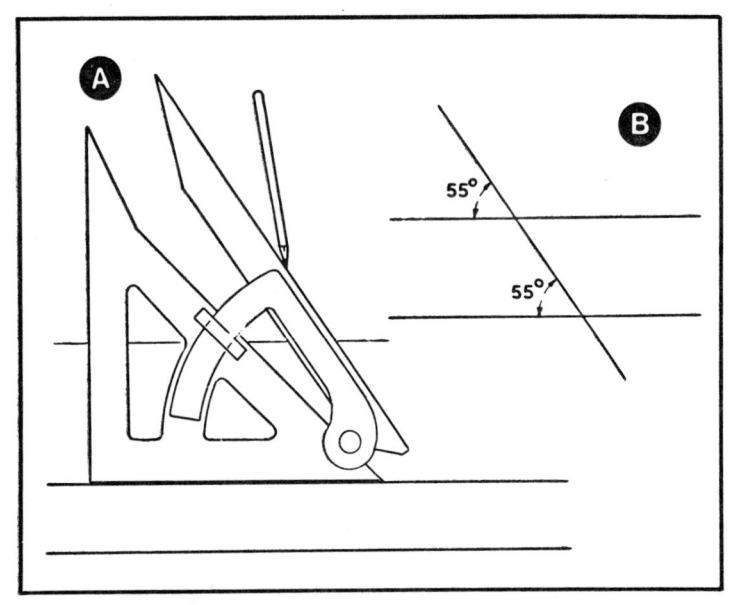

Fig. 6-8. An adjustable triangle allows nonstandard angles to be marked directly.

difficult to positively identify when you take the protractor away. An alternative is to make a "peck" (Fig. 6-7G) freehand, with its point where you will put a straightedge to draw the angular line (Fig. 6-7H).

Because of its method of construction, the moving arm of an adjustable triangle does not meet either of the other sides, although it can be adjusted to a range of angles to them. It is usually possible to draw a line through a point in one action by moving the straightedge far enough away after drawing the baseline, providing it is kept parallel (Fig. 6-8A). A line at a particular angle to one line will be at the same angle to another line parallel to it (Fig. 6-8B).

USING CURVES AND TEMPLATES

There are templates made for shapes and symbols to suit various branches of specialized drafting. There are French curves available in an enormous range of sizes and shapes. A draftsman starting to collect instruments could accumulate far more of these than he really needs. Many beginners can manage without any. It is unwise to become dependent on special aids for drawing shapes and patterns. It is better, at first, to draw what is required with ordinary instruments. Symbols and special character templates

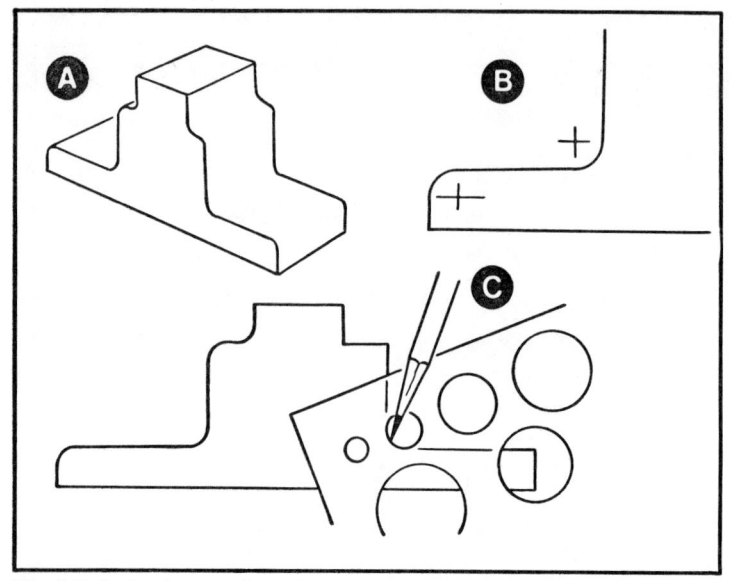

Fig. 6-9. A circular template allows curves to be marked without compasses.

may find uses when you have established what type of drafting you will be doing regularly; then some of them will speed your work.

A few simple templates may be time savers. Suppose you have to draw engineering castings with fillets (roundings) in all corners (Fig. 6-9A). You could draw each of the small curves representing fillets with compasses after locating centers (Fig. 6-9B). If you have a template with many holes of different diameters, a suitable hole can be put over a corner and the quarter circle for the fillet drawn through it (Fig. 6-9C).

French curves are mainly of use in lining in outlines that have been first drawn by other means. You may be able to adapt a curve to suit part of a French curve, but it would be wrong to try to force a shape to it if that means distortion or a departure from the necessary size.

Suppose you have located several points on a curve and have to draw a fair curve through them (Fig. 6-10A). If it is a fairly large drawing, the best way is to bend a spline through the points and draw around that (Fig. 6-10B). There is then no doubt that the shape is fair (meaning smooth and without kinks or distortion). If it is a smaller drawing with curves too tight for a spline to follow, you have to use other means. An example is a section of a molding (Fig. 6-10C). It may help to lightly pencil in freehand what you expect the curve to be (Fig. 6-10D). You can see by standing back and looking at it where alterations are needed to make it fair. Try parts of French curves and lightly pencil around them until you can make up a satisfactory shape (Fig. 6-10E). A common beginner's fault is to try to go too far around a French curve. Once it is obvious that the curve is running away from the intended line, stop there and look for another part to follow on. A good guard against distortion in a line is to see that adjoining curves obtained from French curves overlap a short distance smoothly.

If French curves are used in drawing symmetrical outlines, care is needed to get the same shapes for each side. With the usual transparent French curves, you can use pencil marks for locating when you turn the instrument over. Suppose a part of a French curve suits a part of a tool handle drawing. Pencil on it where its

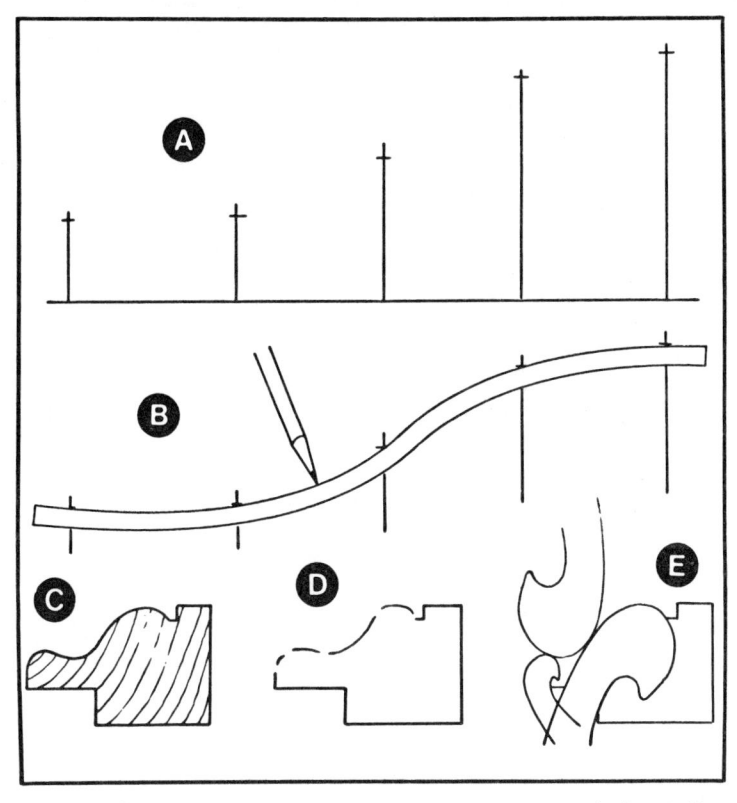

Fig. 6-10. Curves that cannot be drawn with a compass can be located by points, and then a spline is bent through them (A and B). Small curves can be drawn around parts of French curves (C, D, and E).

75

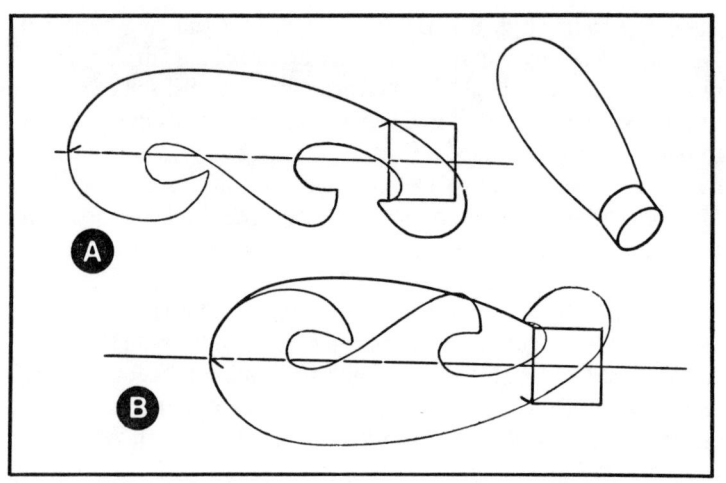

Fig. 6-11. For a symmetrical design, positions on a French curve should be marked temporarily.

limits are when drawing one side (Fig. 6-11A); then use these as guides as you turn over to draw the other side (Fig. 6-11B).

When starting to use French curves, it is very easy to waste a lot of time trying to find the curve you need for a particular part of a drawing. It is actually better to have a few French curves with which you are familiar than to have to search through a large collection of curves that you do not really know. It is surprising how you can get to know particular French curves so you visualize the outline and turn to one for a shape which will match. Start your French curve collection with one or two general purpose ones. You will probably never need any more.

This chapter is mainly concerned with the instruments used for pencil drawings. Ink instruments are only mentioned where their use is related. Detailed instructions on the use of instruments with ink are given in Chapter 14.

Chapter 7
Drafting Geometry

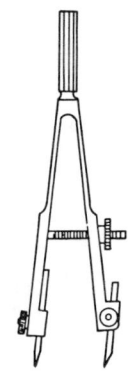

Much drafting can be done with the usual equipment without consciously considering geometric construction, but some knowledge of geometry is valuable even in routine mechanical drawing. There are some specialist applications where geometry may provide the only solution to a particular problem. There is no need to delve deeply into the complexities of advanced geometry, but there are many geometric constructions that can make successful drafting easier.

Geometrical constructions are particularly useful for very large drawings where you have to deal with sizes outside the range of your normal T square, triangles, and other instruments. If you want a line perpendicular to another, it is no use drawing with a 9-inch triangle if the line has to be extended to 36 inches. It could finish a degree or so out at its limit, due possibly to just the thickness of a line at 9 inches. This applies to any very large drawing. Construction should be as big as the final drawing or bigger; then any risk of error is minimized.

RIGHT ANGLES

For very large construction, there is the 3:4:5 method to mark 90°. This uses the property of a triangle with sides in those proportions and having a right angle contained between the two shorter sides (Fig. 7-1A). To set it out, choose a scale that will produce lines large enough for the final drawing. On a baseline,

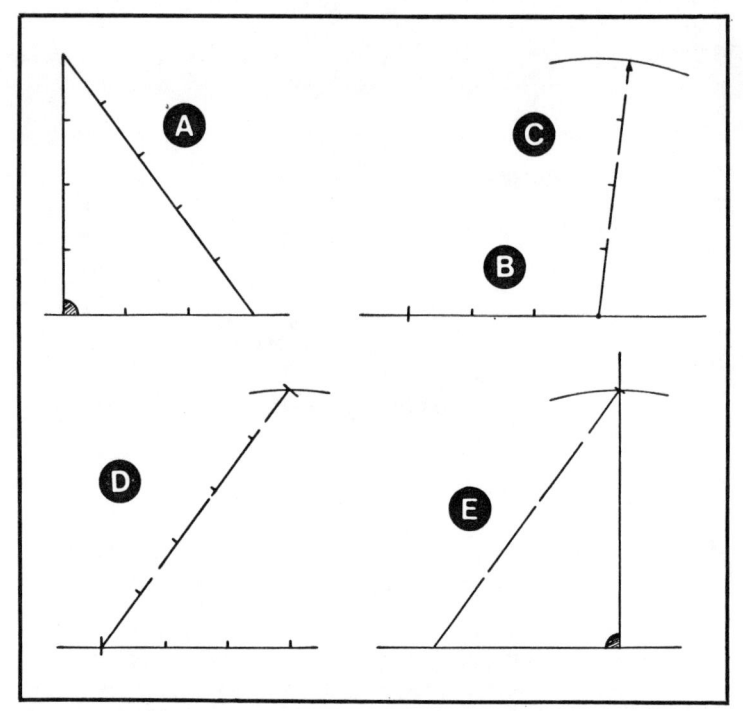

Fig. 7-1. A right angle larger than can be marked with instruments may be drawn by using the 3:4:5 method to draw a triangle with sides in those proportions and a right angle between the shorter sides.

mark three units from the point where the perpendicular line is to come (Fig. 7-1B). From that point, draw an arc of four units radius that will obviously cross a perpendicular line (Fig. 7-1C). From the other point measure five units to a point on the arc (Fig. 7-1D). If you erect a line from the point on the baseline through this spot, it will be perpendicular to the first line (Fig. 7-1E).

Another simple method of drawing a large right angle or long perpendicular line uses the fact that in a semicircle, two lines drawn to meet on the circumference from the ends of the diameter will be at right angles to each other (Fig. 7-2A), whatever position they meet.

From the place where the perpendicular line is to come, locate a point in space, for convenience at about 45° to the baseline. This is not critical, and the angle need not be measured. Use this distance as a radius to swing rather more than a semicircle to cross the baseline (Fig. 7-2B). Put a straightedge through the center and the crossing of the baseline. Mark where it crosses the other side of the arc (Fig. 7-2C). A line through that point will be perpendicular to the baseline (Fig. 7-2D).

BISECTING

The geometric way to find the center of any line (to *bisect* it), or to get the center between two points on it, is to swing two arcs of the same radius from each end, large enough to be obviously more than half the distance, so they can cross each side (Fig. 7-3A). A line through these crossings will break the line at its center (Fig. 7-3B). It will also be perpendicular to the baseline. Similar way of erecting a perpendicular line to a baseline, is to mark two points equidistant each side of the position (Fig. 7-3C). Then strike arcs of the same radius from these points (Fig. 7-3D). A line through these crossings will be perpendicular (Fig. 7-3E). For the least risk of error, choose spacings and radii that let the arcs cross at about right angles.

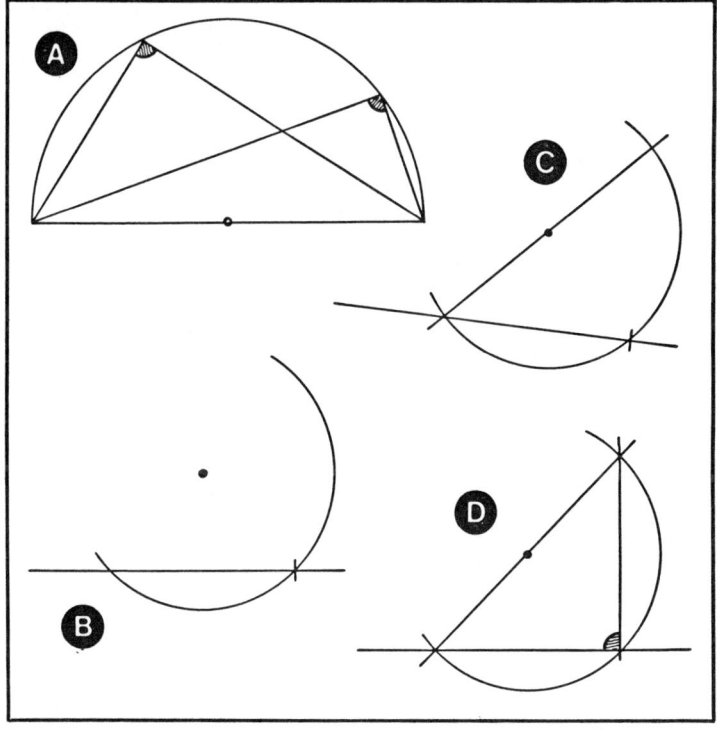

Fig. 7-2. When a triangle on a diameter touches the circumference, the angle there is 90°. This property can be used as another way of drawing a large right angle.

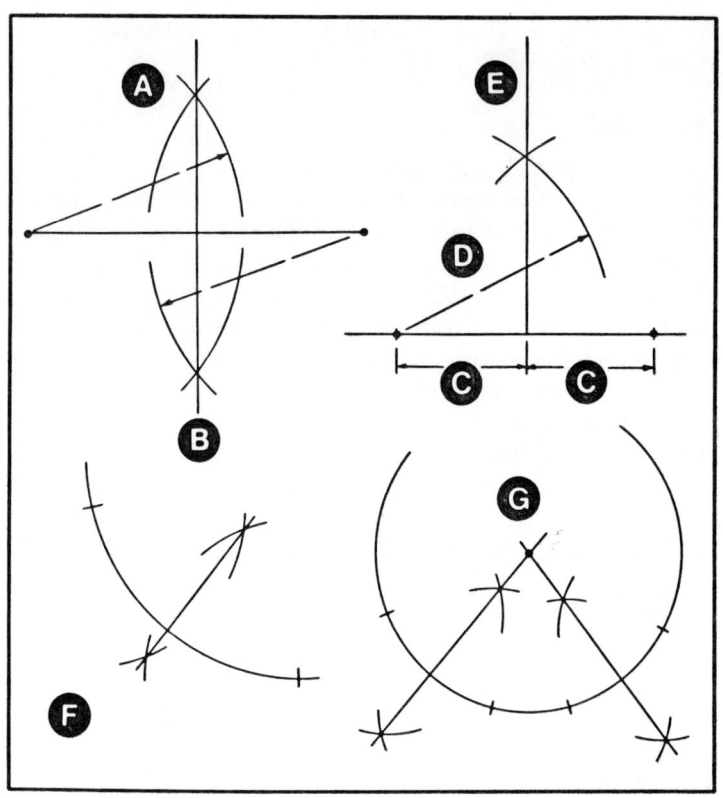

Fig. 7-3. A bisecting line is also square (A to E). This also applies across an arc (F), and two such bisecting lines will cross at the center of the circle (G).

The method of bisecting with crossing arcs can be used for other things than straight lines. The center between two points on an arc can be found (Fig. 7-3F). The bisecting line will pass through the center of the circle if it extends far enough. This property can be used to find a center. If you make two bisections at different places, the lines will cross at the center (Fig. 7-3G). This may be useful if you have to make a drawing of an existing solid object, and the center of a curve is not obvious on it. A third bisection will provide a check, if that crosses the other two at the same place.

A line may have to be divided into a particular number of equal parts. This is easy enough if the length matches the graduations on a scale, and these are simple to divide. If the length is not easily measurable or is a distance that is not easy to divide arithmetically, another method has to be used.

Suppose the space between two parallel lines has to be divided into five equal spaces. Put a scale across the lines. Tilt it until you have tow marks on it representing a distance that will divide into five over the lines (Fig. 7-4A). Use a pencil to make dots at the appropriate points (Fig. 7-4B). Remove the scale and draw lines through the dots parallel with the other lines (Fig. 7-4C).

A similar method is used to divide a line, or marks on it, into any number of equal divisions. Suppose it has to be divided into seven parts. Draw another line from one end of it at no particular angle (Fig. 7-4D). Along this line mark seven equal divisions (Fig. 7-4E). Join the end mark to the other end of the line (Fig. 7-4F). With a straightedge and triangle, draw more lines parallel with this from each of the other points to the first line (Fig. 7-4G). They will meet the line so as to divide it into seven equal parts. This works at any angle, but the risk of slight errors is reduced if the projection lines do not come too acutely. If the line being divided leaves the other at about 45°, and the divisions along it total not much more or

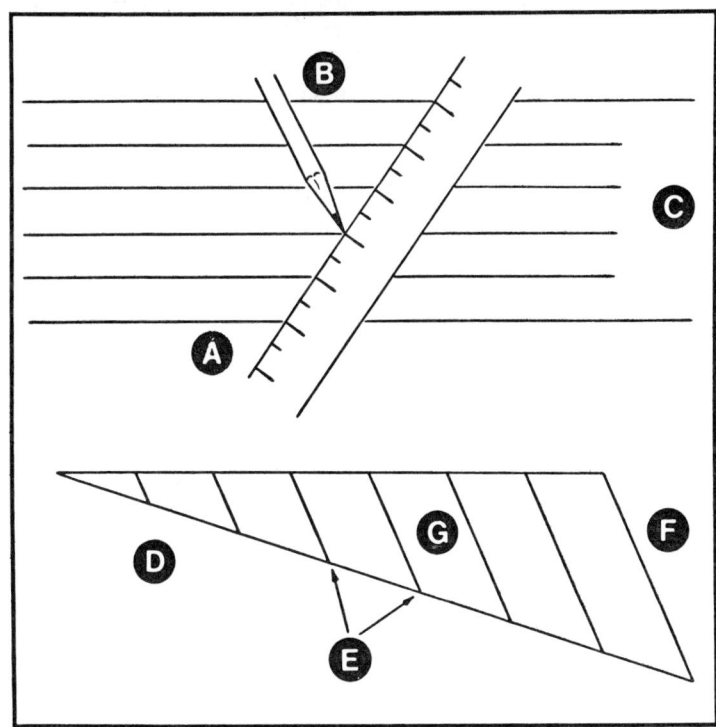

Fig. 7-4. Equal divisions can be marked by tilting a rule (A and B), or divisions on an adjoining line may be projected (D to G).

less than the original line, the projection lines cross reasonably squarely. There is less risk of deviations than if they had to be taken at a fine angle.

If two lines are not parallel, but other lines have to be flared equally between them, the previously discussed two methods can be combined. Put lines across the ends (Fig. 7-5A). Using the method just described, divide these lines into the number of spaces needed (Fig. 7-5B). Then join them (Fig. 7-5C).

These methods can be used to make the divisions in other proportions. Suppose the line has to be divided into one-third and the remainder in one-sixths. Draw the second line and divide it as required (Fig. 7-5D), using any scale which will conveniently allow that. Project from the end. Then make other projections parallel to it (Fig. 7-5E) to get the original line correctly divided.

TRIANGLES

The three angles at the corners of a triangle always total 180°. It is possible to draw triangles if you know one or more lengths of sides and two angles. With the lengths of two sides and the angle between them, you have all the information you need. If you know the lengths of all three sides, you do not need to know any angles. A triangle is a shape that cannot be distorted. Many constructions are triangulated for the sake of rigidity. A square frame would fall out of shape, but a diagonal across it forms two triangles. Its shape is then unalterable.

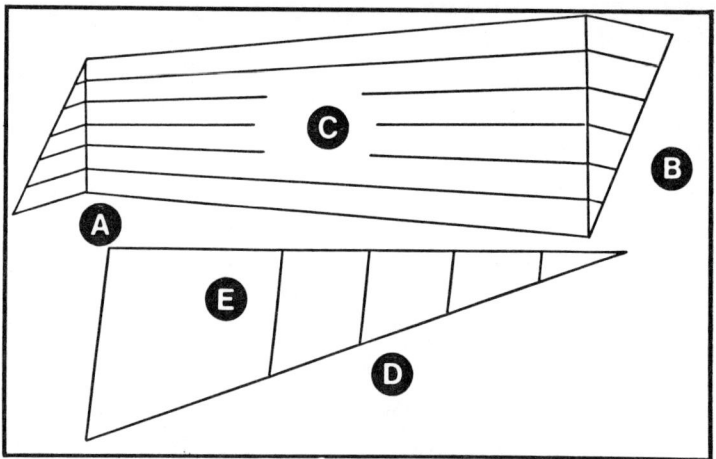

Fig. 7-5. Radiating lines may be marked between measured ends (A to C). Unequal divisions may be projected (D and E).

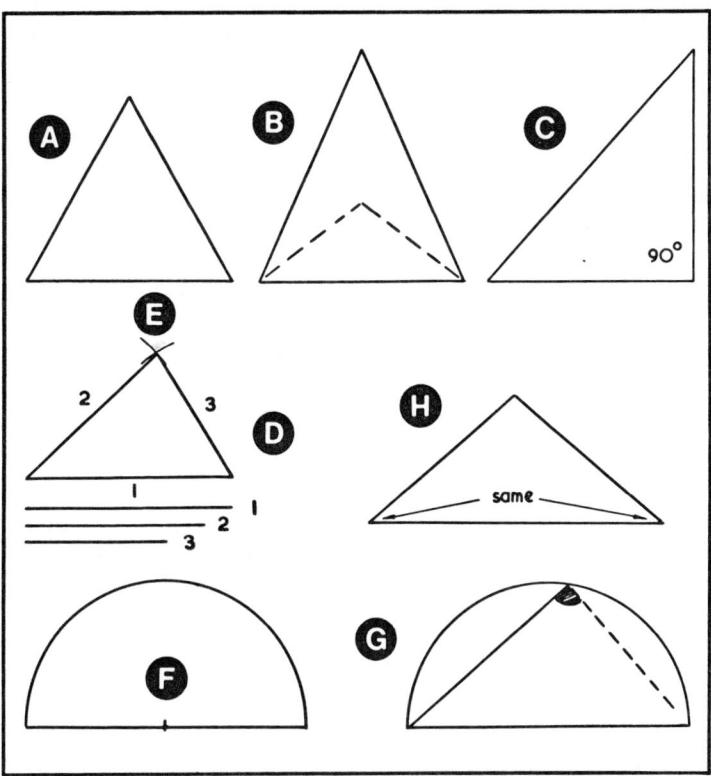

Fig. 7-6. Triangles may be equilateral (A), isosceles (B), and right (C). If three sides are known (D and E), the triangle can be drawn. A right triangle can be drawn on its hypotenuse by using a semicircle (F and G). An isosceles triangle must have two equal sides and two equal angles (H).

The names of triangles should be understood. If the three sides are equal lengths, their corners must also be equal angles (60°). That triangle is an *equilateral triangle* (Fig. 7-6A). If two sides are equal, the other may be longer or shorter, but in any case it is an *isosceles triangle* (Fig. 7-6B). If one corner is a right angle (90°), it is called a *right triangle* (Fig. 7-6C).

When the lengths of sides are known, one can be drawn. Then arcs with a radius the same as the other sides may be drawn from its ends so they cross at the point where the two sides should meet (Fig. 7-6D). The corner furthest from the base is the *apex* (Fig. 7-6E) in any triangle. A right triangle can be drawn with instruments. If it has to be drawn on a *hypotenuse* (the longest side) of a known length, that can be used as the diameter of a semicircle (Fig. 7-6F). If the length of one other side is known, an arc of that radius

83

is struck across the circumference (Fig. 7-6G). If that is joined to the ends of the diameter, the angle at the circumference will be 90°.

In an isosceles triangle two angles as well as two sides must be the same. If the length of the base is known and the angle at the apex, the other angles must be 180° less that divided by two. Draw the base and put lines at those angles from the ends (Fig. 7-6H). They will cross at the apex at the correct angle.

Sometimes it is necessary to draw a circle inside a triangle. To get its center, bisect the angles at the corner of the triangle (Fig. 7-7A) so the bisecting lines overlap. Two would do it, but the third serves as a check. A circle on that point touching one side should also touch the other two (Fig. 7-7B).

Putting a circle around a triangle so as to touch all three points involves a variation of the method of finding the center of a circle. Bisect two sides so the bisecting lines cross (Fig. 7-7C). That is the center of a circle that will touch the triangle points (Fig. 7-7D). Bisecting the third side will check the accuracy of your marking out. With a rather flat triangle, the center of the circle may be outside of it.

POLYGONS

Figures with many sides (*polygons*) may have to be drawn, either independently or in relation to some other construction on a

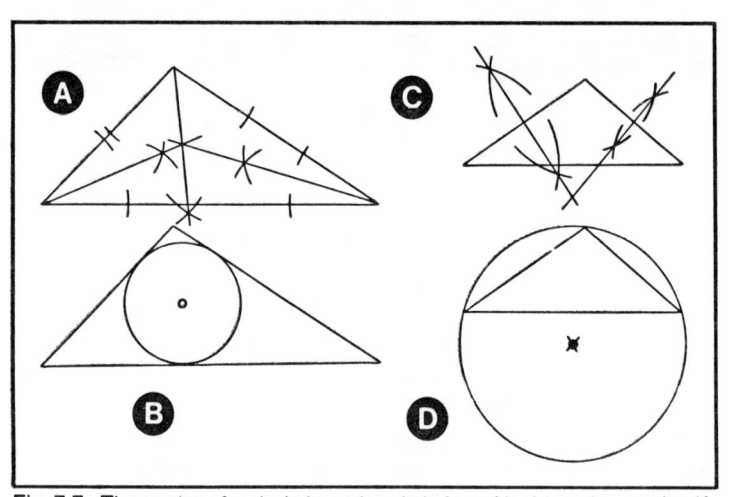

Fig. 7-7. The center of a circle in a triangle is found by bisecting angles (A and B). The center of an enclosing triangle is found by bisecting sides (C and D).

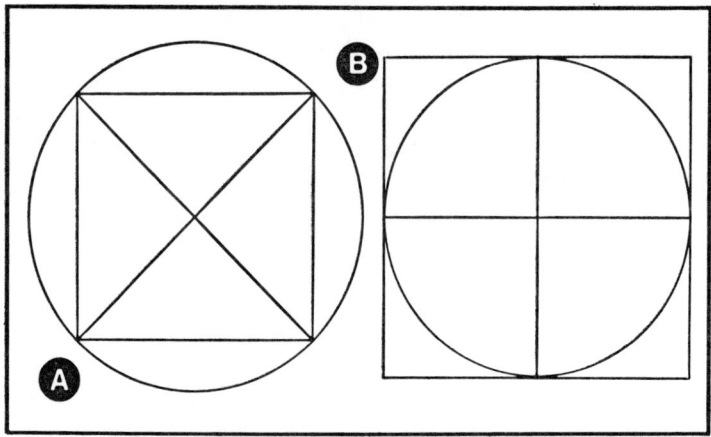

Fig. 7-8. The center of an enclosing circle around a square is found by drawing diagonals. An enclosing square has sides at tangents to the circle.

drawing. Some, such as squares and hexagons (six sides), are easily drawn with instruments. Other shapes, particularly with an odd number of sides, call for special methods. The shapes may have to be related to a circle, as with a hexagonal nut or bolt head. Circles are used in many constructions to make the shapes regular. A polygon does not necessarily have to have all sides the same length. When they are the same length, it is correctly called a *regular polygon.*

There is no difficulty in drawing a square by measurement and using a drafting machine or T square and triangle. For uniformity it is better to set the length of the sides with dividers, so there is no risk of the slight error that might come from marking each way against the edge of a scale.

If a square has to be drawn inside a circle with its corners touching the circumference, draw the circle and make diagonals through the center at 90° to each other. Join where the diagonals touch the circumference to make the enclosed square (Fig. 7-8A). The diagonals may be drawn with the 45° triangle.

If a square has to be drawn outside a circle and touching it, make the circle and draw two diameters at 90° to each other. Draw lines at right angles to the ends of the diameters. They will be tangents to the circle and will form a square (Fig. 7-8B).

The information that there are 360° in a full circle and the angles in any triangle always total 180° can be used to construct a polygon of any number of sides. If all of the corners of the polygon are joined to the center of an enclosing circle, a number of isosceles triangles will be formed, each with its apex at the center. All of those apex angles will total 360°, so the size of each can be calculated by dividing. If that size is deducted from 180°, you get the total of the other two angles in the triangle. As the triangle is isosceles, each of those angles is half of that total.

As an example, take a nine-sided figure (Fig. 7-9A). Each of the triangle apex angles will be 40° (one-ninth of 360°). Subtracting 40° from 180° and dividing by two gives 70° for each of the other angles in each triangle (Fig. 7-9B). The corner of the polygon with nine sides has angles of 140° (Fig. 7-9C).

To draw this shape when the length of a side is given, you could use a protractor at the ends of a line and mark the angles of two more sides, and so on, but that could lead to slight errors. It would be better to find the center of a suitable circle that all the corners would touch. Draw one side and use a protractor to draw lines at 70° from each end. Where they cross is the center of the circle (Fig. 7-9D). Draw the circle and step off the length of the first side around it. If you have worked accurately, it should go exactly nine times (Fig. 7-9E).

Some common polygons such as *hexagons* (six sides) and *octagons* (eight sides) are more easily drawn. If the method just described is worked through for a hexagon, the triangles formed will be equilateral so all their corners are 60°. The base of each triangle is the same length as the radius, and this goes six times around the figure.

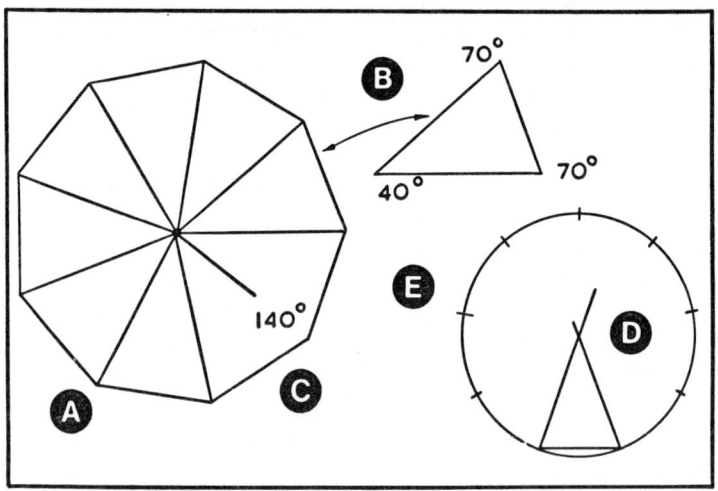

Fig. 7-9. A regular polygon is made up of a number of isosceles triangles.

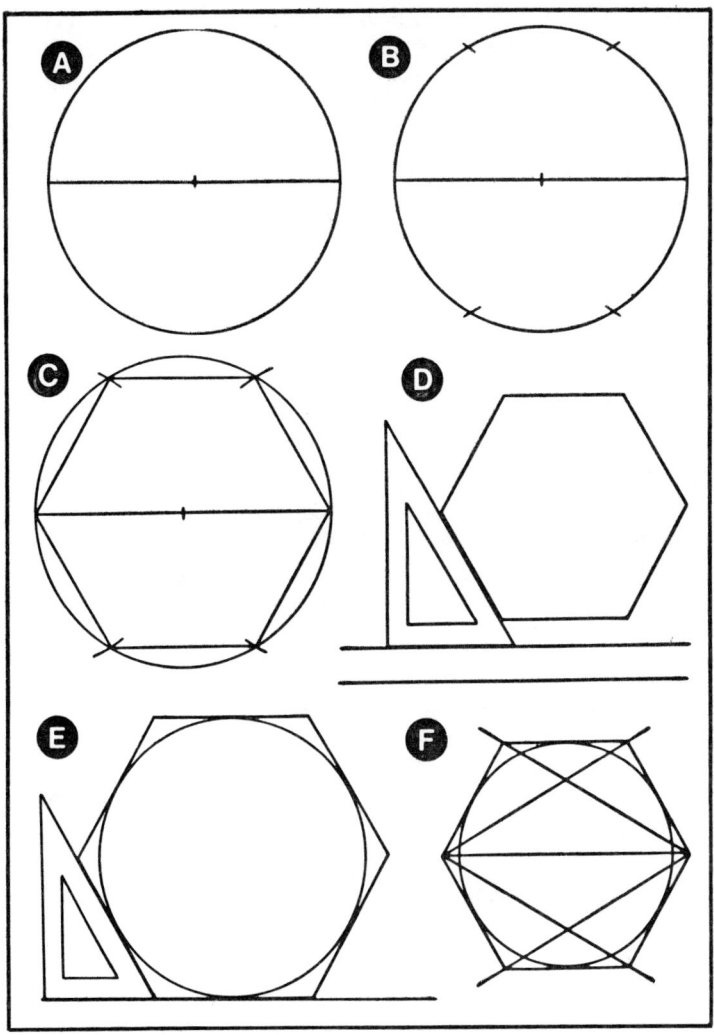

Fig. 7-10. A hexagon can be drawn by stepping off around a circle with a compass or by using a 60° triangle.

Drawing a Hexagon

If the given size is across the corners of a hexagon, draw a circle of that size. If one side of the hexagon is to be horizontal, draw a diameter with the T square across the circle (Fig. 7-10A). Using the compass still set to the radius, or dividers set to the same size, step off around the circle, starting from one end of the diameter line (Fig. 7-10B). You will get the same spacing by

87

stepping off from both ends. Join these marks to form the hexagon (Fig. 7-10C). A similar result can be obtained by using the 60° triangle from the ends of the diameter (Fig. 7-10D) and joining where these lines cut the circle. If one side of the hexagon is to be vertical, start with a vertical diameter and step off from its ends.

That method can also be adapted for drawing a hexagon when the given size is across the flats. Draw a circle of the size across the flats and use the 60° triangle and the T square to make tangents to the circle (Fig. 7-10E). Check the accuracy of your drawing by setting dividers to the length of one side and see if it also matches each of the other sides. A help in getting the meetings of the lines correct is to make some light construction lines at 30° from both points and below a diameter line. They will pass through the corners of the hexagon (Fig. 7-10F).

Drawing an Octagon

The angles at the corners of an octagon are 135°. The complementary angle of that is 45°, so a 45° triangle can be used in drawing that shape. If an octagon has to be drawn outside a circle, make the circle and draw lines tangent to it with the T square and 90° corner of a triangle (Fig. 7-11A). Use the 45° triangle to draw the other four diagonals to complete the octagon (Fig. 7-11B).

That circle indicates the distance across the flats. If it is the distance across the points that is important, the octagon has to be drawn inside a circle. Draw the circle and use the 45° triangle to draw four diameters at 45° to each other (Fig. 7-11C). They divide the circumference into eight equal parts, so joining these points makes a regular octagon (Fig. 7-11D).

Another convenient way of making an octagon, when the size across the flats is given, starts with a square. Draw a square with sides of the length the octagon has to be across the flats (Fig. 7-11E). Draw two diagonals. Set a compass to half the length of a diagonal. Swing it from each corner in turn so the arc cuts a side of the square each way (Fig. 7-11F). There is no need to draw the full arcs, it is where they cut the lines that counts. Dividers could be used to mark the distances. If the points on the sides of the square are joined with lines, they will be found to be the same lengths as the spaces between them, so you have a regular octagon (Fig. 7-11G).

Drawing a Pentagon

A *pentagon* (five sides) is sometimes needed. There is a variation of it to form a star with five points, which is more

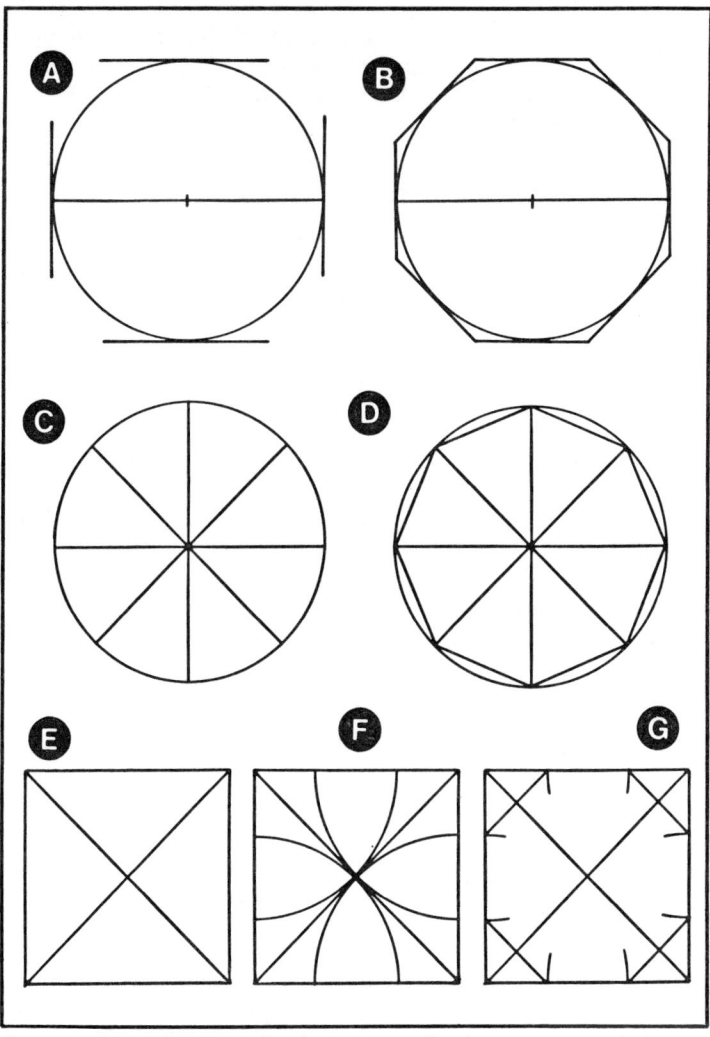

Fig. 7-11. An octagon may be drawn around or in a square or circle.

attractive than one made with any other number of points. One way of making a pentagon is to start at the center of a circle, dividing 360° by five (72°) for the spacing of radius lines (Fig. 7-12A). Draw a circle and join where the radius lines cut it to form the pentagon (Fig. 7-12B). That is a satisfactory method if it is the size of the enclosing circle which is important.

If it is the length of one side that is given, it is possible to find the center of a circle with a protractor from its ends, using the

89

method described for a polygon with any number of sides. In this case, with a 72° apex angle, the other corners of the triangle must each be 54° to make a total of 180° (Fig. 7-12C). Lines at this angle from the ends of the side will cross at the center of a suitable circle. Draw it and step off the length of the first side around it to provide points to join and make a pentagon (Fig. 7-12D).

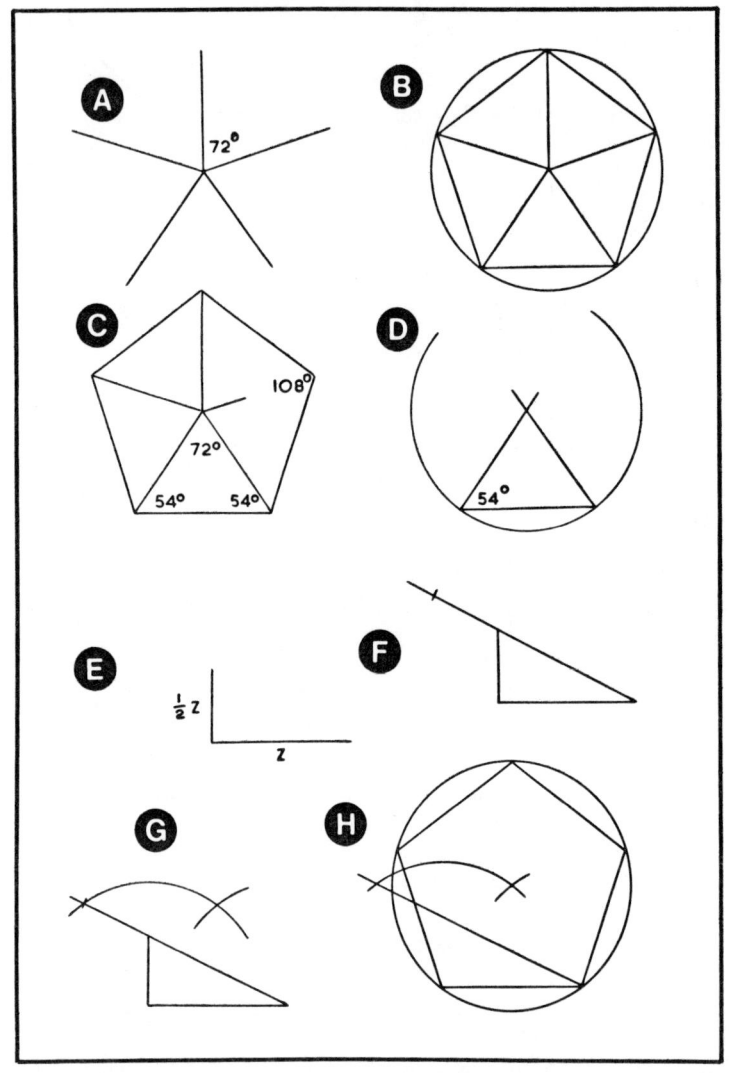

Fig. 7-12. A pentagon may be drawn with spaced radii for an enclosing circle (A to D) or by starting with one side and locating a center (E to H).

Another way of finding the center of an enclosing circle does not involve measuring angles. Draw a perpendicular line to one end of the first side, and mark on it half the length of the side (Fig. 7-12E). From the other end, draw a line through this point and extend it so you can mark on it a length equal to the height on the perpendicular line (Fig. 7-12F). If you set a compass to the distance of this position from the end of the first side and swing it across, it will pass through the center of the circle. Use the same compass setting to draw an arc from the other end of the line. Where they cross is the center (Fig. 7-12G). Draw the circle. Step off the length of the first side around it to get the points of a pentagon (Fig. 7-12H).

If the size of an enclosing circle is known and a pentagon has to be drawn to fit it, make the circle with a diameter line and a radius line square to it (Fig. 7-13A). On the diameter, halfway between the center and the circumference, put the point of a compass and set this to where the radius line touches the circumference (Fig. 7-13B). Swing this to cut the diameter line (Fig. 7-13C). Move the compass point to the end of the radius and set it to where that arc cuts the diameter; then turn it to cut the circle (Fig. 7-13D). That marks the length of one or two sides of the pentagon (Fig. 7-13E), so its length can be marked all round to draw the complete shape.

If alternate points are joined, a pentagon construction can be converted into a star (Fig. 7-13F). The star lines cross to form an inverted pentagon, which may be of use in some decorative drawing.

TANGENTS

A *tangent* is formed when a line, which may be curved or straight, meets a circle or other curve. In many drawings a sufficient degree of accuracy can be obtained by putting a straightedge or French curve against the circle and drawing against it, just as described in using triangles against circles in forming polygons. In some constructions, however, there has to be maximum accuracy. It is important to know the exact point of contact. This becomes particularly apparent when finishing a drawing with ink. A line has to blend into a curve smoothly without the change of direction becoming obvious.

If a tangent has to pass through a particular point on the circumference of a circle, draw a line radiating from the center through that point (Fig. 7-14A). Mark points on it equidistant from the crossing with the circumference. Use these points to swing arcs each way to cross each other (Fig. 7-14B). A line through these crossings will form the required tangent (Fig. 7-14C).

A similar result can be obtained by using a triangle on a straightedge. Put the hypotenuse of the triangle through the point and the center of the circle (Fig. 7-14D). Bring a straightedge up to its outer edge. Then draw along the triangle, hold the straightedge still, and turn the triangle over on it (Fig. 7-14E) so you can draw a tangent line through where the first line crosses the circumference.

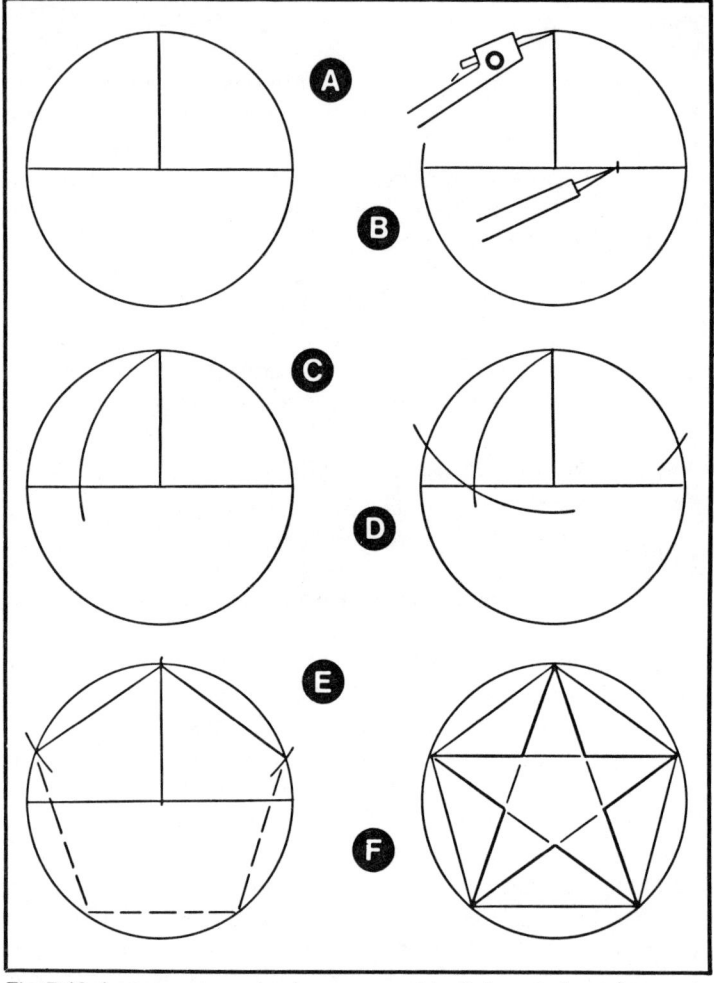

Fig. 7-13. A pentagon may be drawn geometrically in a circle and converting to a star.

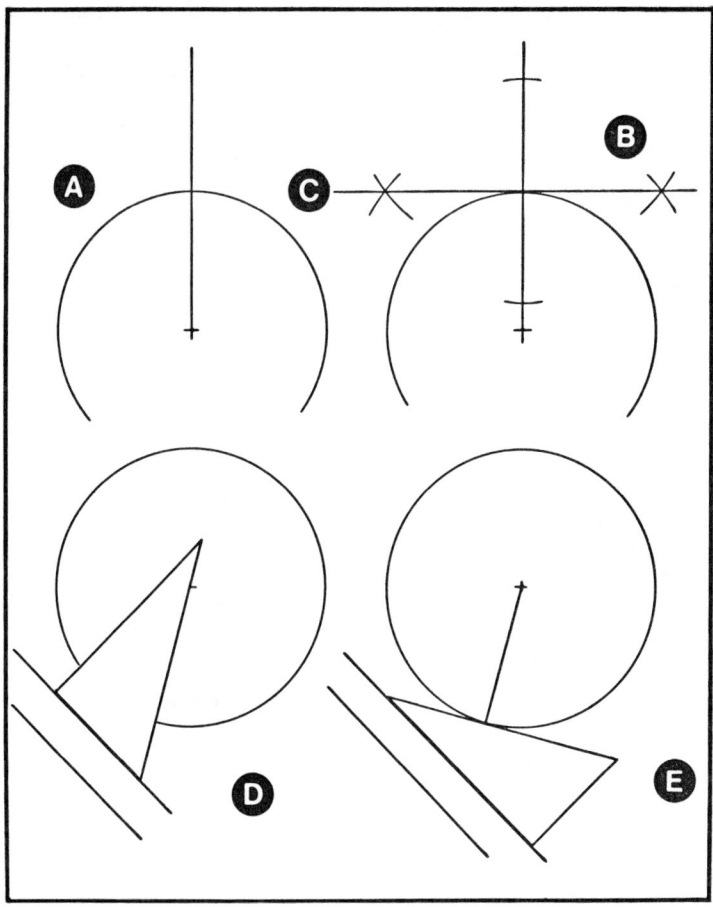

Fig. 7-14. The point of contact of a tangent is at the end of a radius line.

Rounded corners are common needs for drawing tangents. With a square corner, draw lines parallel to the main lines at a distance equal to the intended radius (Fig. 7-15A). Where they cross is the center of the circle or part of a circle (Fig. 7-15B). The exact points of meeting are where the extended lines meet the main lines (Fig. 7-15C). This method also applies if the corner is acute (Fig. 7-15D) or obtuse (Fig. 7-15E). The contact points are perpendicular to the lines.

In some engineering constructions lines have to be drawn between circles or arcs of different sizes, and these lines may be straight or curved. The problem is to find the exact points of contact.

93

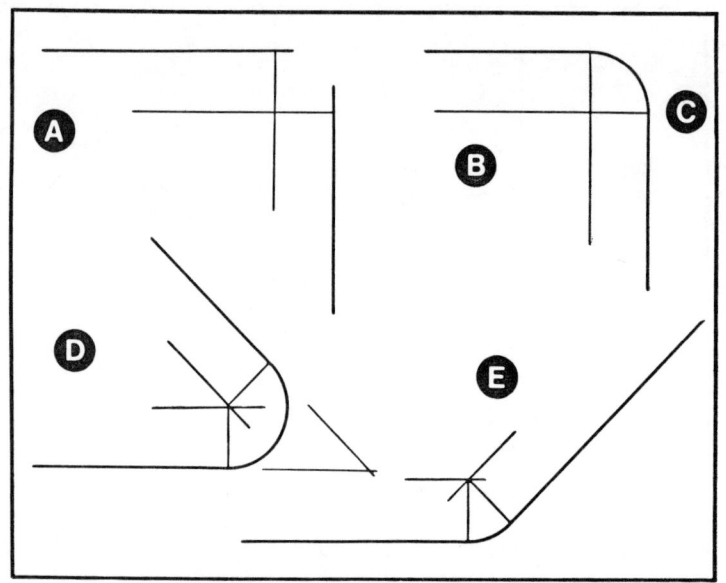

Fig. 7-15. Find the meeting points between straight and curved lines by drawing lines at the radius distance from the straight lines and others square to them at the curve centers.

Suppose tangents have to be drawn to two circles of different sizes and a short distance apart. Find the difference between the two radii, and draw a circle with that radius with the same center as the large circle (Fig. 7-16A). From the center of the small circle, draw a tangent to this new circle (Fig. 7-16B). Through the center of the large circle, draw a line at right angles to this tangent, extending through the circumference of the large circle (Fig. 7-16C). That is the point of contact of a tangent to the large circle. Erect a line at right angles to the construction tangent at the center of the small circle. That cuts the small circle circumference at the point of contact of the tangent (Fig. 7-16D). Draw the tangent.

If the tangent has to come diagonally between the circles and the points of contact are to be found, add the radius of the small circle to the radius of the larger circle (Fig. 7-17A) instead of deducting it as in Fig. 7-16A. Construction is very similar from this point. Draw a tangent from the center of the small circle to this construction circle (Fig. 7-17B). Draw a line at right angles to this line to the center of each circle (Fig. 7-17C). These lines cross the circumference at the points of contact. Draw in the tangent (Fig. 7-17D) and repeat the construction if another tangent is required the other way.

94

In some mechanical equipment the tangent between the circles is curved. The sizes of the two circles as well as the radius of the tangent arc are known. From the centers of the two circles, strike arcs. The radius of each arc should be the radius of that circle, plus the radius of the tangent arc. Where these arcs cross will be the center about which the tangent arc will be drawn (Fig. 7-18A). Draw lines from this crossing to the centers of the circles. Where these cross, their circumferences will be the points of contact of the tangent arc (Fig. 7-18B). Draw the tangent arc and repeat the construction at the other side if necessary.

ELLIPSES

An ellipse is a flattened circle or the view obtained by looking diagonally at one. It is the shape obtained by making a diagonal cut across a cone or cylinder. From the draftsman's point of view, it is a nuisance. A true ellipse cannot be drawn with the normal instruments. There have been special compasses with a sliding part so they can be used tilted. Then the circle drawn is flattened to an

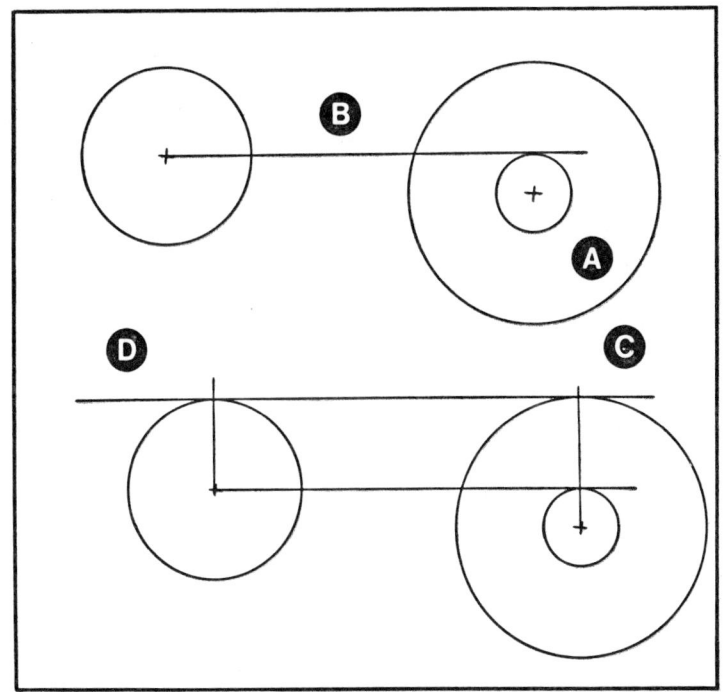

Fig. 7-16. Contact points for tangents between circles of different size are found by relating the two radii.

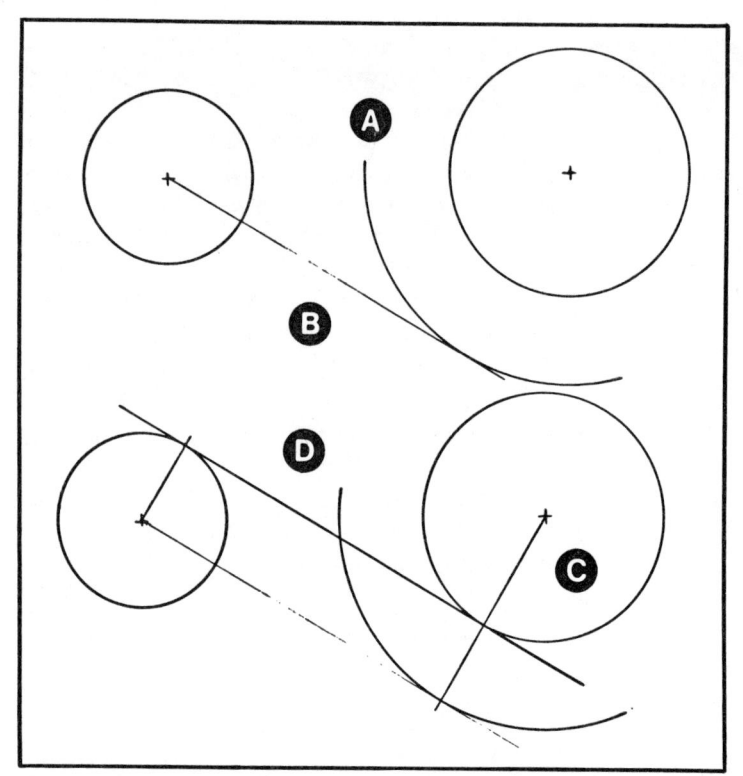

Fig. 7-17. The contact points for a tangent diagonally between two circles are found by adding the smaller radius to the larger.

ellipse similar to a diagonal cut across a cylinder at the same angle, but there are problems in using these compasses with precision. The trammel arrangement (Fig. 3-16) is the nearest thing to a tool that traces an ellipse, using first principles in its formation.

An ellipse is a closed curve traced out by maintaining a constant sum of distances from two fixed points (called focus points). There are two axes (plural of axis) at right angles across the ellipse. The major axis traces the longest distance across the ellipse and has the focus points on it. The minor axis is on the smallest distance across the ellipse.

The foci are like twin centers of circles. If they are brought so close together that they coincide, the ellipse becomes a circle. The further they are apart on the major axis, the more slender will be the ellipse.

A practical way of drawing an ellipse illustrates the principle, but it is not much use for the small ellipses required on a drawing,

96

although it works well for a large shape on the floor. Use two *awls*, pushed in as foci. Have a loop of cord around them. Push a pencil into the end of the loop and keep it tensioned (Fig. 7-19A) as you pull it around (Fig. 7-19B). Providing a steady tension is kept on the cord, the shape traced by the pencil will be an ellipse. The lengths of the two arms of cord extending from the two awls always equal the same distance from the awls, as required by the definition of an ellipse.

The two awls are the foci. The limits of the major axis are controlled by the length of the cord loop when stretched along the awls. The length of the minor axis is controlled by the distance the awls are apart (Fig. 7-19C). Try moving the awls in or out and see what effect that has on the shape of the ellipse produced.

Usually the major and minor axes are known, or the ellipse has to be drawn to fit within certain boundaries (Fig. 7-20A). It is possible to find the positions of the foci by calculation, but it is easier to set them out geometrically. Set dividers to half the length of the major axis (Fig. 7-20B). Have one point at the limit of the minor axis and mark with the other on the major axis, both ways (Fig. 7-20C). These points are the foci.

For drawing an ellipse on paper, where the cord and awl method would not be practical, it is possible to obtain a large number of points that will fall on the ellipse; then join them to get the shape. This is a variation of the trammel method. Draw the major and minor axes, with the limits marked at x and y with the crossing o (Fig. 7-20D). Use a straight edge of a piece of card and

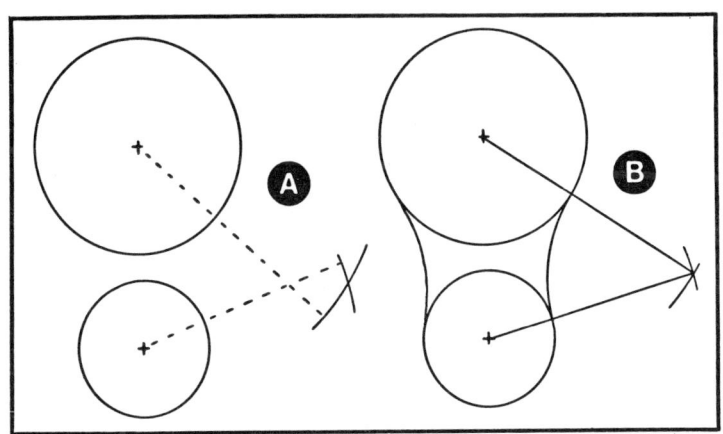

Fig. 7-18. The meeting points for curves meeting are on the combined radius lines.

97

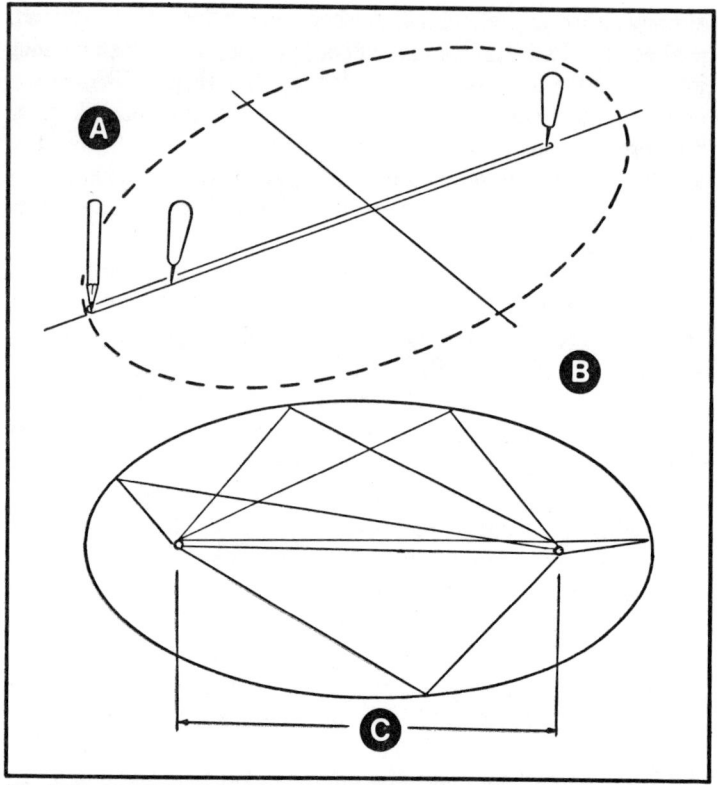

Fig. 7-19. One way of drawing an ellipse uses cord looped around two awls at the focus positions.

put a mark on it near one end to represent o, and from it pick up the drawing distances to mark x and y (Fig. 7-20E).

Use this strip with o as the position for marking points on the ellipse. Position the strip so the other two points come exactly on the axes (Fig. 7-20F). Make a mark opposite point "o" on the strip. Continue to do this at many positions, always taking care to keep the two inner marks exactly on the lines of the axes. This will build up a pattern of marks to be joined, with a spline for a large shape, or with parts of French curves for smaller shapes.

A beam compass with three heads could be used instead of the card strip, with a greater degree of accuracy. These are not usually supplied with more than two heads.

It is sometimes necessary to draw one ellipse inside another, as in marking the inner and outer limits of an arch. They have to be treated as separate shapes. The inner ellipse cannot be made about the same foci as the outer one. Draw rectangles to enclose each ellipse, and draw each shape without reference to the other one.

There are several ways of drawing shapes that approximate closely to true ellipses, but their outlines can be made with compasses. In some circumstances, such a shape may be just as satisfactory as a true ellipse. The method enables you to produce a neat drawing entirely with normal instruments. For one method, use lines at right angles for the major and minor axes, with distances marked (Fig. 7-21A). Draw a semicircle to one side with the compass point at the crossing and the major axis as the diameter. Divide the space between this and the minor axis mark into three, and make one more division of the same size inside the minor axis mark (Fig. 7-21B).

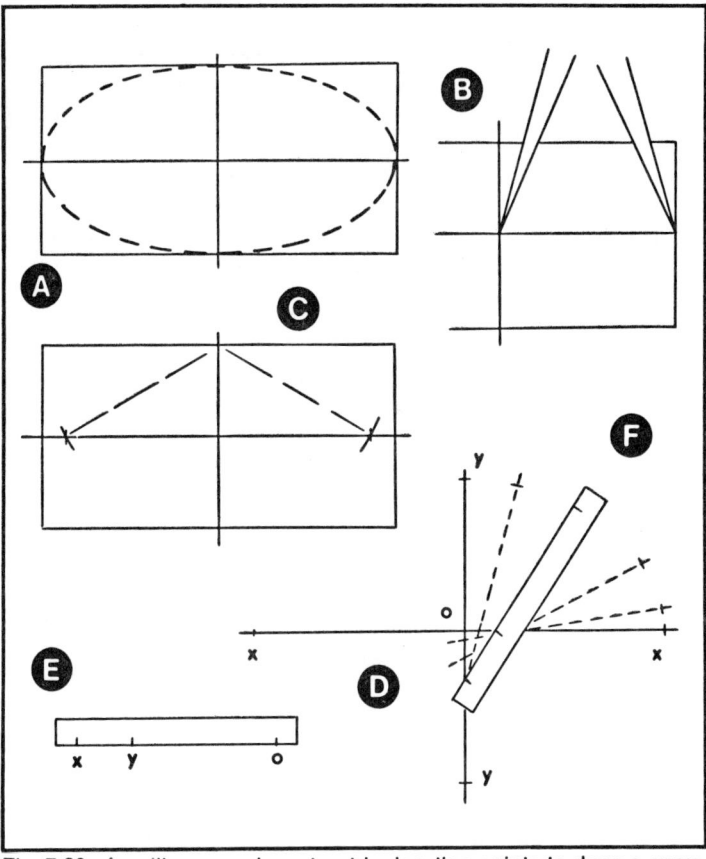

Fig. 7-20. An ellipse may be set out by locating points to draw a curve through.

99

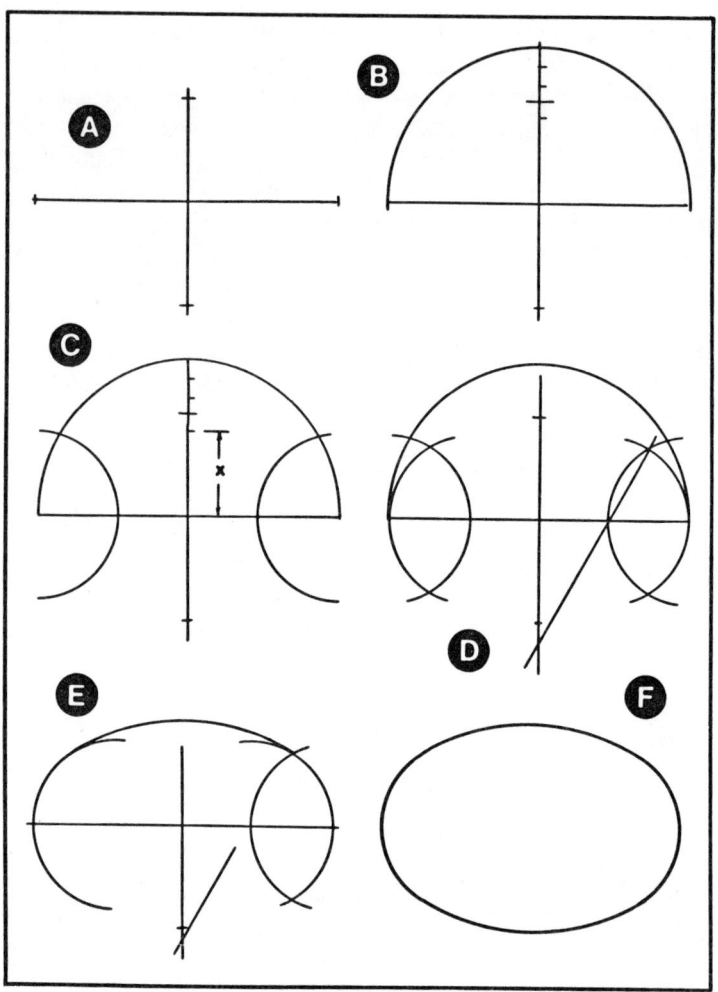

Fig. 7-21. An approximate representation of an ellipse can be made by combining curves drawn with a compass.

Using the ends of the major axis as centers and the compass set to x as radius, draw arcs almost as large as semicircles (Fig. 7-21C). From the places where the arcs cross the axis line and without altering the compass setting, draw more arcs to cross them. Draw a line through an arc crossing and where one of these arcs crosses the major axis, far enough to reach the minor axis line (Fig. 7-21D). With this as a center, draw an arc through the minor axis point to join the arcs at the ends (Fig. 7-21E). Do the same at the other side to complete the approximate ellipse (Fig. 7-21F).

Chapter 8
Lettering

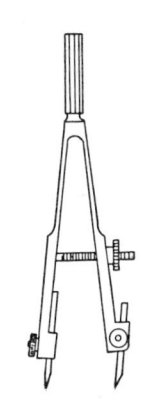

Nearly all mechanical drawings carry letters and numbers. In some cases the only figures are dimensions, and the lettering forms just the briefest notes or references. In other drawings there may have to be a considerable amount of explanation, with lines of text on the drawing. Doing satisfactory lettering can be time-consuming. If much description and explanation is needed, this is better done on any accompanying sheet, typewritten or reproduced in some other way.

You might think that quickly written words would be satisfactory on a drawing, but they look untidy and are not as clear as carefully hand-printed letters. Written words may be acceptable on a preliminary rough sketch, but at further stages it is better to take some trouble over any lettering required. In principle, it is good policy to keep lettering to the minimum. A good drawing should provide all the information needed graphically, as far as possible. Even then there will have to be a few words in various places to indicate what the view represents, possibly the sizes of holes and similar things that can be covered by just a few words. It is usually possible to scheme a drawing so any lettering can be confined to brief notes. If you find that what you have in mind will require several lines of explanation at some point, ask yourself if there might not be a better way of dealing with the drawing that would convey what you intended with far fewer words.

Figures will mostly be for dimensions. They need to be as clear as possible so there can be no risk of error, so they should be complete and plain. When you write figures in notes or sketches, you tend to skimp on some of the lines, lifting the pen or pencil before getting to what should be the limit of the figure. You will probably read it correctly later because of the context. Figures on a drawing should be to standard shapes and drawn more deliberately than you might write them. More information on the use of figures for dimensioning is given in the next chapter.

Letters for drawing are often described in printers' terms. Capital letters are called *upper case*. Small letters may be called *lower case*. Both terms come from the relation of the cases of type in a printing work. For much lettering on a drawing upper case letters are clearer, but if there has to be a lengthy note, it is easier to read in lower case letters. A long passage all in capital letters is taxing on the eyes and not as easily comprehended.

STYLE

Everyone has a different style of writing, so it is possible to recognize a familiar hand without the need to refer to a signature at the end. In the same way people do not all favor the same style of printing on drawings. Lettering on drawings comes as a new activity. It is unwise to think about writing styles and to concentrate on good plain lettering. After some practice there may be few personal touches that creep in, but they have to be very minor to be acceptable. The main differences come in the choice of upright or sloping letters. Upright are preferable, unless you find you can maintain a steady slant in a line of lettering. Sloping lettering that varies in its angle looks very untidy.

Lettering on the majority of drawings is done freehand. It should be kept small, but it obviously must be legible. Overly large lettering looks ugly and brands the work as amateurish. There may have to be some larger lettering for titles. You may have to draw a notice with large letters. If letters larger than the usual descriptive notes on a drawing have to be made, instruments should be used; otherwise, freehand curves will look very uneven and untidy. When learning drafting lettering, it may help to make some letters, at least, to a large size, using instruments to get the shapes and proportions right.

UPRIGHT LETTERING

Upright lettering with the minimum of lines is generally termed *Gothic*. It is sometimes called *Roman*, but that is more correctly applied to upright lettering with lines of different thicknesses and serifs—points extending from the corners (Fig. 8-1A). This can be very attractive when printed or carved on signs, but it would be too laborious for use on a drawing. The usual Gothic, which might be distinguished by being called commercial Gothic, may be upright with the larger letters occupying approximately square outlines (Fig. 8-1B). M is a square and O is a circle. Matching lower case letters have similar proportions for the parts that do not extend upwards or downwards (Fig. 8-1C).

Some draftsmen prefer condensed or compressed upright lettering. The M is then not as wide as it is high, and the O is elliptical (Fig. 8-1D). Care is needed with some lower case letters to make them clearly. Increased spacing may be needed so there is no confusion when the lettering is small.

If you prefer sloping lettering, which may also be called *slanting* or *inclined* lettering it is possible to keep letter forms almost as wide as they are high. This form does not allow the O to be a circle, if all of the lettering is to match. It has to be an inclined ellipse (Fig. 8-1E). More information on the forming of sloping letters is given later in the chapter. *Italic* lettering also slopes, but this is a type of form used in printing for contrast or emphasis. It has no place in drafting. It is incorrect to use the name for ordinary sloping lettering.

Fig. 8-1. Open letters with serifs may be used in headings (A), but normal lettering is kept simple (B to E).

Several systems have been devised to provide guides for ink lettering, either in the form of stencils to be used with a special pen, or as a form to be followed by a guide while a pen draws the lettering remotely. There have been pencil guides made. Although these may have uses for large display work, where a pencil outline can be filled in with paint or ink, guides for a pencil to make lettering as small as required on the usual drawings are not very effective. It is better to learn to do that lettering freehand.

Gothic lettering is sometimes described as *single-stroke, one-stroke,* or *down-stroke.* When forming the letters, the pen or pencil does not go over any part of the letter twice, and so far as possible the strokes are from top to bottom or left to right. A few letters illustrate the technique (Fig. 8-2A).

Notice that in all letters, if it is possible to use a straight line, that is chosen instead of a curve. There is no place in this style of lettering for flourishes or the wavy lines that may come naturally in your handwriting. For practice, it is worthwhile using graph paper or drawing grids on which to lay out letters in the standard form (Fig. 8-2B). Do the whole alphabet if you wish, but there are many letters of basically similar form such as O, Q, and C (Fig. 8-2C). There are letters that would look top-heavy if their center part came exactly at half height. E and F should have their stroke above the middle (Fig. 8-2D). M may occupy a square, but N should be

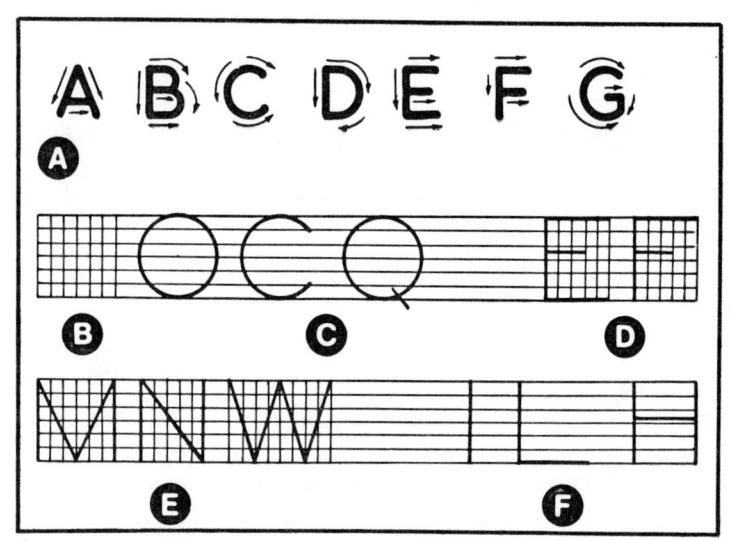

Fig. 8-2. Letters are made with down strokes (A). Grids may guide in learning shapes (B to F).

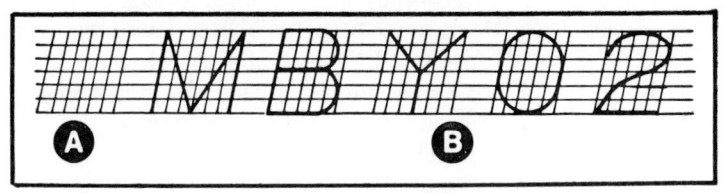

Fig. 8-3. Sloping lettering should be evenly angled.

narrower. W needs to be wider to look right (Fig. 8-2E). The straight line letters, such as I, L, and H are simple to form, but care is needed when working freehand to keep the lines straight and square.

PROPORTIONS

For a proportion that looks right, the letter S and those figures with curves in them, such as 8 and 3 should have upper curves slightly smaller than those below. If they are equal or larger, the construction looks top-heavy. An exception is 9 where its curve may come at or below the center.

If you decide on sloping letters, you must determine the amount of slope that suits you. It may be up to 15° from vertical, but it is unwise to go further than that. All of the notes concerning upright letters apply, but you have to keep a constant angle. You will not be able to get graph paper with lines inclined correctly, but you can draw grids to suit your practice letters (Fig. 8-3A). As with upright letters, M can have the same number of squares each way, but because of the slope O will be elliptical in the same grid (Fig. 8-3B).

GUIDELINES

Capital letters are all the same height. This makes for neatness and is one reason why they are favored for most lettering on drawings. To do Gothic lettering, it is advisable to draw a pair of guidelines (Fig. 8-4A). For general lettering on most drawings, letters should be not much more than ⅛ inch high. An experienced draftsman might manage with just a bottom line, but he would never do the lettering without that guide. If lettering is not horizontal or follows a curved course, it is very obvious to any viewer. Make the guidelines extremely light. You should be able to see them while you are lettering, but they should be as inconspicuous as possible afterwards. Keep the lettering to the guidelines. Even if there are slight inconsistencies in the way you have formed some letters, keeping the height uniform gives an overall appearance of neatness. Details are less obvious.

Lower case letters need more guidelines. There will be at least one upper case letter with them, so allowance has to be made for that as well. There is no insistence on particular proportions, so you may eventually choose an arrangement of guidelines that suits your style. For early experiments, this is a reasonable proportion. Assume the body of a lower case letter is a square, indicated by a grid of four small squares each way. Then the letters that go downwards extend to a depth of two squares, and those that go upwards extend two squares, which can also be the height of any upper case letters (Fig. 8-4B). Make lower case letters by the single-stroke method. Those that lend themselves to the treatment can be based on circles (Fig. 8-4C), drawn in two parts from top to bottom at each side. Then any adjoining lines are added. Take care that lines extending upwards or downwards stop always level on the guideline.

If only one word has to be drawn, it may be possible to manage with just a pair of guidelines. Slight variations in the extensions of letters above and below will not then be sufficient to show. If there is a line of lettering, use four guidelines (Fig. 8-4D). If you are using sloping lettering, it may be worthwhile drawing some guidelines at intervals (Fig. 8-4E) so you can compare your freehand letters with them.

The guidelines can be drawn each time by measurement, but there is then a risk of slight differences. It would be possible to cut

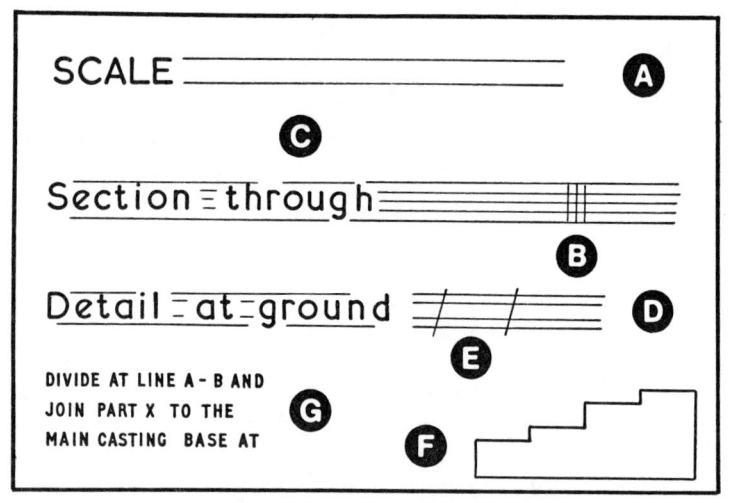

Fig. 8-4. The number of guidelines depends on the lettering (A to E). A template can be made for spacing (F). Keep notes evenly spaced (G).

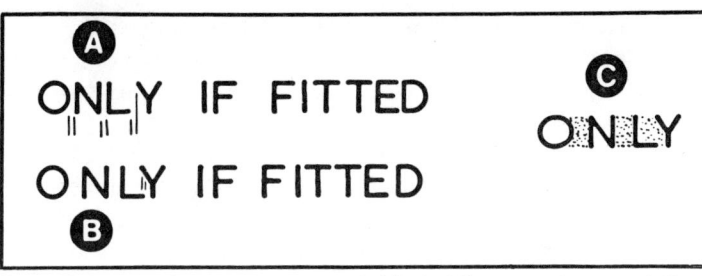

Fig. 8-5. Linear spacing is not enough (A), but the areas between letters should be about the same (B and C).

a piece of plastic in steps, so it could be drawn along the T square with a pencil (Fig. 8-4F). More sophisticated aids can be bought. One is a triangle, with many holes through it, arranged squarely to the hypotenuse. By choosing the correct holes, a pencil can be inserted in them in turn. The whole thing drawn along the T square to make the guidelines. Another guide, which uses a similar arrangement of holes, has them in a part that can be rotated in the main assembly so as to vary the spacings without altering the proportions.

If there have to be two or more lines of lettering, the spacing between the lines should be enough for clarity. Somewhere between one and two times the heights of the letters should be satisfactory. Too wide a spacing may lead the reader to assume they are different notes. For neatness, start all lines vertically under each other (Fig. 8-4G). The other ends of the lines come as they will, although take care to keep lines approximately the same length.

One problem in lettering when you are aiming for neatness is in the spacing between letters. At first glance it may seem that you must have the same space along the line between each pair of letters. That would do with letters that approximately fill a rectangle, but you meet difficulty when two letters with open sides are adjoining. The worst examples are L followed by Y. What seems a reasonable horizontal spacing between many other letters is then excessively large (Fig. 8-5A). L is particularly bad in relation to many other letters.

You must aim at an equal area of space between letters, not an equal horizontal distance. Consider what lettering has to be done. If you have an open combination such as L and Y, allow for them coming as close as you want and even overlapping horizontally (Fig. 8-5B). Then visualize the spaces between the other letters to

107

get a comparable amount of area each time (Fig. 8-5C). Don't decide on a close spacing between normal letters, and N then get to an open arrangement where the minimum that could be managed is not as small as you have been using in other places.

The spacing you allow between letters in a word and between words depends on the amount of paper available. Compressing too much can be confusing, while very wide spacing looks ugly and is difficult to read. Spaces between words should be about the width of a letter M, with a little more between sentences. Do not put commas or periods too close to the words. If you are making a note of three or four words, there is no need to put a period at the end. That would only be necessary if another sentence followed.

FRACTIONS

One problem that arises with numbers is how to deal with fractions. If decimals can be used, there is little difficulty. The figures before and after the decimal point are the same (Fig. 8-6A). Make the decimal point clear. It is better at half the numeral height than at the bottom, where it could be mistaken for a period. On many drawings decimals are used, but there is a difficulty when fractions have to be used. One way is to use figures of the same size as whole numbers, but with a diagonal stroke between them (Fig. 8-6B). There is always the risk that they will be read as whole numbers.

Another way is to use figures of the same size as whole numbers, but above and below a horizontal line. This arrangement is less likely to be confused, but the large figures look ugly (Fig. 8-6C). It is better to use smaller figures, but they must still be large enough for easy reading. That means the overall height of a fraction is usually more than the height of whole numbers (Fig. 8-6D).

Larger lettering for titles and display is better kept plain so it still has a mainly Gothic form, rather than being arranged in any ornate way. If it is large enough to justify using instruments, letters are best drawn in skeleton form and filled in by shading or with ink. There are several ways that the letters can be laid out. One simple method uses a grid of five squares high and three squares wide for the majority of letters (Fig. 8-7A). The plain letters such as E, F, L and I then follow grid lines (Fig. 8-7B). Those with diagonal lines such as M, N, W, X, and Y can use the grid squares (Fig. 8-7C). For those letters with curves, the compass point goes into the center of a grid square. O has a curved top and bottom with straight sides

$12 \cdot 375$ **Ⓐ** **Ⓑ** $15\ 13/32$

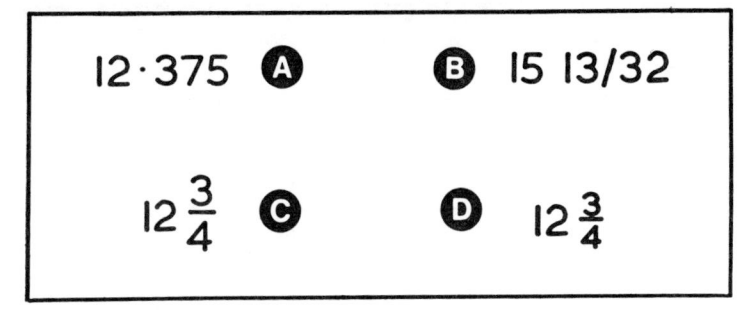

$12\dfrac{3}{4}$ **Ⓒ** **Ⓓ** $12\dfrac{3}{4}$

Fig. 8-6. Decimals are easy to mark (A), but common fractions can be drawn in different ways.

(Fig. 8-7D). C, G, and Q follow the same pattern (Fig. 8-7E). Numbers follow a generally similar pattern (Fig. 8-7F).

A basically similar form can be drawn using a grid of seven squares vertically and five squares horizontally to produce slightly more open and slender letters and numbers.

Another form that avoids curves in the letters is a block style, where what would be curves are parts of octagons. As well as avoiding the use of compasses, all of the lines can be drawn with triangles on a T square or with a drafting machine. The basic letters are better made slightly wider than in the above examples, or the inner angles replacing curves in some letters do not become practical. A grid of seven squares high and six squares wide for most letters will be suitable.

A letter O will show the technique. Draw diagonals of the corner squares, with lines parallel to them at the same distance as the width of a square (Fig. 8-8A). Line this in and fill in the letter to see the effect (Fig. 8-8B). Letters without curves are made in the usual way (Fig. 8-8C). Those which would have had smaller curves such as B, D, and P have flat horizontal bars and diagonal corners (Fig. 8-8D). S falls conveniently into this pattern (Fig. 8-8E), as do 8 and 9 (Fig. 8-8F).

For most drawings, any larger lettering is there so as to be more prominent than the other letters. If the eye is drawn to it, it has served its purpose. If a title is lengthy in relation to the available space, the grids need not be based on squares, they can be rectangles, narrower than they are high. Similarly, if the title would look better stretched to fill a space, the basic rectangles could be horizontal to make wider letters.

A point to watch in designing large lettering is not to make the area to be filled in too large. Shading in pencil neatly can be

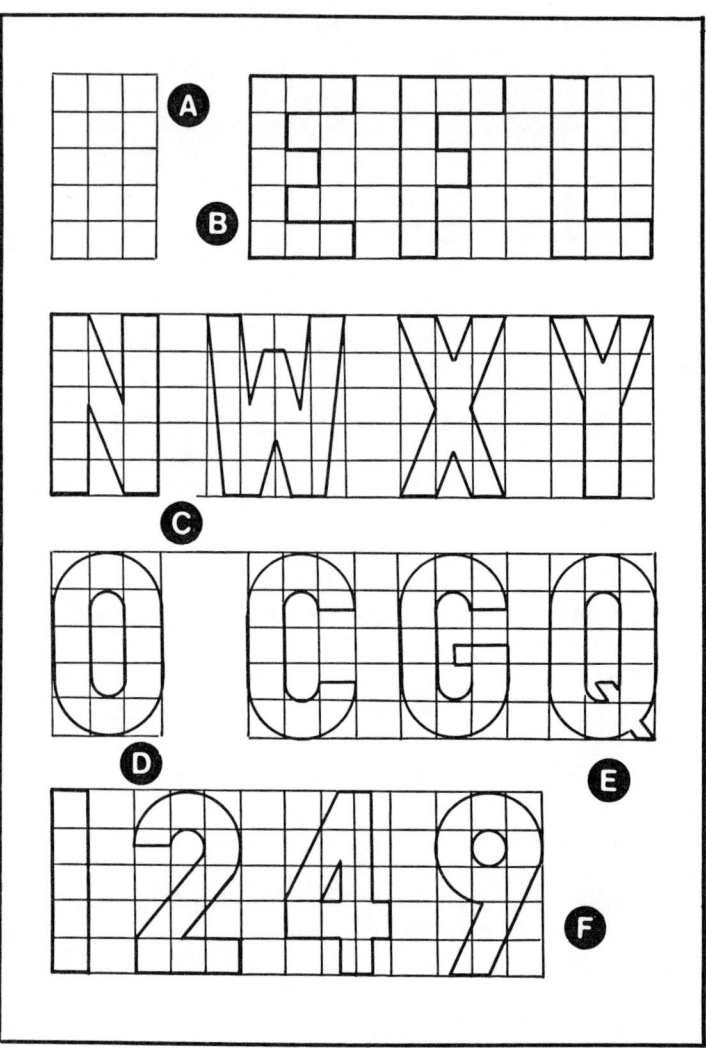

Fig. 8-7. Samples of large letters drawn by using a grid of squares.

tedious, while a large amount of Indian ink may take a long time to dry. It will tend to buckle some types of paper. An alternative to filled lettering is to complete it fairly heavily in skeleton form, while any construction lines are erased so the outlines are emphasized (Fig. 8-9A). A variation on this is to make the ends of the letters semicircular (Fig. 8-9B) instead of square.

Yet another variation is to convert the outlines into shadow letters. Assume there is a light shining from the top left corner.

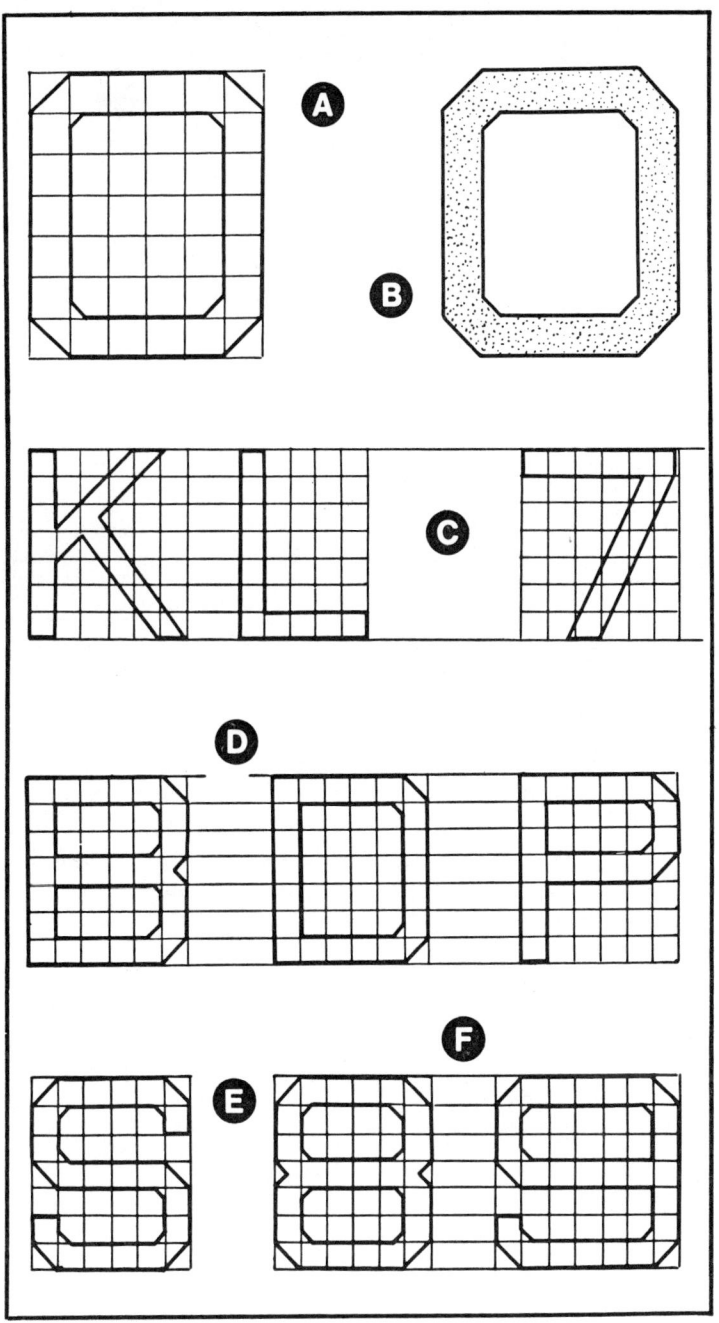

Fig. 8-8. Another form of letters without curves, based on a grid of squares.

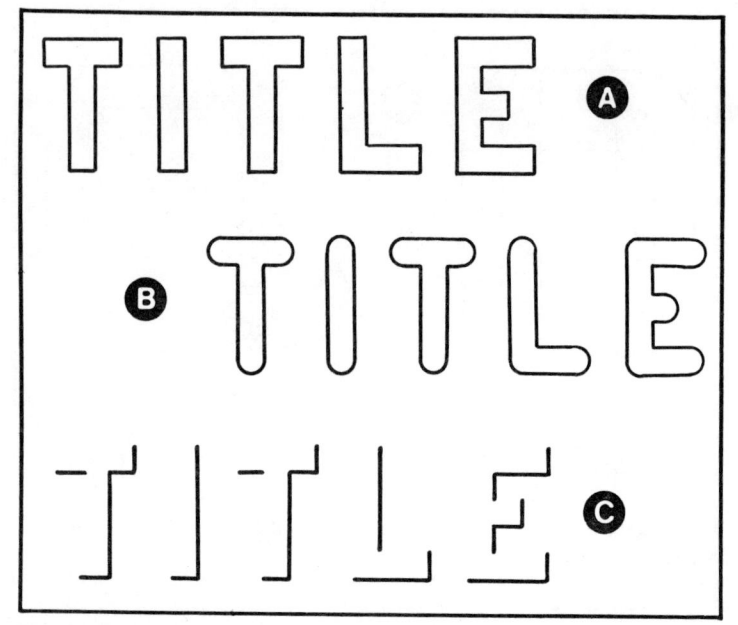

Fig. 8-9. Emphasis can be given to a title by using open letters (A and B) or by only drawing shadows (C).

Draw in what would be the shadows if the letters were solid (Fig. 8-9C). The straight line letters are particularly effective and easy to draw, but where there are curves you have to allow for the shadow changing sides. In doing this sort of lettering, make sure there is a shadow one side or the other of every part of the letter. In titles, it is best to avoid lower case letters for the shadow effect, although they can be drawn.

Chapter 9
Dimensions

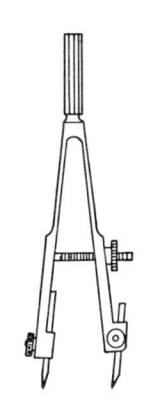

The main purpose of most drafting work is to produce drawings that show others the shapes and sizes of things that either exist or have to be made. The various views may be to scale or full-size. Although the reader might get the sizes by measuring directly on the drawing, it is more helpful to provide dimensions for all the parts to which it is expected he will need to refer. This means that besides the lines that show details of the subject of the drawing, there will be other lines and numbers indicating sizes.

The alternative which is sometimes used is to draw a scale on the sheet, long enough for the reader to be able to refer any measurement he needs with dividers. To a certain extent this simplifies the work of the draftsman, as he does not have to provide dimension lines. It could be a nuisance to the user of the drawing, as he has to set dividers every time he needs a measurement instead of reading it directly from the figure in a dimension line. A scale is useful when there is considerable detail on a fairly small drawing, and it would be impossible and confusing to try to dimension every part. This would apply to a general drawing of a piece of antique furniture with elaborate moldings and carvings. This would be drawn as accurately as possible, maybe with a few dimension lines showing overall sizes. Then a scale is put near the border at one side and as long as the greatest dimension of the piece of furniture (Fig. 9-1). There may have to be other sheets of drawings giving details, but anyone referring to the general drawing can discover unmarked sizes by using dividers on the scale.

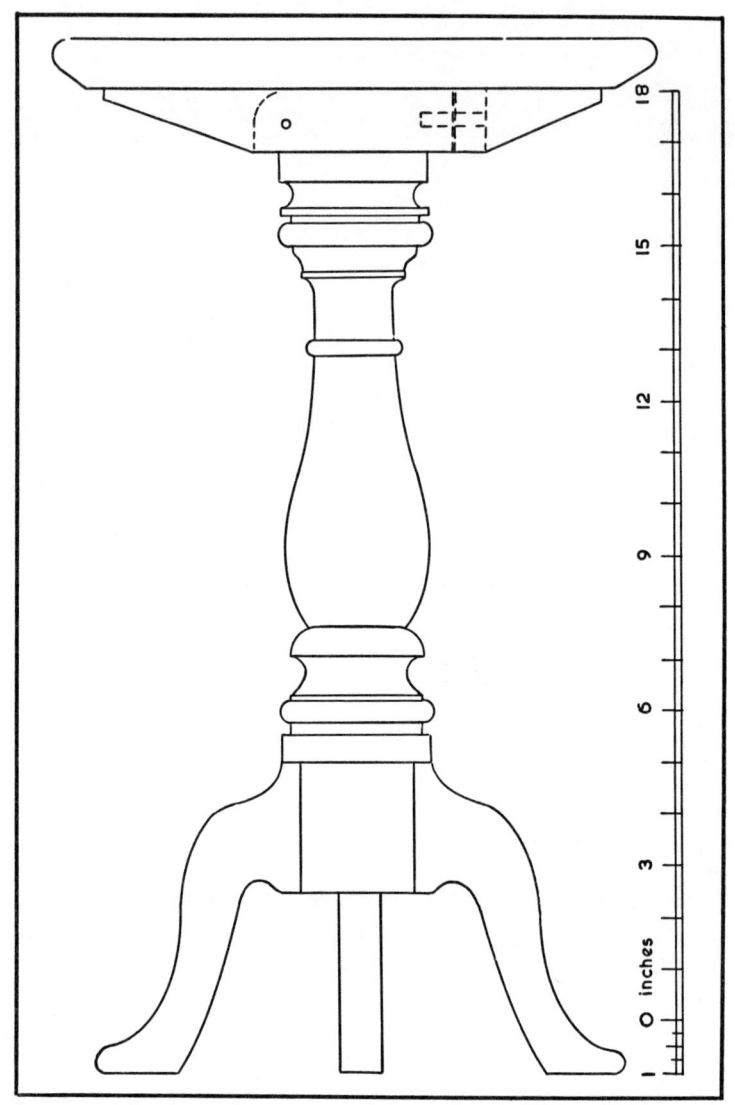

Fig. 9-1. Where a drawing is involved and individual dimensions would be difficult and confusing, it is better to include a scale for reference.

DIMENSION LINES

Dimension lines have already been described in Chapter 5. It is obviously important that there should be no doubts about the points to which they refer. The arrowheads should finish exactly opposite the reference points and be connected to them with fine projection lines. It would be wrong to have the arrowheads short (Fig. 9-2A) or without projection lines (Fig. 9-2B). In nearby all cases the projection lines should be square to the dimension lines and go past the arrowheads (Fig. 9-2C). For neatness they may stop short at the outline, but any gap should be quite small (Fig. 9-2D). Sometimes, as in a pictorial view, it is better to let projection lines continue construction lines.

In nearly all drafting, the figure indicating the size is included in a break in the dimension line (Fig. 9-2E). In some architectural and civil engineering drawing there is a preference for putting the figures alongside the dimension lines which are unbroken

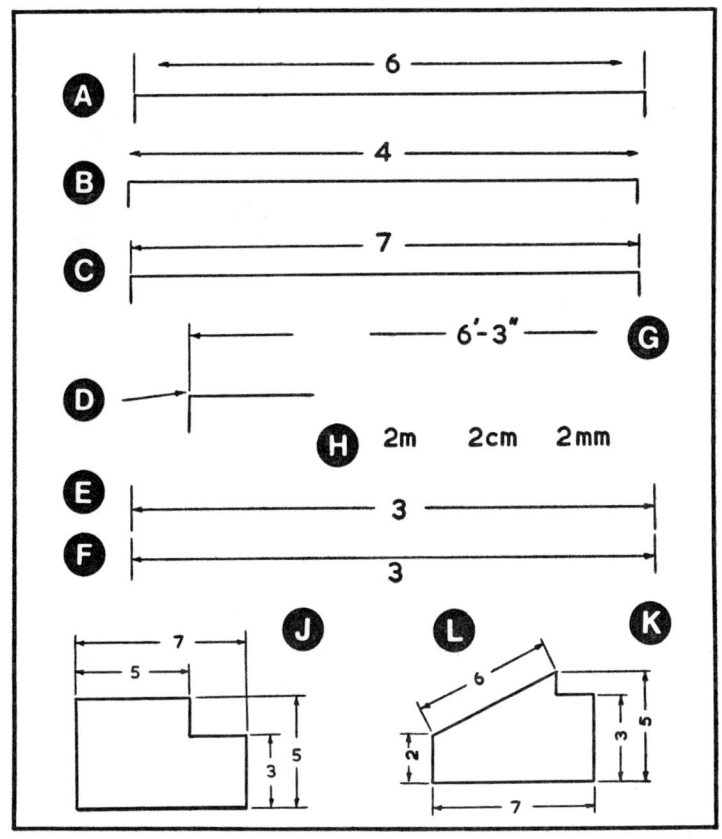

Fig. 9-2. Dimension lines should not be short (A) or without projection lines (B), but should leave no doubt (C). Projection lines need not quite touch the main lines (D). The figure may be within or alongside the dimension line (E and F). Mark the system used (G and H). Make dimensions clear and outside the main lines (J to L).

115

(Fig. 9-2F). It is important that the meaning of the figures are clear and they are easily read. This is particularly important with fractions (see Chapter 8). If there can be no doubt what the figures represent, there may be no need to indicate feet, inches, or millimeters. If all the sizes shown are in inches, there could be a note on the drawing, probably under or near the title, saying, "All sizes are in inches." This would remove any doubt and risk of error. If feet and inches are mixed on the drawing, use the appropriate symbols, with a dash between feet and inches (Fig. 9-2G). If the drawing is metric, use lower case letters without stops—m for meter, cm for centimeter, and mm for millimeter (Fig. 9-2H). Note that the letters are close together as well as without stops. Where larger distances have to be shown, it may be better to write complete words such as miles.

Assume that a drawing will be kept the right way up for most of the time it is being read, so dimensions should be arranged to be understood easily. Sometimes all dimensions are arranged with the figures upright, whatever the direction of the lines (Fig. 9-2J). Another common practice is to have horizontal dimension lines with the figures to be read that way, while vertical dimension lines are arranged as if the reader views them from the right-hand side (Fig. 9-2K). Any diagonal lines then have the figures square with the lines (Fig. 9-2L).

LOCATING DIMENSIONS

The main dimensions of most things are *length, breadth,* and *thickness.* There may be many details within these limits. The actual outlines may be anything but square, although for setting out they can be regarded as enclosed by rectangles. The first priority in planning dimension lines is to allow for measurements overall in these three directions.

If it is a pictorial or isometric view, you can put all the main dimensions on the one view (Fig. 9-3A). With the usual three views, you have a choice of where to put some of the dimension lines. It is a good rule to never repeat a dimension. For instance, there is no need to put the width on the plan and the end elevation. It is sufficient to have it on one view. In a simple drawing it is unlikely that there could be a mistake, but many drawings are used to develop designs, with possible alterations along the way. It could be that a dimension on one view is altered, while a duplicate dimension on another view is left unaltered, leading to mistakes and confusion.

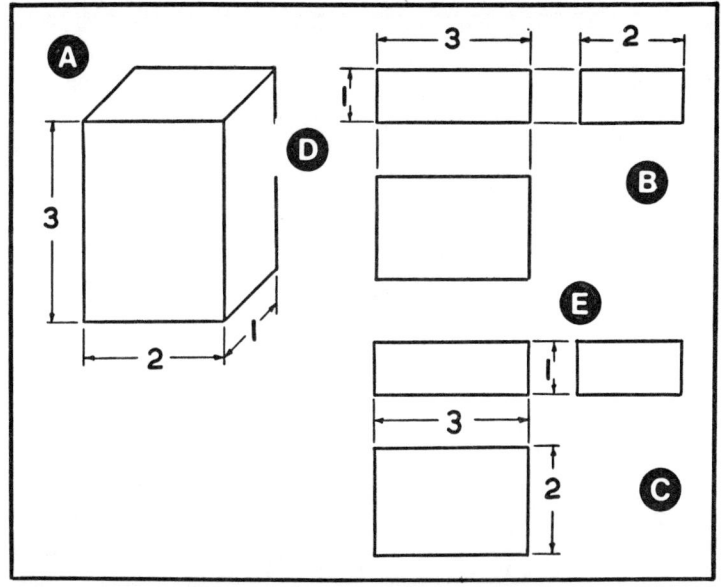

Fig. 9-3. Projection lines for dimensions should be square with or extend from main lines.

With three views, you could put length and breadth in line above the two elevations (Fig. 9-3B), or you might use the projection lines between the side elevation and the plan for the lengthwise dimension (Fig. 9-3C). It would then be neater to put the breadth dimension on the same level at the side elevation. It would not be wrong to put it on the plan view instead. The height might come at one end of either elevation (Fig. 9-3D). If you are using the projection lines between the side elevation and the plan for length, it would be logical and neater to use the projection lines between the two elevations (Fig. 9-3E).

It is bad practice to put dimension lines across the body of a drawing, as that can lead to confusion. In nearly all cases it is better to have them outside a view, although there are a few rare occasions when a dimension on the main outline has to be accepted as the best way.

Keep dimension lines all the same distance from the main lines to which they refer. Let them be far enough away for the figures to be clear of lines and easily read, but not so far that the reader's eye has to follow projection lines a long way to see where the measurements apply. Difficulties come if there have to be several dimension lines applying to the same edge.

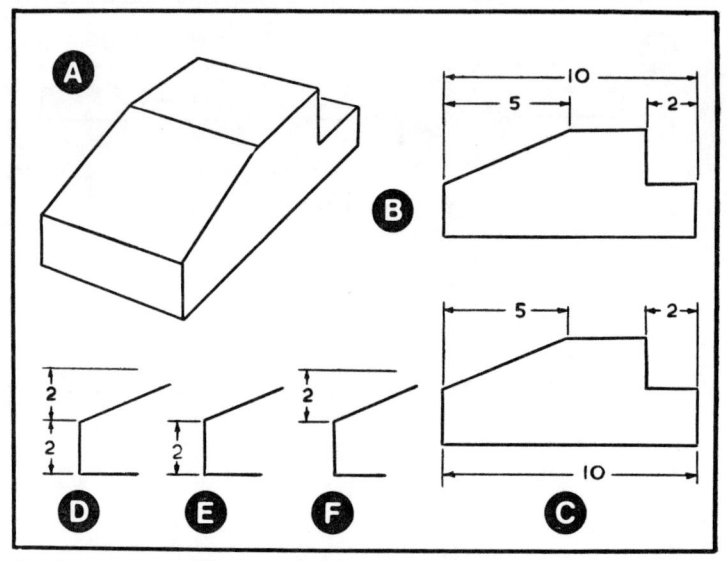

Fig. 9-4. Locate dimension lines so they are all there and do not confuse, while not omitting anything.

Suppose the simple rectangle is stepped and beveled at one end (Fig. 9-4A). Besides the overall length, you have to show the dimensions of the step and bevel. You can have their lengths in line with each other and the overall size outside them (Fig. 9-4B). Another way is to put the overall length at the other side of the view (Fig. 9-4C). This also applies to the depths of the step and bevel. Give either the depth of the solid part or the depth of the cut part, but not both. Sometimes it is necessary to show both (Fig. 9-4D), but unless there is a special reason for doing so, show the overall size and one of the part depths (Figs. 9-4E and 9-4F). Which of the two is chosen depends on which you regard as more important in the particular item.

With three views it may be better to put some of the dimensions on one view and some on another. Spreading dimensions around the views often leads to clarity and may make a tidier drawing, but it is the usefulness to the reader which counts. Which arrangement will be most easily understood and be least liable to error?

SMALL DIMENSIONS

One problem that has to be dealt with is the dimensioning of small measurements. A point will be reached where putting arrowheads at each end of a short dimension line does not leave space for figures. One treatment has the size at the side of a short dimension line, even when elsewhere the figures are within longer lines (Fig. 9-5A). Another treatment is to reverse the arrowheads so they come outside the projection lines (Fig. 9-5B). The figure goes in the space if there is room or outside if it is very close. A further problem comes if you have two small measurements adjoining. With this method of dimensioning, the lines have to come at two levels on the projection lines (Fig. 9-5C).

Another way of dealing with short dimension lines is to arrange a leader from the center of the line to the figure (Fig. 9-5D) or from one end to it (Fig. 9-5E). It is good practice to take the leader away at an angle that cannot be confused with other lines, but put a short horizontal end on it towards the figure (Fig. 9-5F).

CIRCLES

The reader needs to know the location of the center for any circle. If the circle represents a hole that can be related to two edges which are square to each other, two center lines (long and short dashes) can cross at the center and extend far enough to be used as projection lines for dimensions (Fig. 9-6A).

The size of a circle can be shown with projection lines from the circumference and the usual dimension line (Fig. 9-6B). Sometimes, particularly with a large circle, there can be a dimension line across a diameter (Fig. 9-6C), preferably at an angle different from any other line on the drawing so it cannot be mistaken. Another

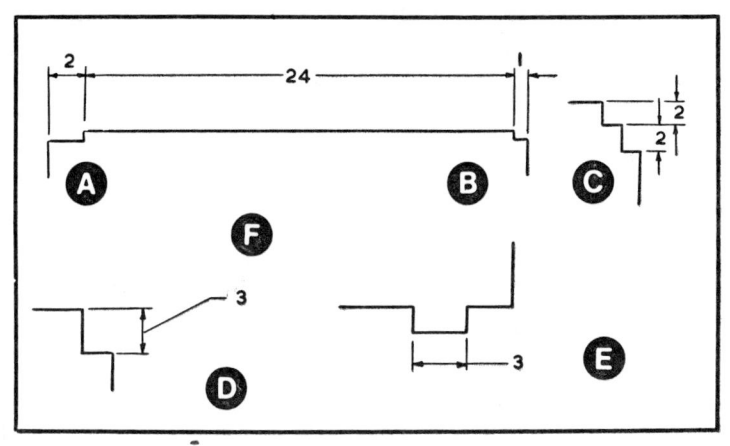

Fig. 9-5. Small dimensions may have reversed arrows, and figures may have leaders.

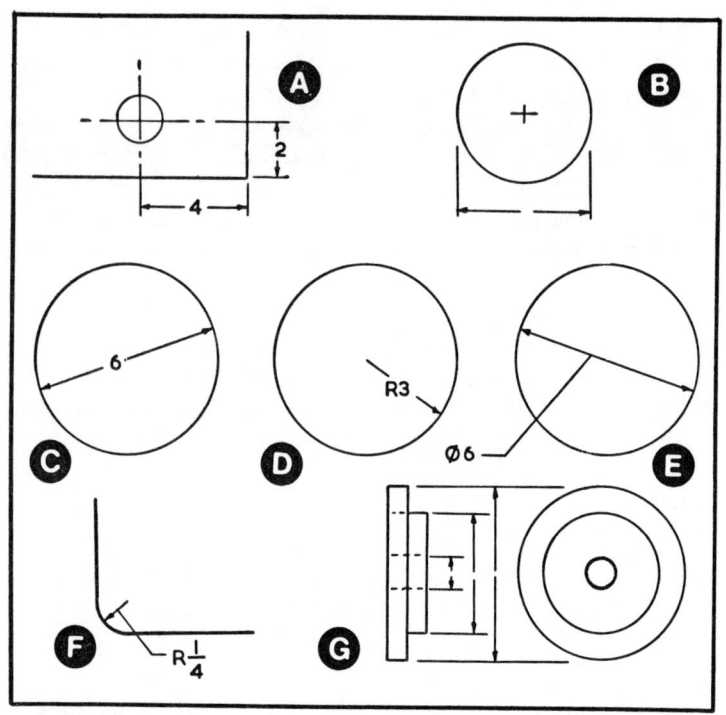

Fig. 9-6. Circles are located by their centers and dimensions may be outside or inside. Dimensions common to two views may come between them.

way is to show the radius, with a letter R alongside the figure (Fig. 9-6D). The dimension can be the diameter, which is indicated by the letter D or a symbol having a diagonal stroke across a circle. This might be taken outside the main outline with a leader (Fig. 9-6E). Using the letter R and the figure would be the best way of indicating the small radius required at a corner (Fig. 9-6F). For neatness, keep this sort of thing level and possibly in line with dimensions. If what you are drawing is round in one view and with parallel lines in the other view, or the circle is represented by a straight line in the second view, put your diameter dimensions on the straight line view (Fig. 9-6G) for preference.

It is usually possible to arrange a series of dimensions so the shorter is near the outline. Then they get progressively longer and no projection lines cross (Fig. 9-7A). It would be wrong to have them the other way so projection lines cross dimension lines (Fig. 9-7B). On rare occasions crossing may be unavoidable. In that case, break the projection line, not the dimension line (Fig. 9-7C).

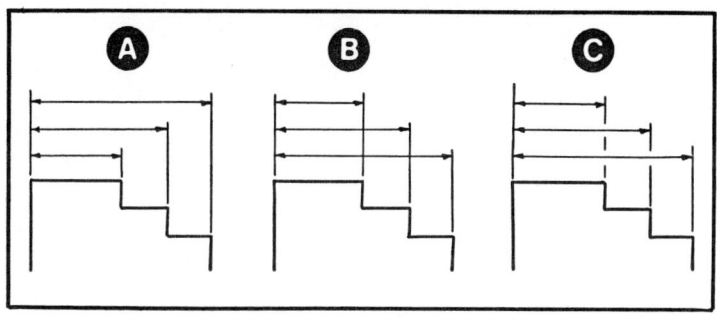

Fig. 9-7. Keep longer dimensions outside shorter ones (A), but if they have to cross (B), break the lines (C).

ANGLES AND CENTER LINES

Dimensions for angles usually have to be directly at the angle, which may be external to the drawing (Fig. 9-8A), but still there when that puts the dimension within the drawing (Fig. 9-8B). The dimension line is made with a compass having its point at the angle and far enough away from it to admit the figure and angle symbol (Fig. 9-8C). Generally, it is better to keep the angle dimension close in, although for a small angle it may have to be some way from the corner to provide space for the figure. In an extreme case the figure may have to be put outside (Fig. 9-8D).

Center lines are important and should be included to indicate that a thing is symmetrical, as well as to locate the centers of circles. Quite often symmetry is the most important feature of the object being drawn, so many dimensions are better made from the center line than from edges (Fig. 9-9A). Where circles are involved, as in a crank with both ends rounded, the center line shows the object is symmetrical. The distance between the centers of the circles is the important one (Fig. 9-9B). The overall size is secondary, as it comes as it will be by drawing the circles (Fig. 9-9C). It may still help the reader to give this size. Another view projected from it shows the thicknesses and other details (Fig. 9-9D). In a case like this where the two bosses project opposite ways, it would be acceptable to put an overall dimension across the drawing (Fig. 9-9E), although some draftsmen may prefer to project clear of the drawing (Fig. 9-9F).

If a part is hidden, it is shown with dotted lines. Avoid giving a dimension to a hidden edge. In another view it is almost certainly shown with a solid line, and that is the one to dimension (Fig. 9-9G). On very rare occasions the dotted line may have to be dimensioned, but always explore ways of avoiding that.

If things are repeated and obviously the same, it is possible to dimension in one place only. Indicate by a note that the others are the same. This happens when holes are repeated (Fig. 9-10A). Keep the figure that gives the number of holes away from that giving the size, so the numbers cannot be misunderstood. "Four holes, ½ diameter," is better than "4½ holes." If that form of note cannot be avoided, put the first figure in a word as "four ½ holes."

A note may take care of a small detail. A chamfer on the end of a rod could be dimensioned and the angle projected (Fig. 9-10B), or there could be a note with a leader giving the details (Fig. 9-10C).

TOLERANCES

In engineering it is possible to work to fine limits, and the draftsman has to take this into account. It is common to work to thousandths of an inch, or three places of decimals. It may be acceptable for something to be made bigger or smaller than the specified size. When working in wood or some other materials, fine limits are impossible, so the draftsman is not concerned with fine dimensions. If it is a precision piece of engineering that he is drawing, he has to take into account *tolerances*.

Suppose a hole has to be made ¾-inch diameter and a ¾-inch rod has to slide through it. If both were made exactly to those sizes, they would not go together. There must be some clearance so they will slide. The amount by which the sizes can vary is the tolerance. It may be decided that the hole could be exactly ¾ inch diameter or up to .005 inch larger. That is a plus tolerance. These dimensions are usually written wholly in decimals taken to three places, even if one or more figures are 0, so ¾ inch is .750 inch. The size would then be shown with the tolerance below it (Fig. 9-10D).

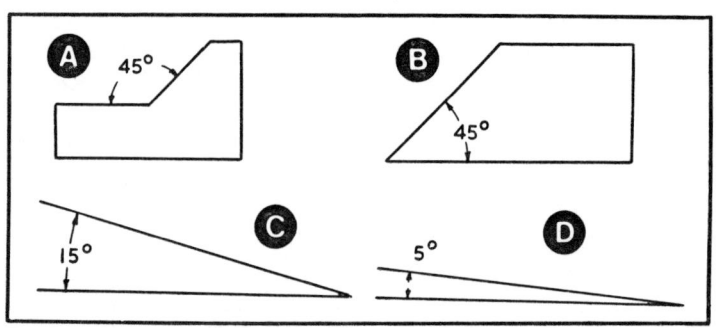

Fig. 9-8. Used curved dimension lines for angles.

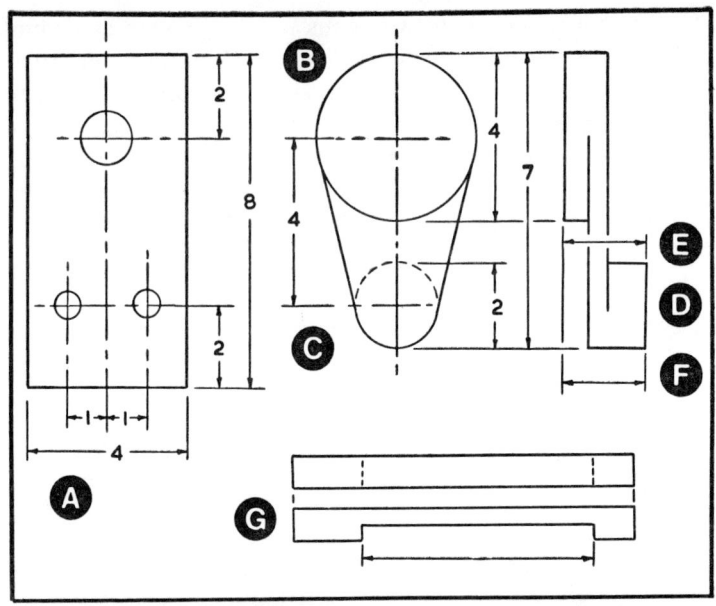

Fig. 9-9. Centers of circles may be located about a center line. Keep dimension lines outside where possible. Put dimensions against solid lines rather than dotted ones.

The diameter of the rod may have to be no more than .005 inch less than the hole diameter, but for its purpose it could be .002 inch less than that. Its size is shown as .745 inch with a minus tolerance below it (Fig. 9-10E).

Some items might be acceptable if made bigger or smaller than the given size. If the tolerance is the same both ways, there can be plus and minus signs before a single figure (Fig. 9-10F). If the tolerances are different, give the two figures with the appropriate indicators (Fig. 9-10G).

If you are dealing with whole figures, as you would be for a large measurement or if you are dealing with millimeters, it is advisable to put 0 in front of the decimal points of any tolerance (Fig. 9-10H). This draws attention to the decimal point and makes it unlikely to be overlooked. A decimal point may not be obvious at any time, and 0 in front of it may be a good idea in other circumstances.

As with decimal points, there can be a risk that the minus sign will not be read as such. It should be the same length as the horizontal bar of the plus sign, and the two bars of that should be the same length (Fig. 9-10J).

Another way of showing tolerances is to give the maximum and minimum sizes. The person working from the drawing does not then have to do any arithmetic, but he knows that any size between the two given will be satisfactory. For a dimension line long enough to allow it, the two sizes are given above and below an unbroken line (Fig. 9-10K). For a smaller dimension the two sizes come outside, above and below a horizontal leader line (Fig. 9-10L).

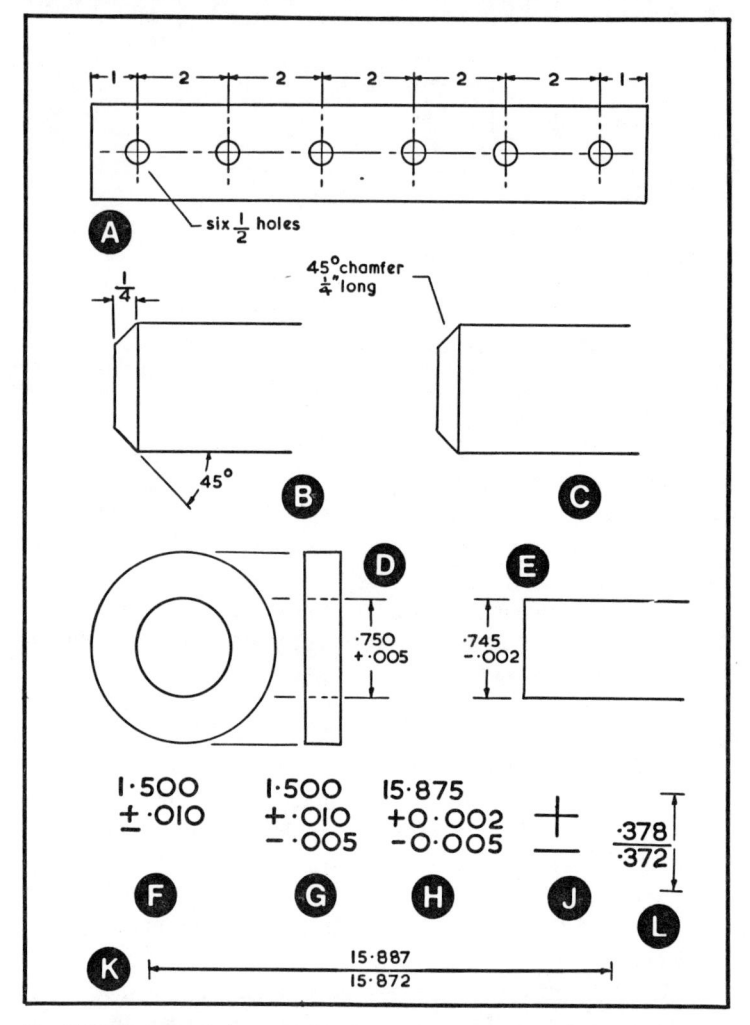

Fig. 9-10. Repeated sizes can be covered by a note (A). Chamfers may be dimensioned (B) or covered by a note (C). Tolerances are indicated in several ways (D to L).

Finish marks are allied to dimensions. These symbols are a sort of graphic shorthand and are dealt with later where they apply.

Dimensioning is not something which can be brought down to rules that can never be altered. The foregoing notes show the acceptable ways of providing dimensions, but some draftsmen have different ideas. Even two draftsmen working in the same office may produce drawings of the same things dimensioned in different ways; yet both are satisfactory. The important consideration is that all the information required is there. The user of the drawing should be able to get all the sizes and details of variations and tolerances from the drawing without having to refer elsewhere for any of them.

A secondary consideration is neatness which must be coupled with clarity. It is no use producing a drawing which has an overall appearance of neatness if this means some dimensions are crammed together and are difficult to decipher. The arrangement of dimension lines can be tried faintly in different ways to get the best overall appearance before going ahead with the final arrangement. With a choice of views which can be dimensioned to give the same information, practice will show how to get the most satisfactory results.

A beginner will find it worthwhile examining drawings by others to see how dimension lines have been arranged. An experienced draftsman usually provides all the sizes needed with an economy of dimension lining; yet he does not leave anything out. Not every draftsman uses dimension lines in the best way, so be critical and consider how a particular drawing might have been dimensioned better.

For neatness, it is important that the thickness of dimension lines all be the same on a particular drawing. Arrowheads should all be the same size, irrespective of the length of lines. It is a common mistake to draw arrowheads larger on longer lines. Similarly, all figuring and lettering should be to standard sizes throughout. Level up dimension lines on different views as far as possible.

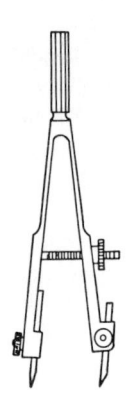

Chapter 10
Sections

Sometimes an internal part can only be shown by dotted lines in the normal views, whatever direction the view is taken. That may not matter with something simple. All the information needed can be extracted from the drawing by the user without risk or error. With other more complicated things, what the dotted lines represent inside the part may not be obvious, and in some cases could be capable of more than one interpretation.

CUTTING PLANE

The way to show something internal or hidden in a form of solid outline that cannot be misunderstood is to provide a section view, usually in addition to any normal exterior views. This is the shape of an imaginary cut across the object, which is indicated by a line across the solid view, often with arrows to indicate the direction of the section view, particularly if the outline would be different if viewed the other way. This line may be drawn like a center line except it is broken with short dashes, or it may be a series of longer dashes (Fig. 10-1A). It is indicating the *cutting plane*. If the cutting plane actually comes on a center line, the normal center line single dot breaks are used.

A simple example shows the method of sectioning. This is a brick with its hollow "frog" on top and a hole through the center. The cutting plane may come across the center and break the hole (Fig. 10-1B). On the side of elevation of the plan, the cutting plane

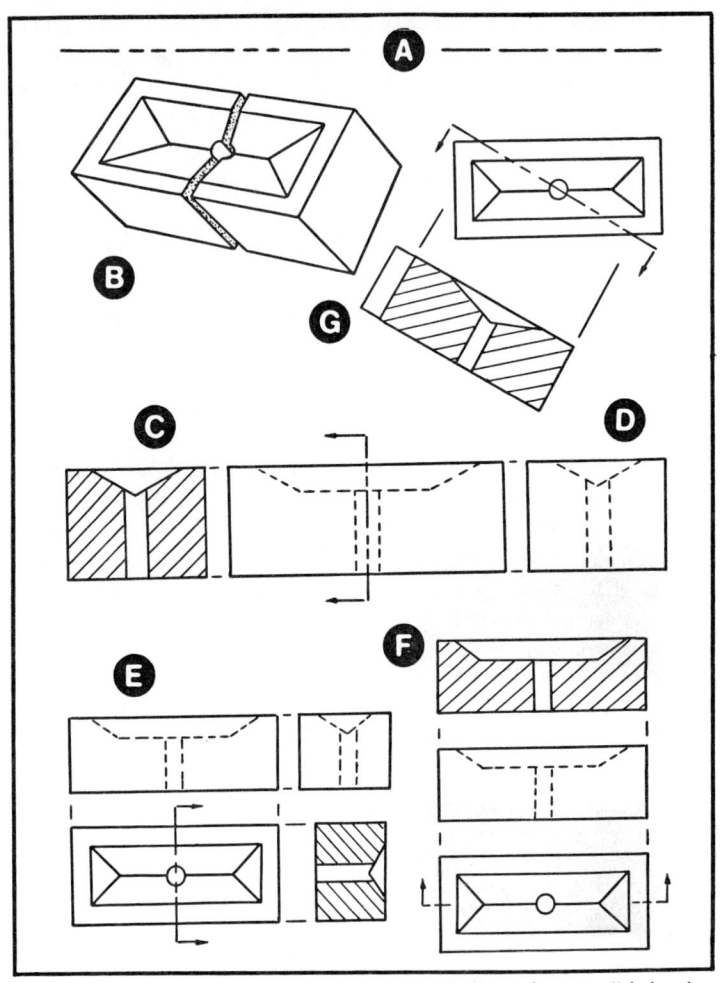

Fig. 10-1. A section is a view at a cut position indicated on a solid view by section lining with arrows to indicate the direction of the view.

is marked and the section is drawn to one side (Fig. 10-1C). If there has to be a normal end elevation, that could be at the other end (Fig. 10-1D), or if because of the particular configuration the end elevation and section have to be at the same end, that is acceptable (Fig. 10-1E).

The cutting plane can be arranged in any direction. The brick could be cut lengthwise (Fig. 10-1F). The section drawing could be put below the plan or above the front elevation, whichever is more convenient or clearer.

127

Sections are nearly always arranged with the cutting plane parallel with one of the edges of the object. In rare cases there may be a need for the cutting plane to be at an angle, as when it may be necessary to know that shape because of its relation with some other part of an assembly. The brick would not be likely to need this, but if it was, the projection would be made parallel to the cut (Fig. 10-1G).

With a round object, such as would be turned on a lathe, a section will give a better picture of any internal work than leaving it to the dotted lines. As the object is round and symmetrical, the cutting plane normally comes across a diameter (Fig. 10-2A). If the section drawing is in the direction of the arrows, what it is should be obvious, but for convenience in laying out the drawing, it may be simpler to put the section elsewhere. It is necessary to indicate it by lettering. Put letters near the arrows (Fig. 10-2B). Then put a title under the section to indicate that it is a view in that direction, using the same letters (Fig. 10-2C). This is important on some drawings where it is necessary to show more than one section view.

CROSS-HATCHING

Cut parts have to be shown by some form of shading known as *section lining* or *cross-hatching*. There are many ways of doing this,

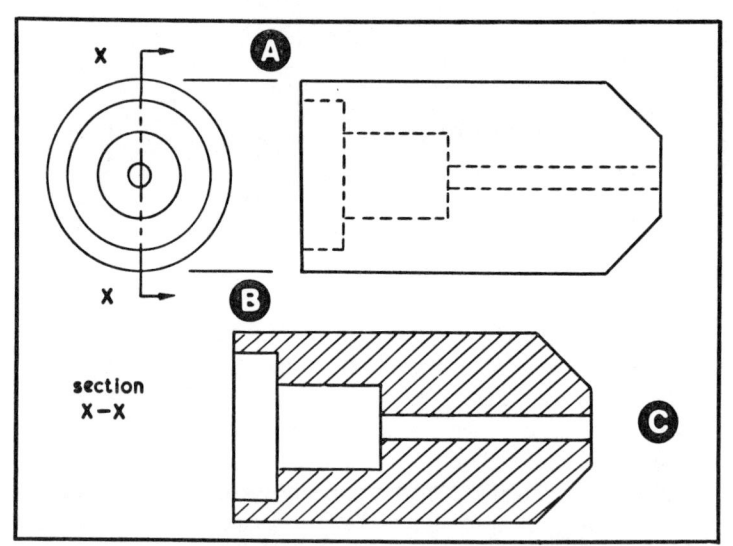

Fig. 10-2. A section may not be directly in line with the arrows, but a caption indicates what it is.

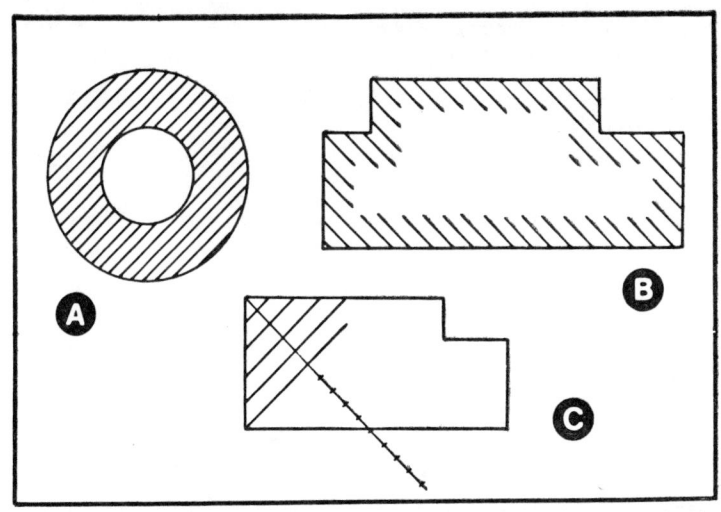

Fig. 10-3. Section lining should be evenly spaced and need not completely cover a large area.

and the shading may be varied to also indicate the material. If color is available, the whole cut area may be given a light color wash. There may be shading with it around the cut edges. This can be done with pencil, giving a gray shade around edges or all over the cut parts. It is more common to use a series of light lines. They should be arranged at an angle that will not confuse any of the outline, and for most cut objects this means at 45° to horizontal.

If there is no call for any other form of sectioning, the cross-hatching should be done with lines neatly spaced (Fig. 10-3A). For smaller parts they may be as close as 1/16 inch, but over a big area drawing them becomes tedious. A spacing nearer ⅛ inch would be acceptable. If there is a large expanse of a section without detail, and where there could be no confusion, the cross-hatching can be restricted to the outlines of the cut (Fig. 10-3B).

With practice it is possible to do cross-hatching without measuring between the lines—merely estimating by eye. For early work, it is advisable to step off with dividers on a faint line drawn at right angles to the section lines (Fig. 10-3C). There are special instruments that move a straightedge by regular amounts between lines. Some of these are quite complex and only justified in offices where many draftsmen frequently have to do section lining. A simpler version is like a hacksaw blade along the T square, and the triangle used for cross-hatching has one or more teeth to engage with it as it is moved along.

SECTIONED MATERIALS

The simple regularly spaced cross-hatching serves for many materials on many drawings, but there are variations on it. Completely different arrangements may be used where other materials have to be indicated. If there is to be differentiation, regular hatching is used for iron only. For steel, the spaces are alternately wide and narrow. For copper and its alloys (brass and bronze), alternate lines are broken into short dashes. Lead, zinc, and the soft alloys have lines both ways. Magnesium and aluminum use crossed lines (Fig. 10-4A).

If wood is involved, a cut across the grain shows annual rings with a few *medullary rays*, while a section taken along the grain shows long grain (Fig. 10-4B). Sand is shown as dots, often closer around edges than in central parts of the cuts, mainly because it is tedious and unnecessary to work closely all over. If it is concrete, larger particles to indicate stones are added (Fig. 10-4C). Dots may be used in technical illustrating for sections of many things or for emphasis by shading, but they are not employed that way on drawings for use in engineering, building, etc.

There are ways of indicating other materials, some of which are vaguely pictorial, while others are similar to some already described for other materials. Anyone dealing with house constructional materials is unlikely to confuse them with engineering sections (Fig. 10-4D).

If the part being sectioned is thin, it may be impossible to draw its outline and then cross-hatch within it. In that case a very thin part may be filled in solidly (Fig. 10-4E). A problem comes when a thin part widens. Filling in the side part solidly might result in too great a mass black, and a change has to be made to section lines (Fig. 10-4F).

MULTIPLE PARTS

A sectioning problem comes when the cutting plane goes through more than one part, so you have sections of two things that are fitted together. When this happens, the cross-hatching must be arranged differently on the two parts. This is simple enough in many things, when the lines go at 45° in one direction on one part and at 45° the other way on the other part (Fig. 10-5A). If the parts are not all basically rectangular, there may be diagonals in the shapes. You must then be careful that the section lines do not come parallel to the outlines of any part. Cross-hatching can be at any angle, so draw the section lines to avoid being parallel to any

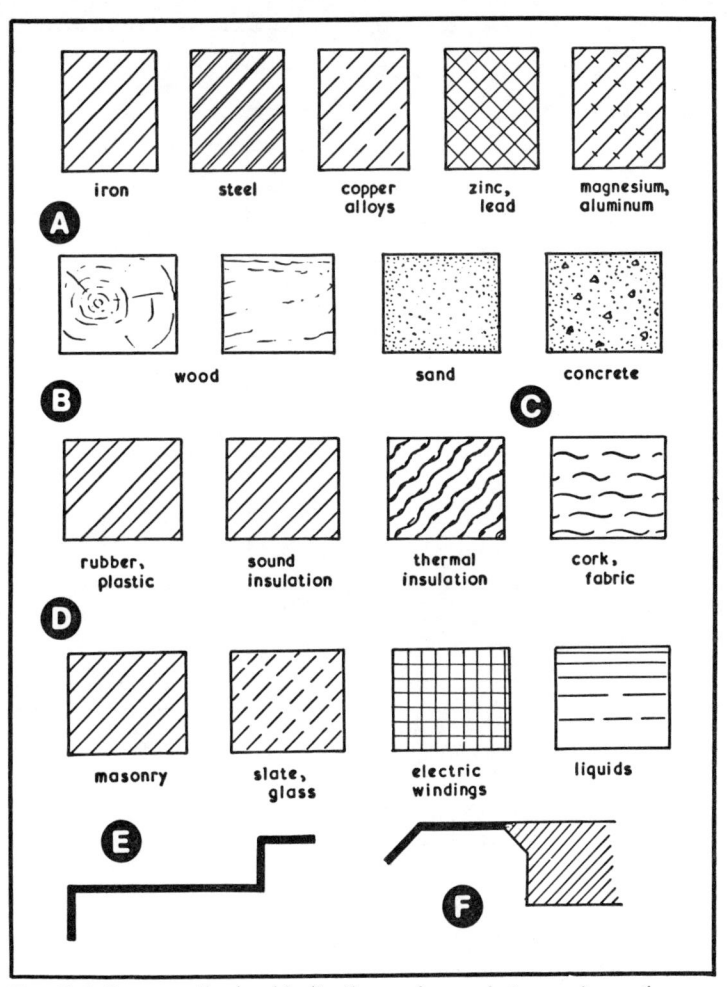

Fig. 10-4. Some methods of indicating various substances in section.

bordering line (Fig. 10-5B). Where the cutting plane goes across a hole or other gap, do not use any section lining (Fig. 10-5C). In a complicated assembly there may be three or more parts to the cross-hatched. If one part separates the others, it does not matter if parts which do not meet are cross-hatched the same way. (Fig. 10-5D). If they do come into contact, you have to choose different cross-hatching angles, and at the same time you must avoid any parts coming parallel with outlines (Fig. 10-5E).

In a section you should show all visible lines beyond the imagined cut. If the cutting plane is through a stepped hole, it

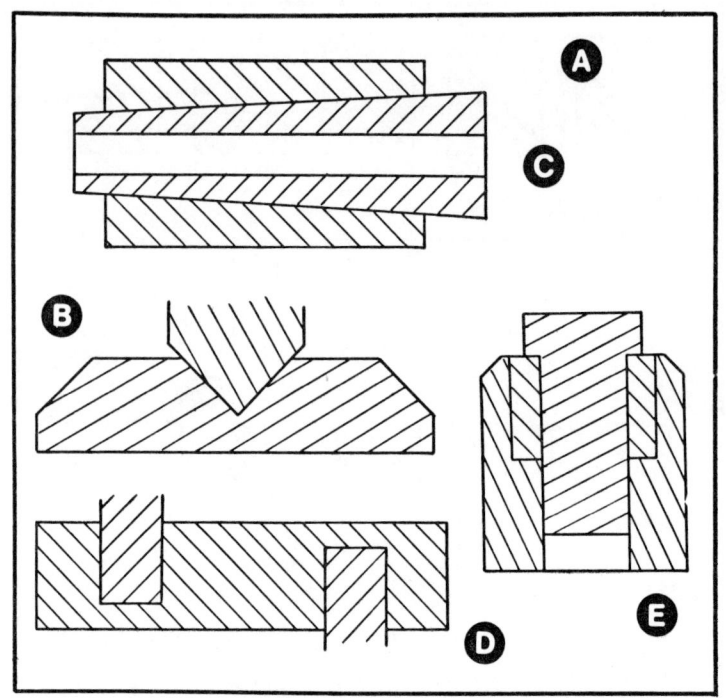

Fig. 10-5. Where a section is through two or more pieces, the crosshatching is arranged differently on each piece.

would be wrong to only show the outline of the hole (Fig. 10-6A). There should also be lines showing the steps at the far side of the hole in the section (Fig. 10-6B).

Although any section should also show solid lines that would be visible, it is better not to include dotted lines to represent hidden parts, unless these are essential to provide some information that cannot be shown better elsewhere.

If a section goes across a bolt, rivet, or shaft which is parallel to the view, those parts are still drawn whole, although what goes around them is drawn as cut. A rivet would be shown with its heads and without section lining across it (Fig. 10-6C). A shaft would be shown in the same way (Fig. 10-6D), while a screwed part would have threads shown in a conventional way (as described later), with the head or nut drawn solid (Fig. 10-6E).

There are some other things that are clearer if sectioning is avoided on them. This happens with ribs. It may be necessary to cut centrally through a hole, and this line splits a rib. Instead of sectioning solidly (Fig. 10-6F), the design is more easily under-

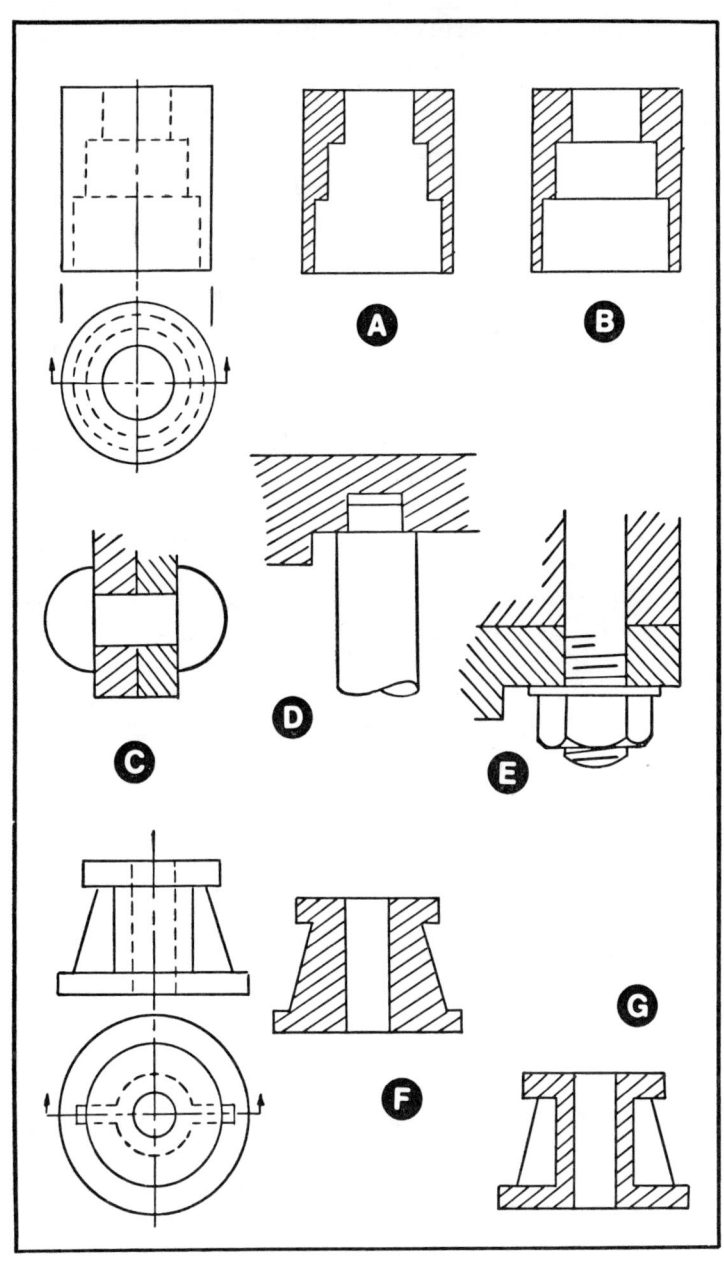

Fig. 10-6. If a section shows internal shaping, do not leave the space bare (A), but show facing edges (B). Round parts fitted into a section (C, D, and E) should not be sectioned. The section of a flanged part should not be fully sectioned (F), but is better with flanges shown whole (G).

133

stood if the rib is not sectioned (Fig. 10-6G). Excessive sectioning of complicated or detailed things may confuse what is supposed to be making the drawing clearer, so experience shows how to arrange sections for clarity and what things like spokes, ribs, shafts, and pins are better not shaded.

The cutting plane is nearly always seen as a simple straight slicing cut. This is the preferable way to arrange it, but suppose there is some detail which does not fall into line. It is then satisfactory to let the cutting plan be stepped as it crosses, so it takes in the offset detail. Steps in the cutting plane should always be at right angles and arranged in a part where the step does not affect the section as drawn (Fig. 10-7A).

It is sometimes possible to draw a section alongside the external view using the same center line if it is a symmetrical object. With a simple drawing, the presence of section lining may be all that is needed to show what the section represents, but if necessary there could be an indication of the cutting plane on another view (Fig. 10-7B).

With some articles, it may only be necessary to section a part to clarify the shape there, while the rest of the object will be clear enough in full views. Such an auxiliary section may be drawn in line with the part cut (Fig. 10-7C). If it has to be drawn elsewhere, it should be indicated as a detail or auxiliary section (Fig. 10-7D). A full section at that point would also show those solid parts further in the same view, but for this purpose these would be an unnecessary complication.

REVOLVED SECTIONS

Another use of sections is to include them actually in the body of a long part. Suppose a long piece has square ends, but is reduced to octagonal for much of its length. This can be shown with break lines each side of a section (Fig. 10-8A). Sometimes a section is drawn on the full view without breaks (Fig. 10-8B), but it is better to cut back the lines so the section is obvious (Fig. 10-8C). At least stop the lines a short way from the section drawing, if there is not much space or as much line as possible is vital (Fig. 10-8D). Such views are at right angles to the parts they represent and may be called *revolved sections*.

Sometimes all that is necessary is a part section that can be combined with a partly completed drawing. In that case a ragged line indicates the division between the two parts. It may be just a piece of casting cut away to show its internal shape, which is

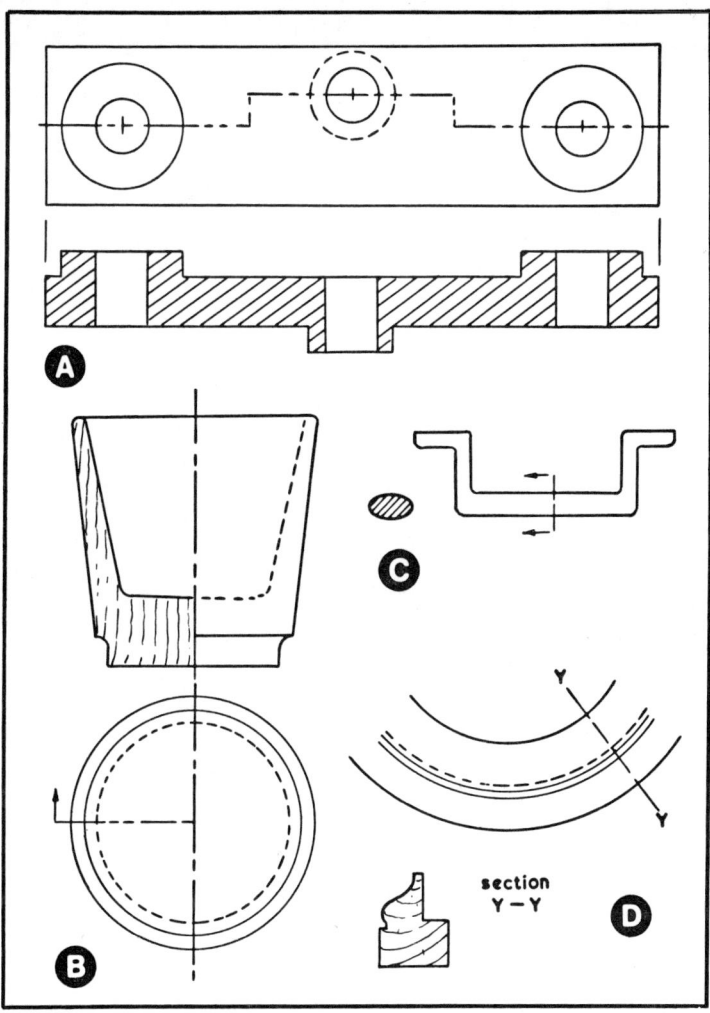

Fig. 10-7. A section line may be stepped (A). A part section can be used (B) or only part of a piece need be sectioned (C). A section across a curved molding makes the shape clear (D).

continued dotted in the other part (Fig. 10-9A). In woodwork it could be a tabletop, shown complete except for one corner where the rails and the top of the leg are visible (Fig. 10-9B). In this case it is only the top which is imagined cut away. The parts exposed are still whole, so there is no need for section lining.

Sometimes a section breaks an object, so a true and complete representation would involve developing the cut line. If a cylinder

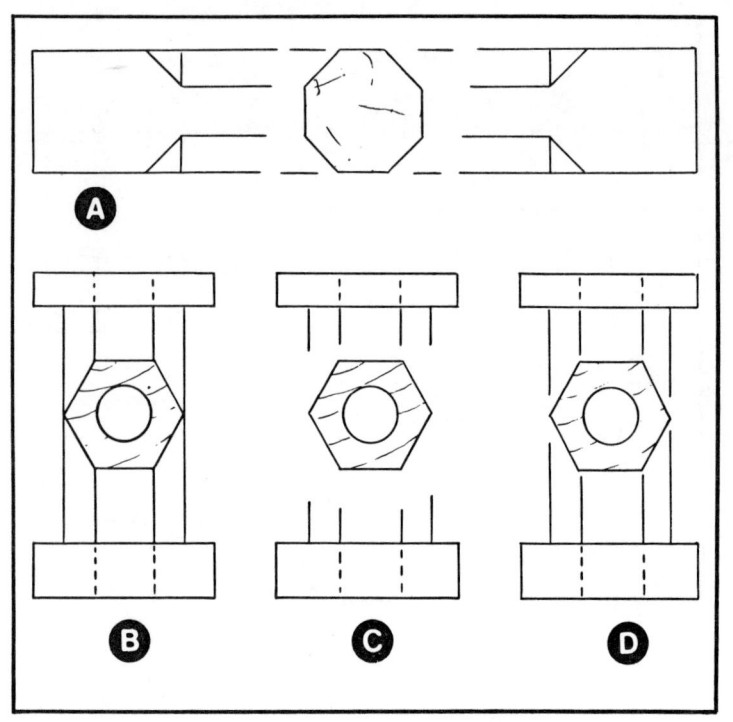

Fig. 10-8. A revolved section can be drawn in position.

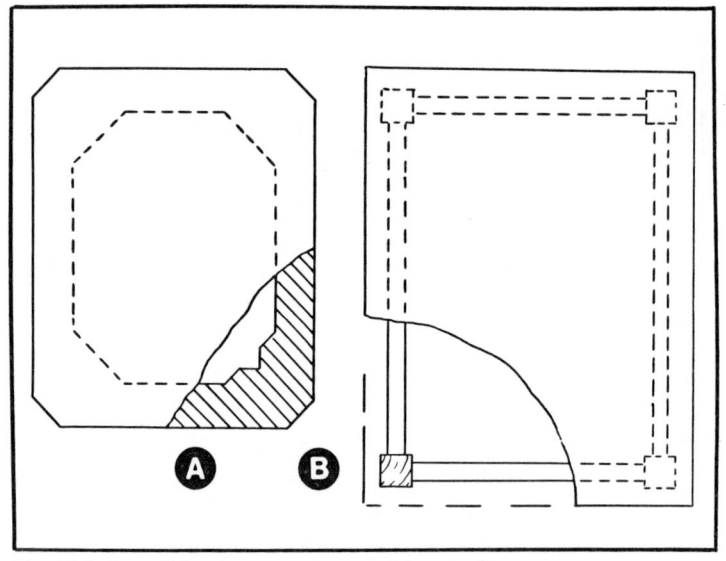

Fig. 10-9. A partial cutaway exposes hidden detail.

136

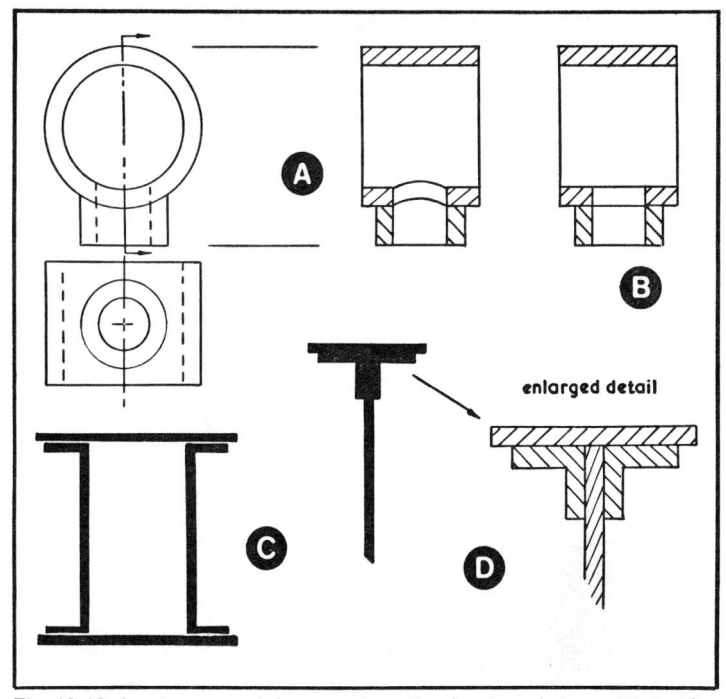

Fig. 10-10. A cut across a tube may show the developed shape (A), but for most purposes it would be drawn straight (B). Thin sections are drawn solid (C), but there may have to be an enlarged detail (D).

comes into another larger one, a section across the entering pipe would give simple lines for the cut cylinders, but the visible solid line indicating the joint would be curved (Fig. 10-10A). It may be possible that getting this shape correctly drawn is important for some reason, but for many purposes the developed shape does not matter as the section has been made to illustrate some other point. In that case a straight cut is preferred (Fig. 10-10B).

THIN SECTIONS

When sections are so thin as to need filling in solidly, there comes a problem where two adjoin, as when building up sheet metal. Leave gaps between them, although in fact the parts would be touching (Fig. 10-10C). If for the sake of correct dimensioning they have to be drawn touching, it may be necessary to draw a constructional detail to larger scale alongside (Fig. 10-10D).

As can be seen, sectioning is used supplementary to orthographic projection. It is something that should not be used more than necessary. When you start to visualize how you will lay out a drawing to make its details clear, get your effect with the three normal views if possible. In the vast majority of cases, everything you want to show can be dealt with. You do not need any further drawings. If there is an internal detail that you think would be easier to understand if it was sectioned, consider if it is necessary to have a complete and separate sectioned view or if it can be shown with a part of one of the normal views cut away. Obviously, section drawings should always be used if they make the work clearer. They must serve a purpose, though, and not have to be used because you did not deal with another view in the clearest way possible.

Chapter 11
Auxiliary Views

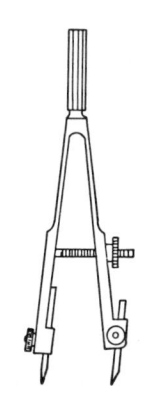

A drawing may consist of many views of the object sufficient to convey all the information a reader needs. For many straightforward things, it will be sufficient to use no more than the usual three views, consisting of two elevations and a plan. These may be supplemented by other elevations and a top and bottom plan, if these views are necessary to show differences. Adding the extra views of this sort is simple for the draftsman, but there are complications if the object will not conform in its entirety to horizontal and vertical planes. There may be no part of the conventional views that shows every detail as clearly as you wish. It may be advisable to draw a further view taken squarely from the part where you want to show that part of the object in a way that will not confuse. This extra is an *auxiliary view*.

EXAMPLES

A simple example is a block with holes drilled in a diagonal face (Fig. 11-1A). The information needed concerning sizes could be shown with a plan and two elevations. For anyone coming to the drawing with no previous idea of what is represented, the arrangement of holes on the diagonal part is not immediately obvious (Fig. 11-1B). An auxiliary view parallel to the diagonal face will allow it and the hole arrangements to be shown full-size or true to the scale of the rest of the drawing.

Widths will be the same as in the end elevation, so draw two construction lines with this spacing in a suitable position and

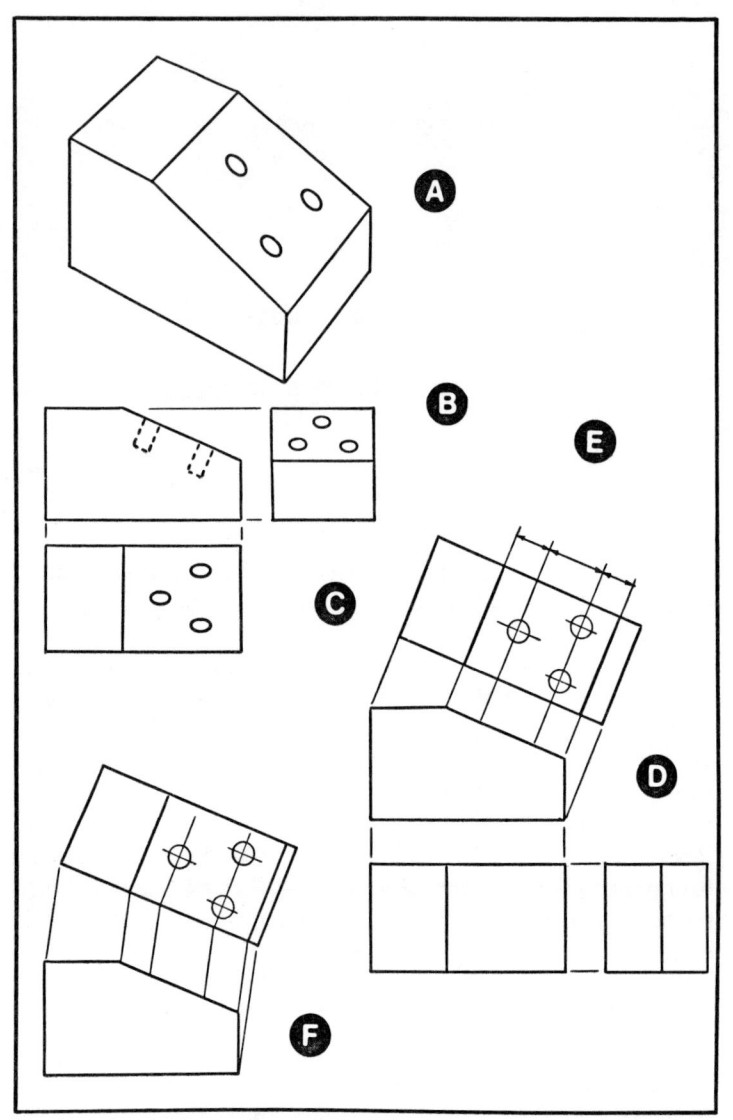

Fig. 11-1. An auxiliary view gives details correct to scale, where they would not be so clear on other views.

parallel with the diagonal part (Fig. 11-1C). Make projection lines squarely with this from the parts of the main drawing which have to be shown, including the hole positions (Fig. 11-1D). Complete the auxiliary drawing including dimension lines, if required (Fig. 11-1E).

140

For convenience in laying out the drawing, the auxiliary view may be better placed by drawing the projection lines at an angle other than 90°. This is feasible, but make sure all the projection lines are parallel. Take them only to the near edge of the auxiliary view, so the lines on the view can be drawn squarely from them (Fig. 11-1F).

Many auxiliary views are symmetrical about a center line parallel with the projected face. That should be drawn first. The other positions are obtained by measuring each side of it on the projection lines. A simple example is a square bar cut diagonally (Fig. 11-2A). The ordinary views do not show clearly the cut face and would be further complicated if there had to be holes or other details on it (Fig. 11-2B). Use an elevation and plan, as viewed from the corner across the cut (Fig. 11-2C). Draw a center line for the auxiliary view far enough away from and parallel to the cut in the elevation (Fig. 11-2D). Project across this line from the corners of the cut. The tips will be on the center line, but use dividers to get the positions of the other corners on the middle projection

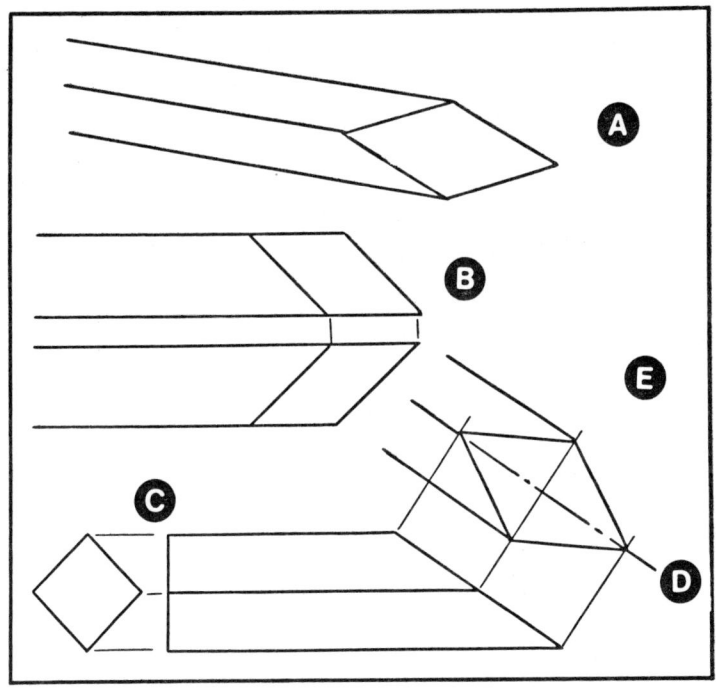

Fig. 11-2. A diagonal cut may not be obvious in an orthographic projection, but an auxiliary view parallel with the cut gives its true shape.

line, using half the diagonal of the plan (Fig. 11-2E). Join these points to complete the shape.

The same sort of projections can be used with diagonal parts on any straight-sided object, whether a polygon (Fig. 11-3A) or an irregular figure (Fig. 11-3B). If the object is considered enclosed in a skeleton form, the various faces can be projected on the outside. The diagonal one will require a diagonal face on the skeleton (Fig. 11-3C). It is not always easy to plan the best way of laying out a drawing so as to produce the most useful auxiliary view. The examples are straightforward and illustrate the point, but actual projects may be more complicated. Remember that you have to draw one side view that includes the true length of the face you wish to project. Other conventional views have to complement this.

With the side view as a base, you can use projection lines to make the auxiliary drawing. It is always simplest to draw the auxiliary view opposite and near the fact it represents, but it does not have to be there and need not actually be parallel to it. For convenience in making the best use of the paper, the auxiliary view might come in any available space. Sizes would have to be transferred by using dividers, and there should be a note indicating what the view is (Fig. 11-3D).

CURVED PARTS

Diagonal cuts on parts with straight edges project easily as points are found which can be joined with straight lines. So long as the cuts are straight, the lines on the auxiliary view will be straight. There are complications if the diagonal part involves curves. For instance, a cut across a round rod produces an ellipse in the auxiliary view (Fig. 11-4A). There is no simple mechanical way of projecting such a curve. You have to make projections that give a number of points on the curve. Then they have to be joined up. The more reference points there are, the more accurate will be the projected shape.

For a cylinder cut across, draw a side view with a plan (Fig. 11-4B). The plan may be a full circle if what you are showing will be included in the finished drawing. If all you are doing is preparing to find points on the auxiliary view, you can draw a semicircle on the side view (Fig. 11-4C). From the base of the semicircle or a diameter parallel with the elevation on the plan view divide the circle into a number of parts. It is usual to step off equal divisions around the circumference, but similar results can be obtained by

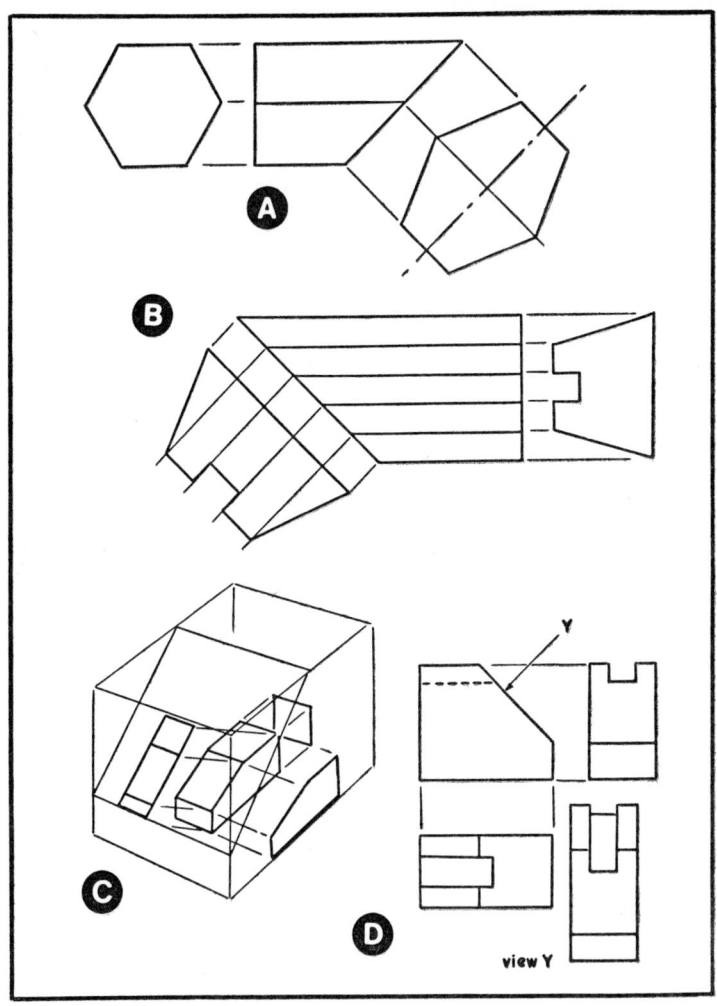

Fig. 11-3. An auxiliary view can be projected (A to C) or drawn in another convenient position (D).

using uneven divisions. As shown, the radius is stepped off in both directions from diameters at right angles to give 12 divisions (Fig. 11-4D).

Project from these points along the cylinder to the cut, keeping the projection lines parallel with the lines of the elevations. From these points project across the center line of the auxiliary view (Fig. 11-4E). It may help in early work to number related parts, but practice will show how to do the work without them.

143

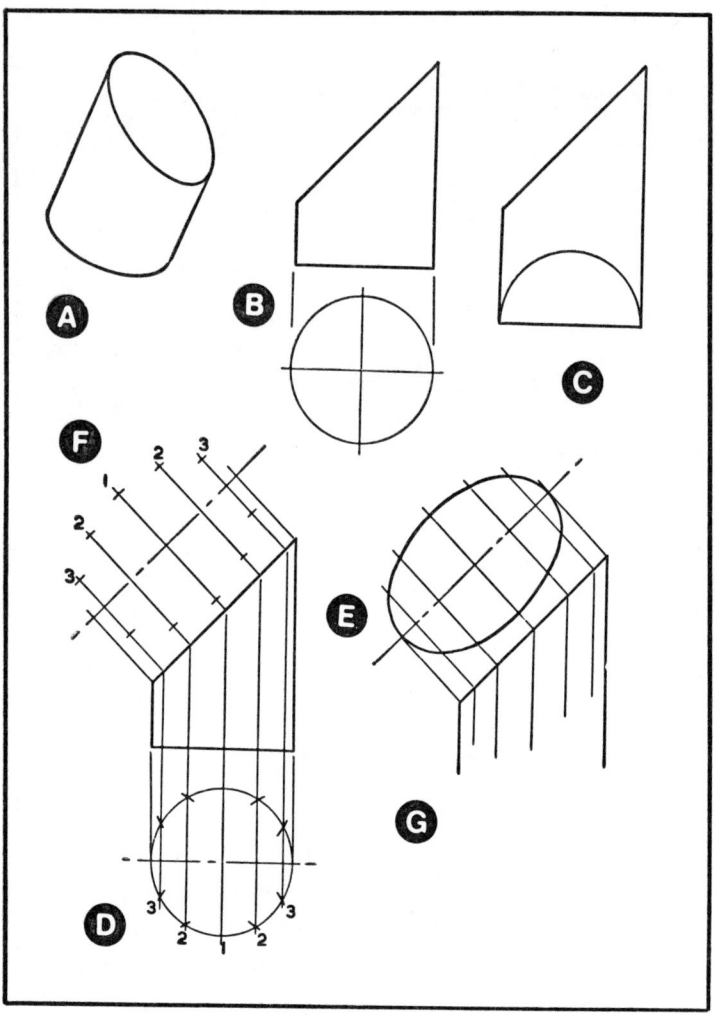

Fig. 11-4. A cut on a cylindrical piece has to be projected to an auxiliary view by getting points on ordinates. Then the result is an ellipse.

Widths on the auxiliary view must now be marked, usually with dividers, to match their corresponding sizes on the plan (Fig. 11-4F). The extremities are obviously on the line, and it helps to next mark the full width at the center. You can then see immediately if you have inadvertently stepped off an intermediate distance on the wrong line. Locate all the positions with short crossing lines. Then draw a curve through the crossing points to complete the view (Fig. 11-4G).

144

Similar methods can be used regardless of the curve involved. It may be a shape combining straight and curved parts. An example is a mitered molding, where the true shape as exposed by the cut is required (Fig. 11-5A). Draw the shape of a cross section taken squarely across the molding, and project a side elevation of sufficient length from it (Fig. 11-5B).

The straight parts can be projected in the way already described without difficulty, leaving a gap to be filled with the curved

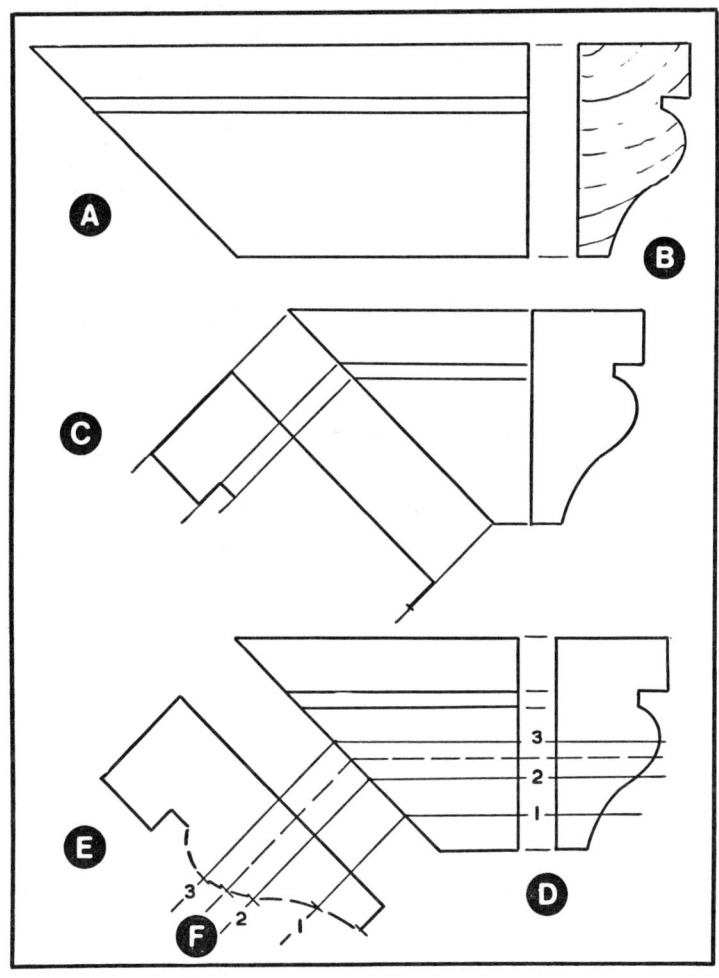

Fig. 11-5. The true shape of a cut across an irregular shape can have the straight lines projected simply. Then the heights of curves are marked on ordinates.

outline (Fig. 11-5C). Draw a number of reference lines across the end view. They need not be evenly spaced, but should cross the curve at parts you think may be difficult to locate accurately on the auxiliary view. Project these lines along the side elevation (Fig. 11-5D). to the cut, and then across to the auxiliary view, where you step off matching heights obtained from the end view (Fig. 11-5E).

When making auxiliary views of curves, you may want to do it with the minimum of projection lines. There have to be enough lines to positively locate points on the curve. If the original projection gives you points that leave you with doubts about how the curve should be drawn at any position, use another line, starting at the plan view and projecting to the cut, and then across to provide an additional reference point on the auxiliary curve (Fig. 11-5F).

It may be sufficient to merely provide the shape of a cut or diagonal surface, but for a full auxiliary view the solid parts visible

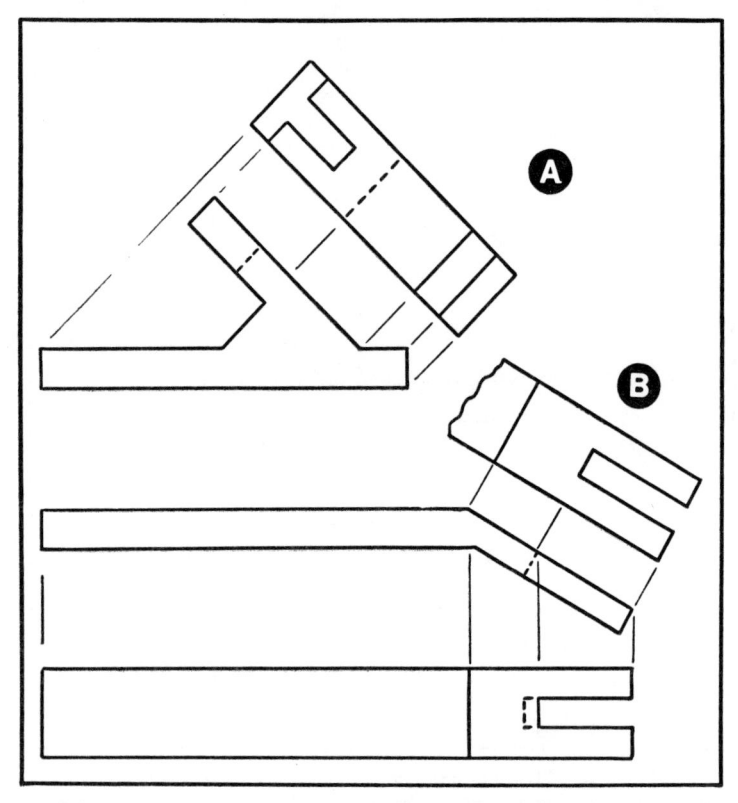

Fig. 11-6. Auxiliary views show the true shapes of parts that are at an angle to the main parts.

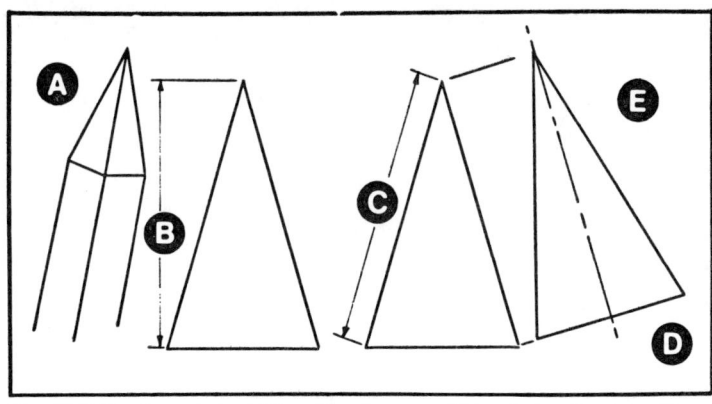

Fig. 11-7. The true shape of the side of a pyramid has to be developed from the center height of a sloping side.

in that direction should also be drawn (Fig. 11-6A). For a proper understanding of many auxiliary views, it is advisable to complete them in this way, including lines for parts hidden in that view and drawn dotted. If only the shape of a cut or diagonal part is shown, for clarity there could be short lines extending into the solid part. Then a break could be indicated (Fig. 11-6B).

CONIC PROJECTIONS

When there are slopes both ways, as in a cone, the true shape of the side is easy to obtain. An example is a regular point in the end of a square rod (Fig. 11-7A). The height of one face is not the vertical height of the cone (Fig. 11-7B), but the height as seen in side view (Fig. 11-7C). A projection to an auxiliary view can have the width of the base marked (Fig. 11-7D) and the points joined to complete the shape (Fig. 11-7E).

Octagon

This would also apply to an octagon, which could be regarded as a square shape with the corners cut off. Its side elevation parallel with one face of the cone will show a side view of another face (Fig. 11-8A). Any other form of cone, where a view square with the side of the base shows a side view of an angle and not another face, needs a different treatment to get the true height and shape of one face.

Draw a plan and side view from parallel with one edge. From the plan, make a projection that will bring the views of the outer surfaces on edge (Fig. 11-8B). Make this the same vertical height

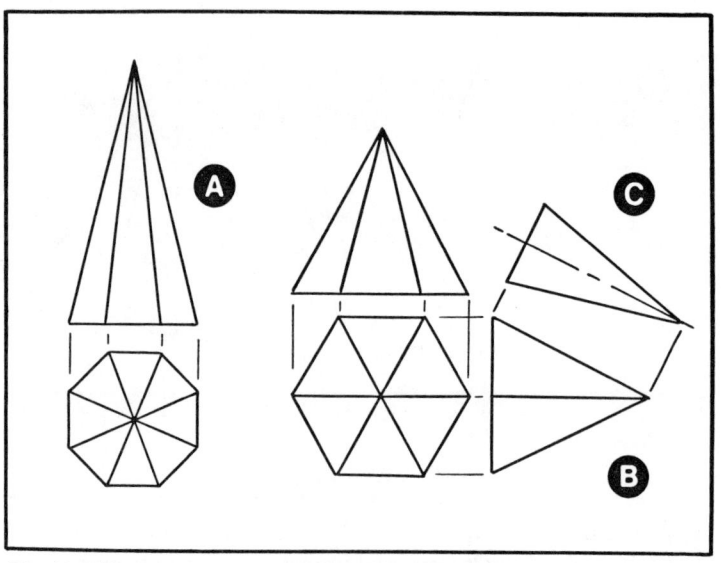

Fig. 11-8. The shape of one side of a cone of any number of sides has to be projected about a center line projection of a slope.

as the first elevation. From the edge of this view, project an auxiliary view, giving the true height and obtaining the width of the base from one side in the plan (Fig. 11-8C).

For any cone with an uneven number of sides, proceed in the same way. Turn the second view so you get a side view of one surface, and then project from that.

Truncated Conic Projections

If you have a truncated square cone cut squarely or diagonally, but parallel with a side, across the top (Fig. 11-9A), you need two side elevations. Each is completed to the full cone, so you have an apex to work from. One elevation has a cut edge parallel with the view (Fig. 11-9B). The other elevation has a corner to the front, so you get an edgewise view of the part to be projected (Fig. 11-9C). Put a plan below one of the elevations (Fig. 11-9D).

Make the projection lines to the center line of an auxiliary view. The extremities of the cut come on this line, but you have to find the width at the center line. The two elevations and the plan are interrelated. Distances on one view can be transferred to another, so it is possible to project from an elevation down to a plan. Draw on it the shape of the cut as seen from above (Fig. 11-9E). The slope there will be foreshortened, but the width will

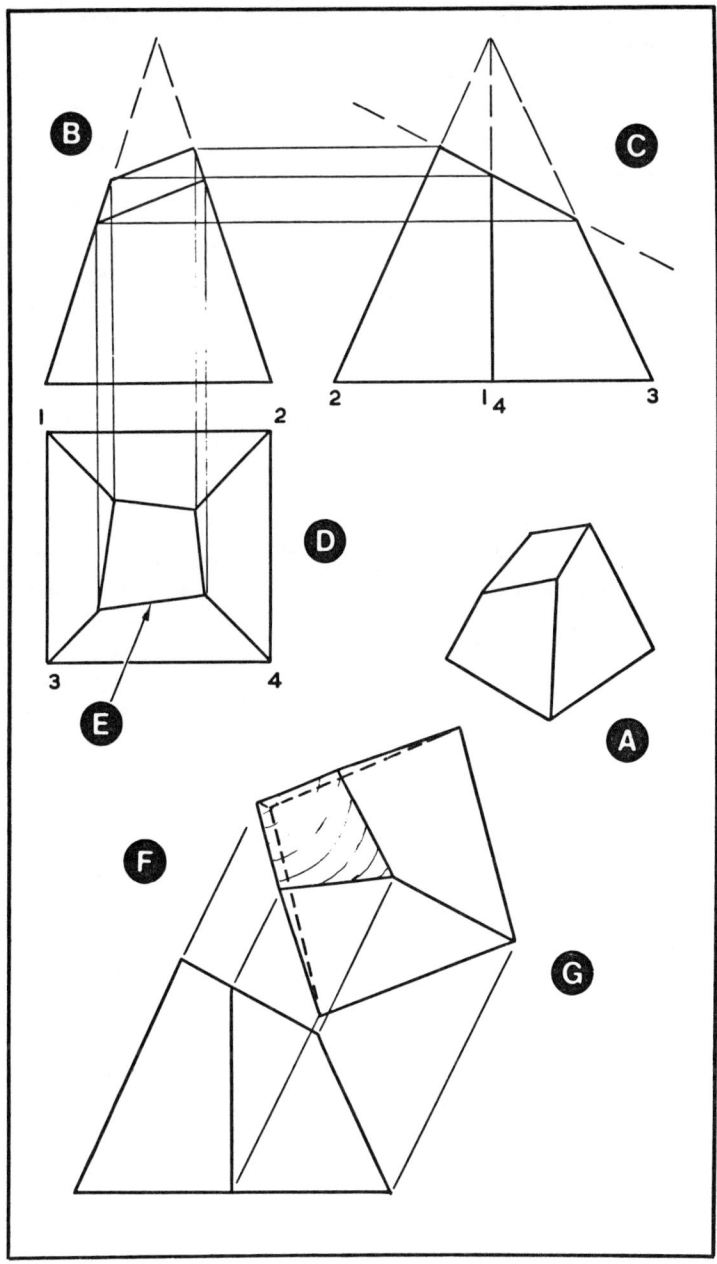

Fig. 11-9. A diagonal cut across a square pyramid (A) can have its true shape projected by using heights and positions projected from four views (B to G).

be correct. Transfer this to the projection to provide points to join and complete the shape (Fig. 11-9F).

That gives the true shape of the cut. If the auxiliary view is to be complete, the solid parts should also be projected to the same center line, using widths obtained from the plan or bases of the elevations (Fig. 11-9G).

If the cone is round, elliptical, or any other curved form, an auxiliary view is made with a combination of the method for a cone with flat sides and that method used for a cut across a cylinder.

Draw a side view edgewise to the cut. Continue lines to the apex of the cone. Draw a similar cone alongside it, as well as a plan view below one of the cones. The second cone will represent an elevation at right angles to the first, but at this stage you cannot show much of the shape on either that elevation or the plan. If you put diameters across the plan and lines on the elevations, you can mark points on the cut on them (Figs. 11-10A and 11-10B).

Divide the circle of the plan view into any number of parts. Only a few are shown to avoid confusion, but in practice there will usually have to be upwards of 12. Project these points to the bases of both cones. Take them to the apexes (Fig. 11-10C). From where the cut on one elevation crosses the lines, project across to the other elevation (Fig. 11-10D). That will give you crossing points representing positions on the curve of the cut in that view. Joining the points gives you the shape of the cut as viewed in that direction. If you need the shape of the cut in plan view, the sizes can be projected to diameter lines between the circumference points on the plan and joined to give the shape (Fig. 11-10E).

Project from all the reference points on the cut to the auxiliary center line. From the widths of these positions as shown in the second elevation, mark locations through which you can draw a curve (Fig. 11-10F) to give the actual shape of the cut. To complete the auxiliary view, project the solid part and complete the shape.

If the cut across a round cone takes in part of the base and the true shape of the cut has to be drawn, the draftsman uses similar methods to those just described. With an angular cone of any number of sides, the method is simpler. The angles can be used instead of projection lines. A vertical cut at any distance across a round cone produces a curve called a *hyperbola* (Fig. 11-11A). If the cut slopes, the shape produced is a *parabola* (Fig. 11-11B).

Draw two similar cones to represent elevations at right angles to each other and a plan below one of them. Show the cut on one elevation and the plan (Fig. 11-11C). There is no need to divide the

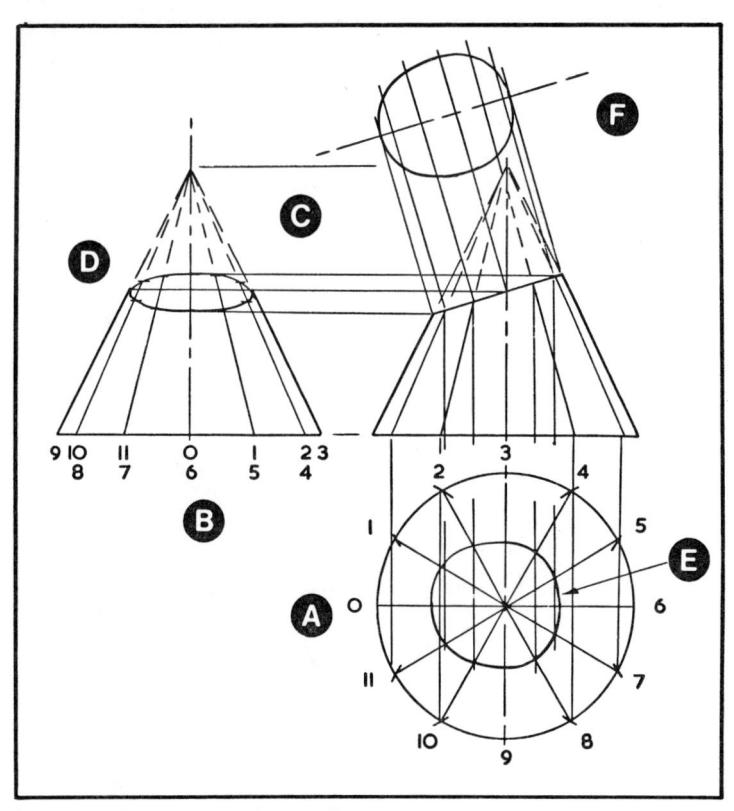

Fig. 11-10. The shape of a diagonal cut across a cone can be found by projecting from ordinates spaced around the base circle and by measuring the widths in other views.

full circumference of the circle in the plan for the vertical cut, as it is only the parts of the cut that need reference points projecting. Make marks around the cut part of the circumference. For a vertical cut, stop at the limits of the cut on the base, but for a sloping cut that crosses to the other side, go all around. Spacing is usually even (Fig. 11-11D), but it may be better to have closer marks towards the sides of the cut, where the flatter curve may need more defining.

Project these lines to the elevation bases. Continue them to the apex (Fig. 11-11E). Note where those on the side view of the cut cross it, and project them to the relevant lines in the other view (Fig. 11-11F). These are points on the curve which can be drawn through them (Fig. 11-11G). This is the true shape of a vertical cut, so there is no need for an auxiliary view.

151

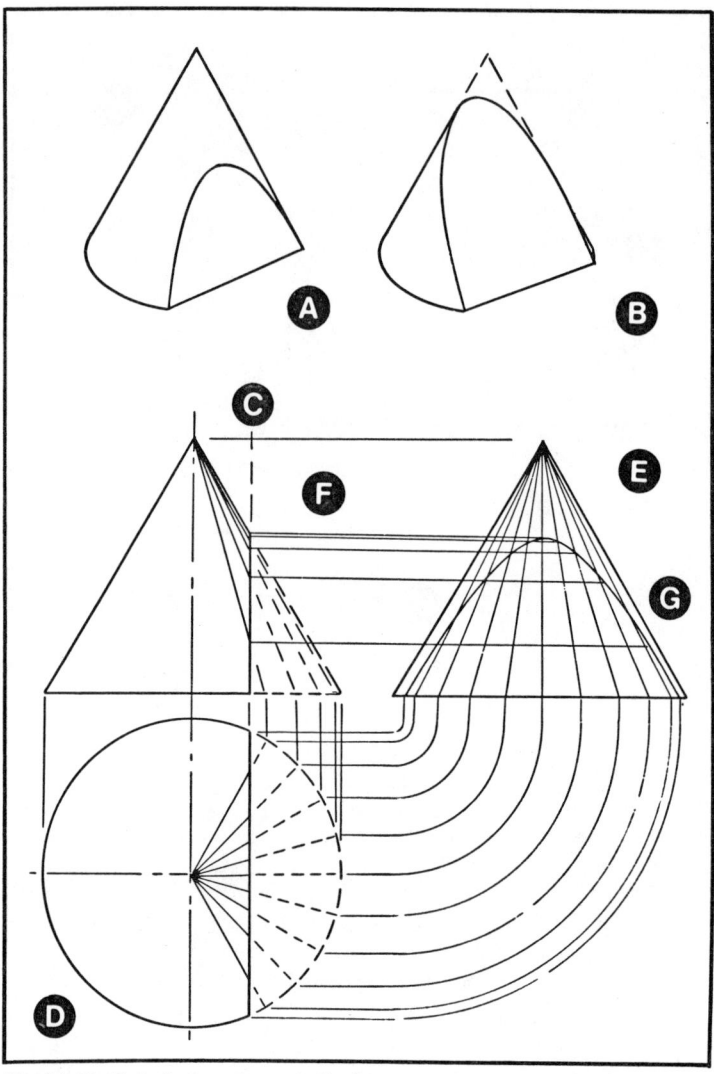

Fig. 11-11. Cuts that go through the base of a cone may be vertical. Their outlines are found by projecting heights of ordinates to a second elevation.

If the cut slopes across the cone (Fig. 11-12A), the method up to this stage is similar, except the curve projected to the second elevation and from that to the plan, if you wish, will have the correct widths. It will be shortened in its depths in both views (Fig. 11-12B).

There has to be an auxiliary view to get the true shape of the cut. Draw its center line parallel with the side view of the cut (Fig. 11-12C). You will have the limits of the cut to mark on it. The width of the cut comes from the plan view, so draw in the base (Fig. 11-12D). Project squarely across where the various reference lines cut the line. Obtain widths of these from the other view or the plan and mark them all (Fig. 11-12E). Then draw a curve through to get the true shape.

REVOLVED VIEWS

It is common to draw views in orthographic projection so edges are vertical and horizontal. Most things are either rectangular or can be assumed to be contained in rectangles, so these views are logical. They may contain all the sizes and other information needed, but the actual shape may not be immediately obvious because none of the views strike the viewer as a representation of the object. A pictorial or isometric view (Chapter 15) might be included to aid in visualizing the project, while leaving the orthographic views to provide sizes and actual shapes.

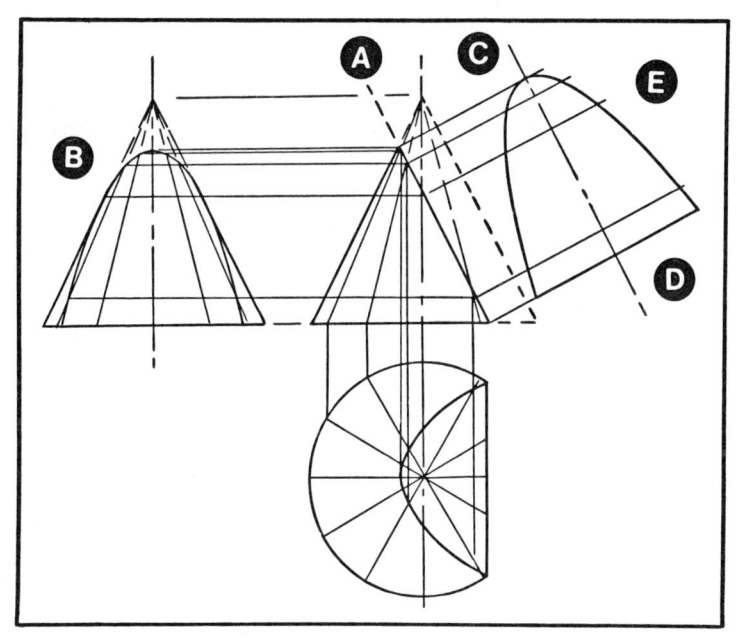

Fig. 11-12. If a cone is sliced across diagonally, positions on the cut are projected between the elevations and plan, and then to the true shape of the auxiliary view.

Another way is to revolve a view. This has the advantage of still being an orthographic view with dimensions that are still true in one or two directions, although one view is diagonal. Measurements that way are probably not of much practical value, unless that way is a true shape of a diagonal or cut face.

The auxiliary views of objects showing the true shapes of faces with the solid parts continued from them are revolved views. The object is pictured with the diagonal face towards the viewer and in correct outline, with the other parts included to the shape and size they are in that view.

The method can also be used for clarity with solid objects. Imagine a solid letter K. Three views would describe it (Fig. 11-13A). If the plan view is tilted to a slight angle, projections of the two elevations will now show a thickness and may give a better idea of the intended shape (Fig. 11-13B). Vertical sizes are the same, while width and thickness can be obtained from the plan by the user.

Revolved views may be used where it is necessary to show a moving part in a second position. In a simpler example, a piece of strip metal may be cranked, so ordinary views do not show the true shape before bending (Fig. 11-14A). The cranked ends can be projected from a side view, so the developed shape is given (Fig. 11-14B).

Revolved shapes have advantages when it is necessary to find true lengths. Many examples are treated in the same way as described for getting actual sizes in auxiliary views. The work is

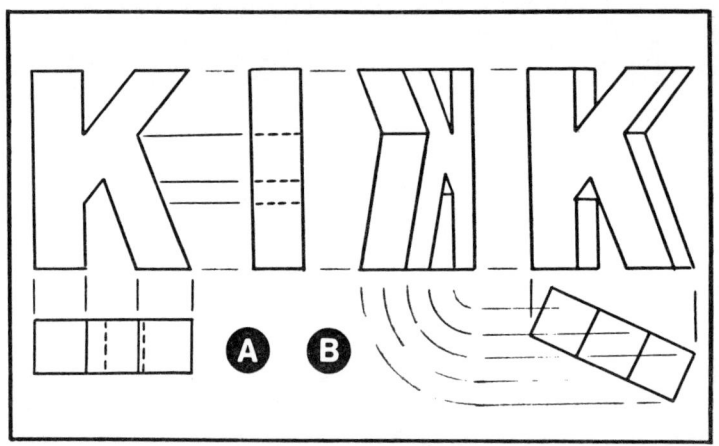

Fig. 11-13. A better idea of appearance may sometimes be obtained by revolving one view, so the other views show the object diagonally.

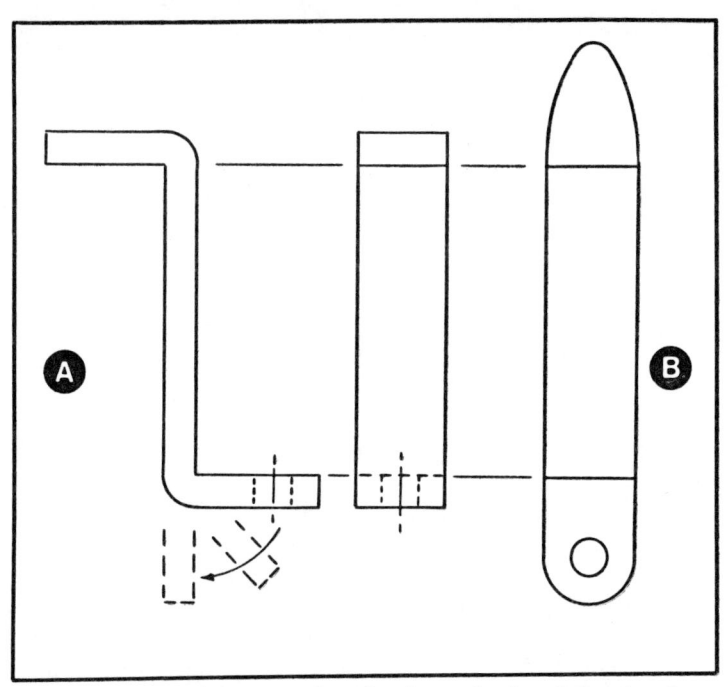

Fig. 11-14. Revolved views can show the shape of a piece before bending.

the application of geometry. The steps are not complicated, but by the time a shape is obtained there may be a confusing collection of projection lines. Do all of the construction work faintly, so unnecessary lines can be erased after the drawing has been completed. The user will be concerned with the result and not how it was achieved, so nothing should be left that might be misleading if a construction line was read as a main line.

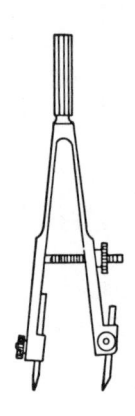

Chapter 12
Developments

If something has to be made from sheet material bent or curved to shape, it is necessary to find the outline of the flat material before bending. This is called its *development.* Sheet metalwork is the obvious application of developments, but there is a need for developments when designing packaging, paper sculpture, and display items made from paper, card, and other sheet material.

Much of the work is related to that described in Chapter 11 for finding true shapes of angular and other cuts. If the item has flat sides and straight folds, development is fairly easy with the minimum of construction lines. Where curves are involved, there have to be many projection lines on which points on the developed curve can be plotted and drawn.

A simple example is an open rectangular box (Fig. 12-1A). It might be made from five parts—two ends, two sides, and a bottom. If it is to be made from material that will not bend, such as wood, that is how it is developed. If it is to be made from something that can be bent, the usual way of making it from one piece is to arrange the folds around the bottom edges (Fig. 12-1B). Allowance would have to be made for joining the corners. If it is a large construction of thick sheet metal, there might be separate strips of angle iron in the corners (Fig. 12-1C). For thin sheet metal or card, there could be flaps to fold in and solder or glue (Fig. 12-1D).

That is not the only way to develop a particular shape. For a special purpose, it may be better to have the upright parts in a continuous length. Then the bottom might be attached to one side (Fig. 12-1E). It might be better to have the bottom separate, particularly if it is required thicker. Then flaps could be on the other parts (Fig. 12-1F). Notice the need to cut miters in the flap corners. It is often advisable to angle the ends of flaps slightly in places where they do not actually have to miter.

TUBES

A rectangular or square sectioned tube is developed like the box. If the top slopes, the heights are projected across or measured with scale or dividers. Simplest is a slope on opposite sides and cut square across the others (Fig. 12-2A). Draw a view from the angled side. Alongside it draw the development, marked out to the total of the four widths of sides. It may help to number the corners. Project across to the relevent fold positions (Fig. 12-2B), and join these points to get the shape of the cut part of the development (Fig. 12-2C).

There is a practical consideration. A short joint is easier to make satisfactorily than a long one. Unless there is a particular reason for making the joint on a long corner, arrange the development to have its ends at a short corner. Then mark a flap there (Fig. 12-2D).

Slightly more complicated is a square section cut diagonally (Fig. 12-2E). The side view has to be drawn diagonally to get the cut to a straight line. First draw a plan with its sides at 45° to horizontal and project upwards, so you can then draw the cut (Fig. 12-2F). Lay out the development as before, with the four sides flat and the corner lines long enough to take the projection lines across. Number the corners if you wish. Project across and mark the positions on the corners. Then join them to get the developed shape (Fig. 12-2G).

The same method can be used to make developments of any flat-sided section. It could be a regular hexagon, octagon, or other multi-sided shape, not necessarily with equal sides. There could be internal folds as well as external ones. In some cases it may be necessary to draw two side elevations to show all the heights of a cut. Whatever the shape (Fig. 12-3A), the development is started by laying out the widths of all the surfaces (Fig. 12-3B), and then projecting the heights of all corners across (Fig. 13-3C). As with the other shapes, allow for the corner joint to be short, if possible (Fig. 12-3D).

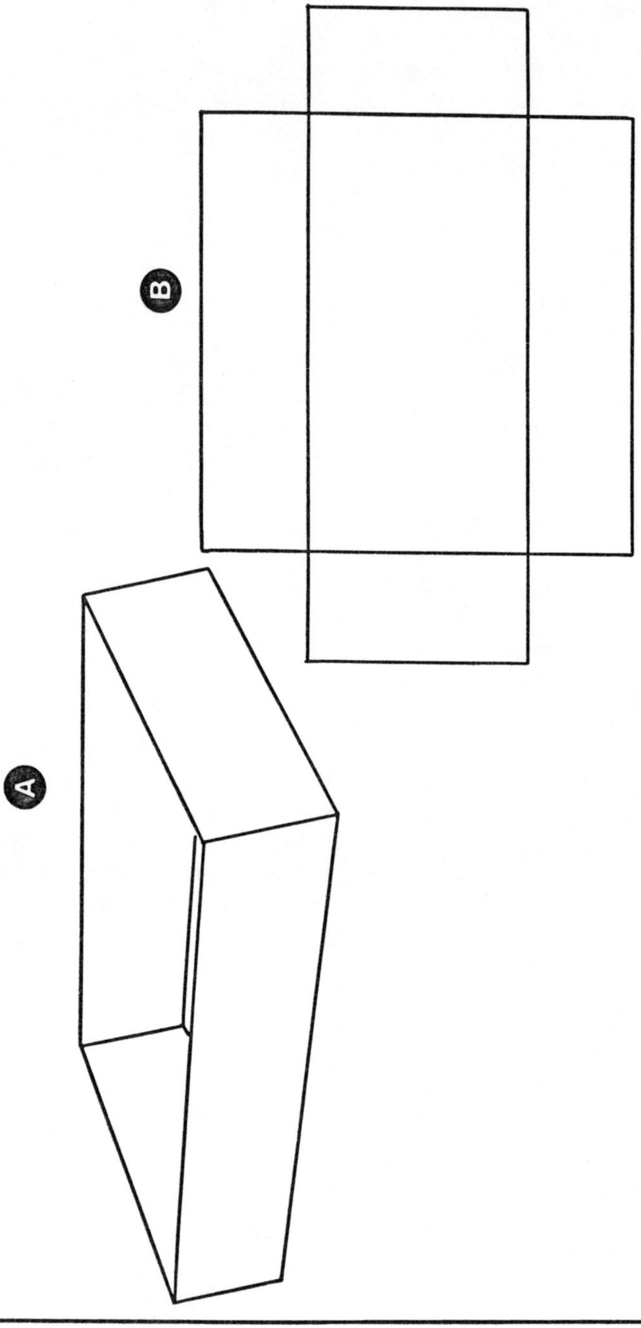
B
A

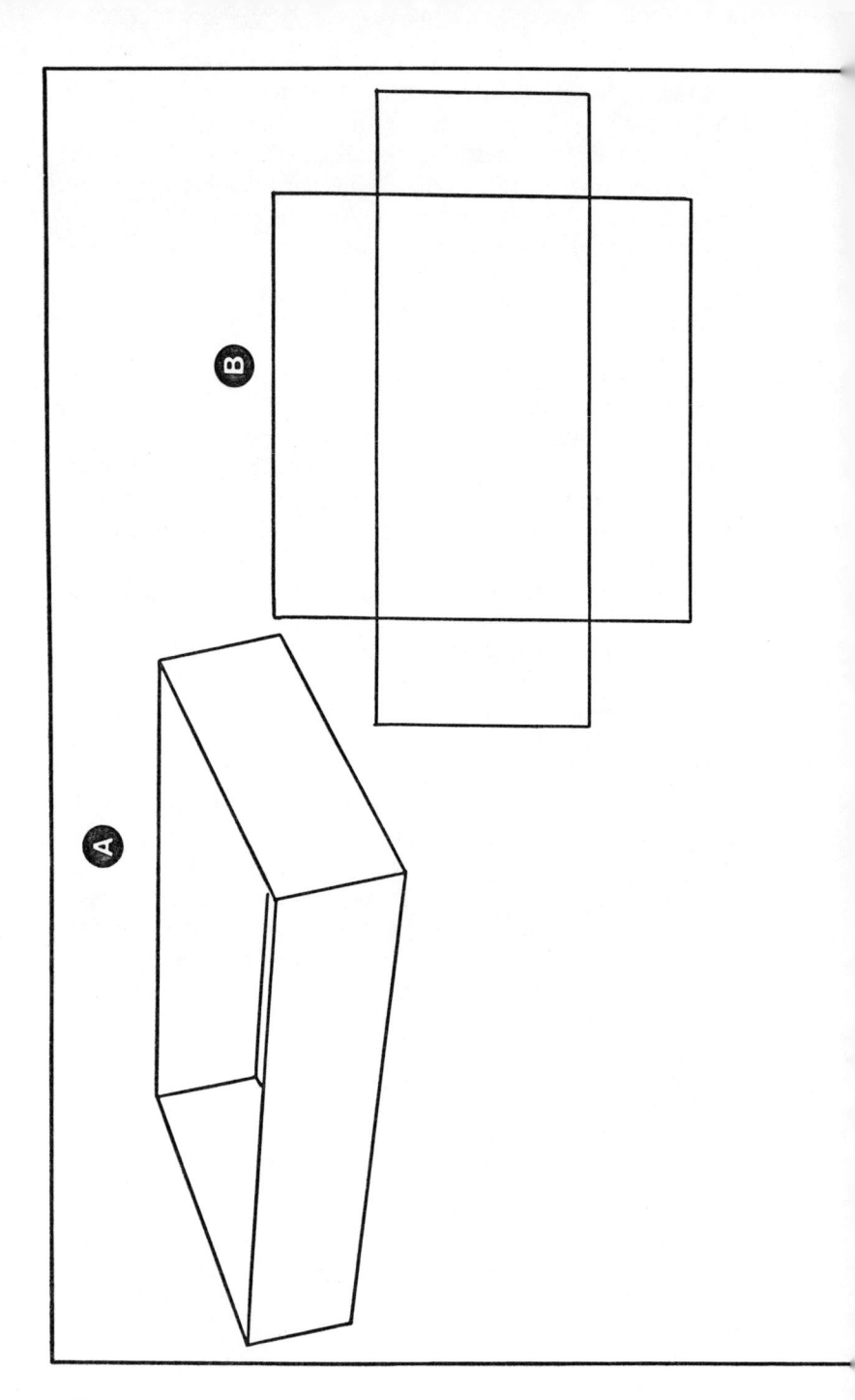

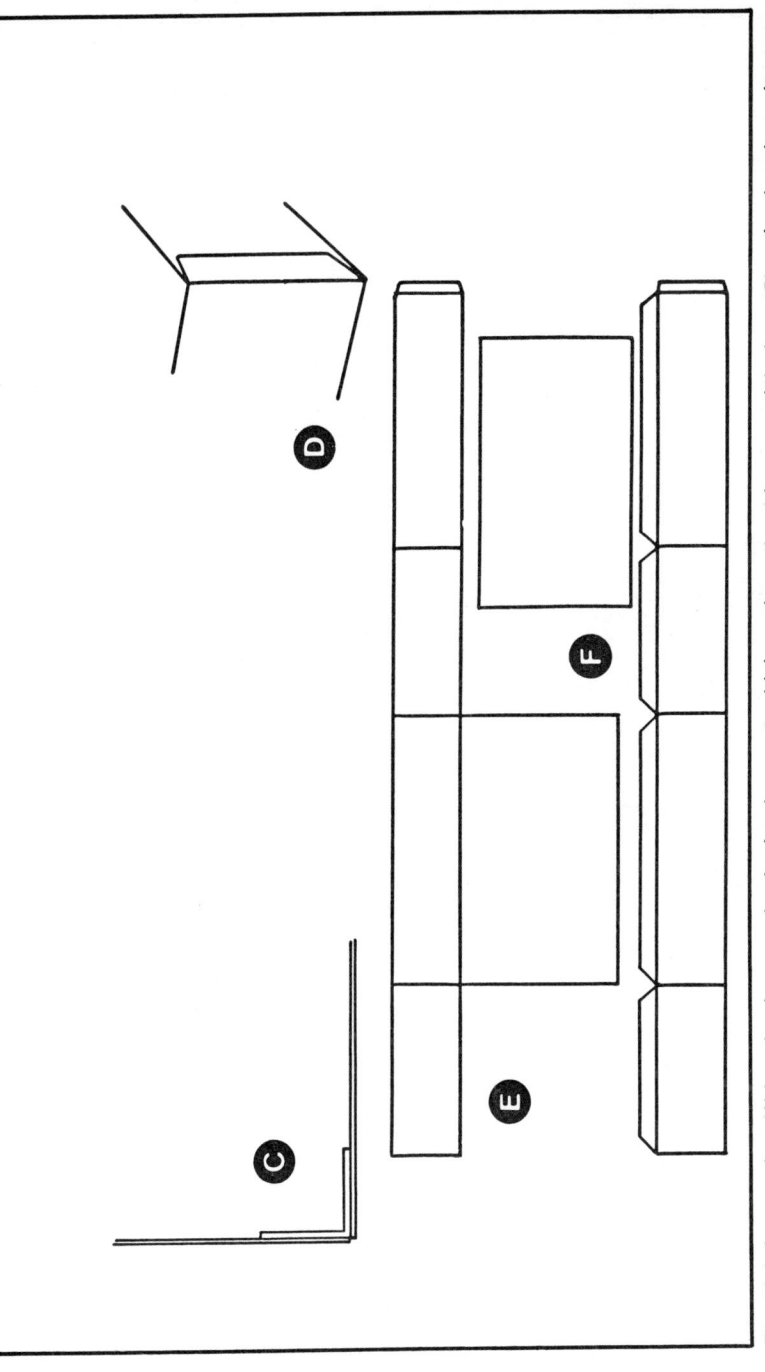

Fig. 12-1. An open box (A) is a simple example of a development, which can have the sides around the base (B) and a piece in each corner (C) or a tab (D). There are other ways of developing to get the same final shape (E and F).

160

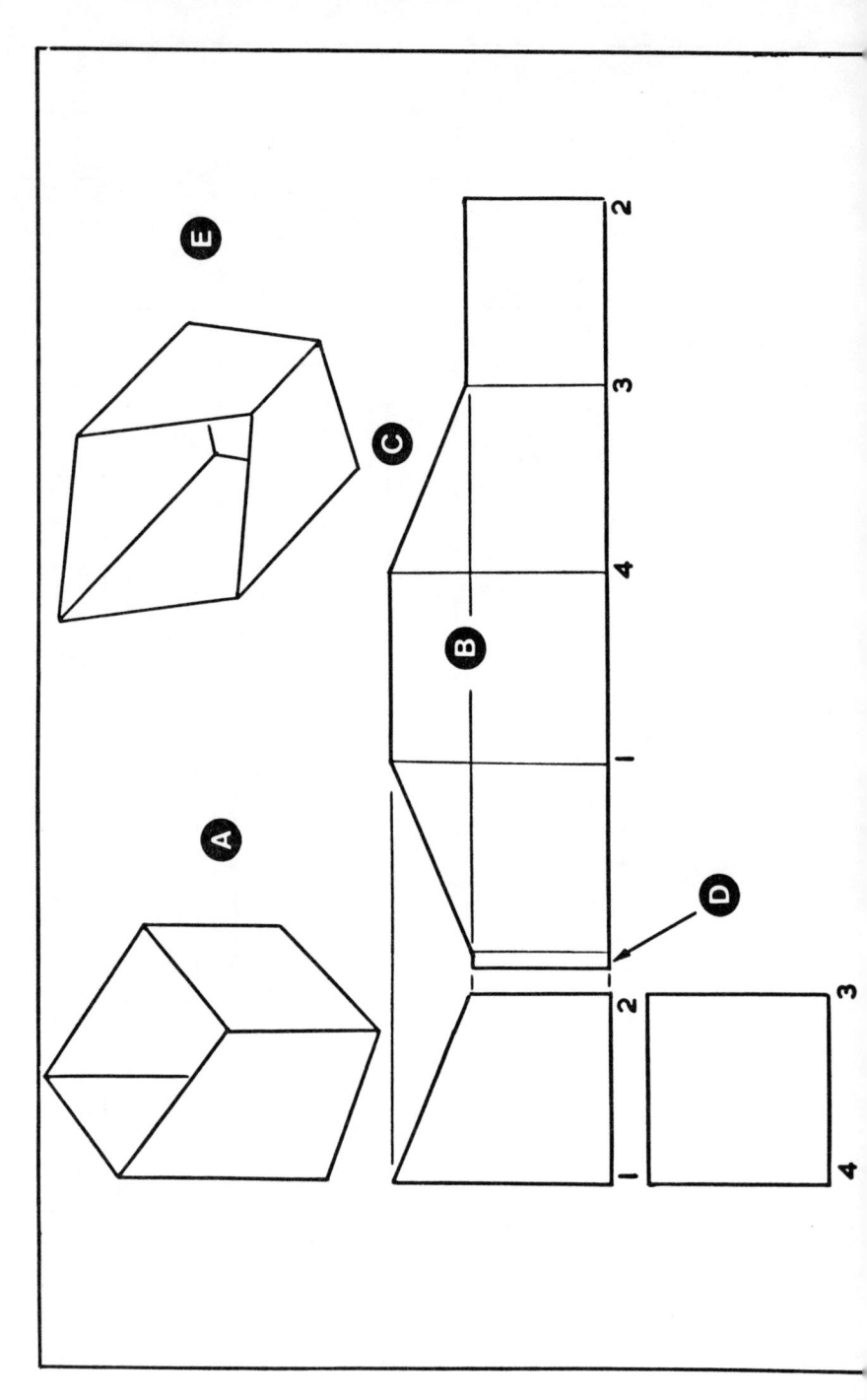

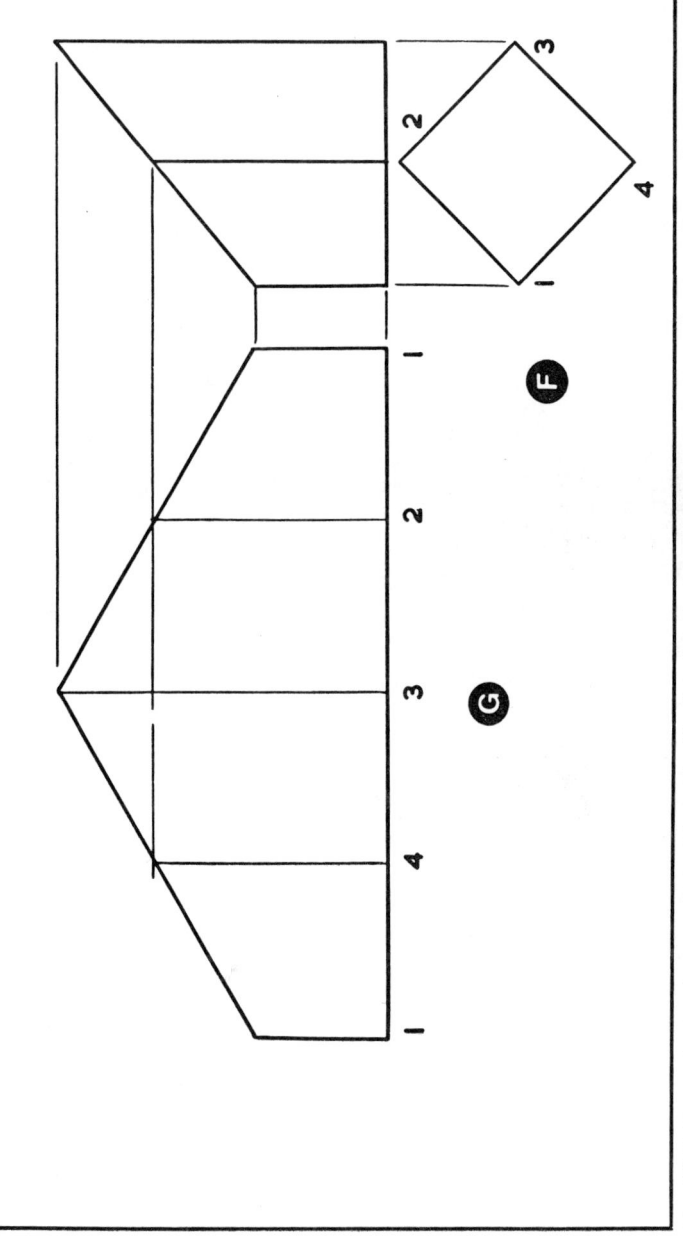

Fig. 12-2. For a sloping top parallel with a side (A), the development is made from a simple elevation (B, C, and D). For a diagonally arranged top, there has to be a cornerwise plan and elevation (F and G).

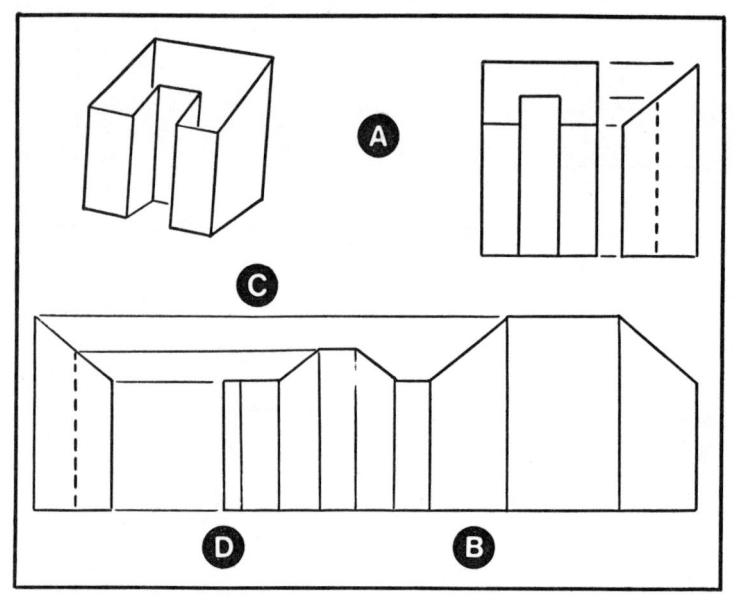

Fig. 12-3. A compound development can be made by imagining the shape opened out in steps.

BEND ALLOWANCES

With paper and card, as well as the thinner gauges of metal, the bend can be assumed to form a sharp angle for which no allowance need be made. The corner shown in the development will be bent on the line, and the sizes of the sides will finish the same for all practical purposes. If thicker metal is bent, the angle will not be sharp but will follow a curve (Fig. 12-4A). Unless it is important that the corner should be as sharp as possible, it is stronger to allow for a moderate curve there, and what this is may depend on the bending facilities available to the craftsman working to the drawing. In this case the draftsman must consult him before deciding on a curve which may not be practical when the drawing gets to the shop.

Metal bends on its *neutral axis*. Metal inside the axis compresses, and that outside it stretches. Only on the neutral axis does its size remain the same. In the case of metal of uniform width and thickness, the neutral axis is in the center of the thickness (Fig. 12-4B). If you draw a center line, with marks at the limits of the curves, the flat part outside each corner will be unaffected and remain the same size. Between the marks is a quarter of a circle that has to be allowed for (Fig. 12-4C). For a simple example, assume the metal is 0.5 inch thick and the radius at the neutral axis is 1 inch (Fig. 12-4D). A full circle of 1-inch radius will have 6.28-inch circumference, so one-quarter of this is 1.57 inches, which is the amount to allow around the bend (Fig. 12-4E). Anyone unfamiliar with obtaining sizes around curves will find instructions in the following sections of this chapter.

Suppose you are setting out a square tube measuring 3½ inches externally (Fig. 12-4F). The distance across neutral axes

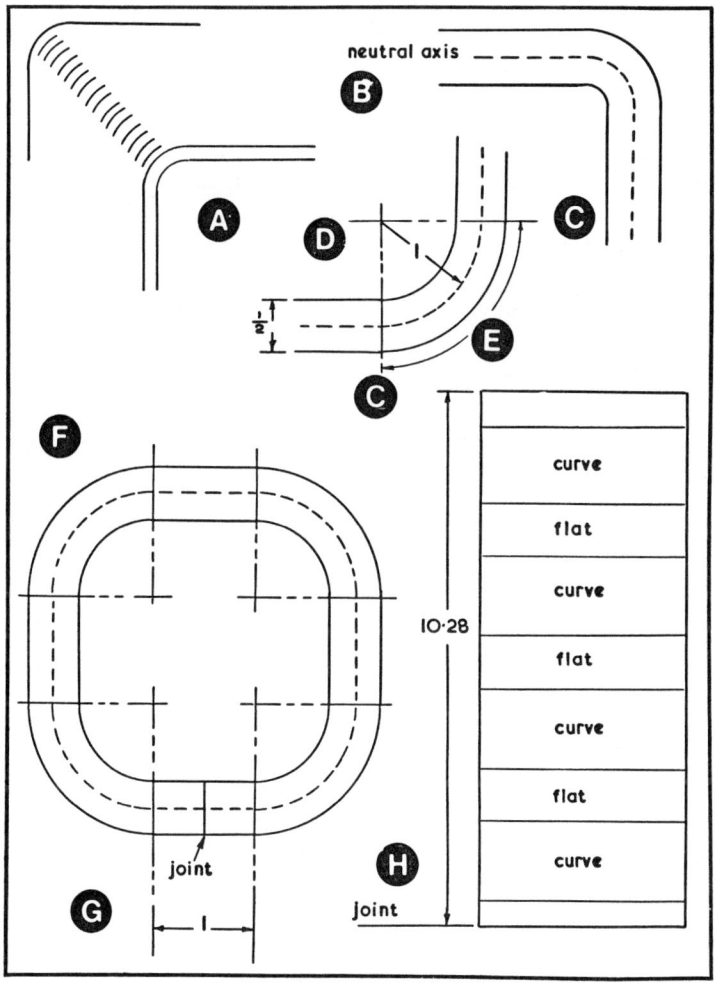

Fig. 12-4. Metal bends to a curve around its neutral axis (A to E). The amount to allow for this can make a considerable difference to the total length of a developed shape (F to H).

will be 3 inches. Then you have to deduct the 1-inch radius at each corner, leaving 1 inch of flat at each side (Fig. 12-4G). If you check the calculations, the overall width of the development is made up of four corners plus four flats, making a total in this case of 10.28 inches (Fig. 12-4H). This is considerably less than if you had used the external widths and assumed sharp bends.

CURVED DEVELOPMENTS

Developing curves involves the calculation of circumference or part of it. Remember that the diameter is twice the radius, and the circumference has a relation to the diameter represented by the Greek letter π. This is a proportion that can be taken to any number of decimal places without reaching finality. For practical purposes, it is usually taken to no more than four places of decimals when its value is 3.1416. Even this goes further than may be required by much precision engineering, and 3.142 is a simpler figure often used. For a very close approximation, you can use 3 1/7. The shape

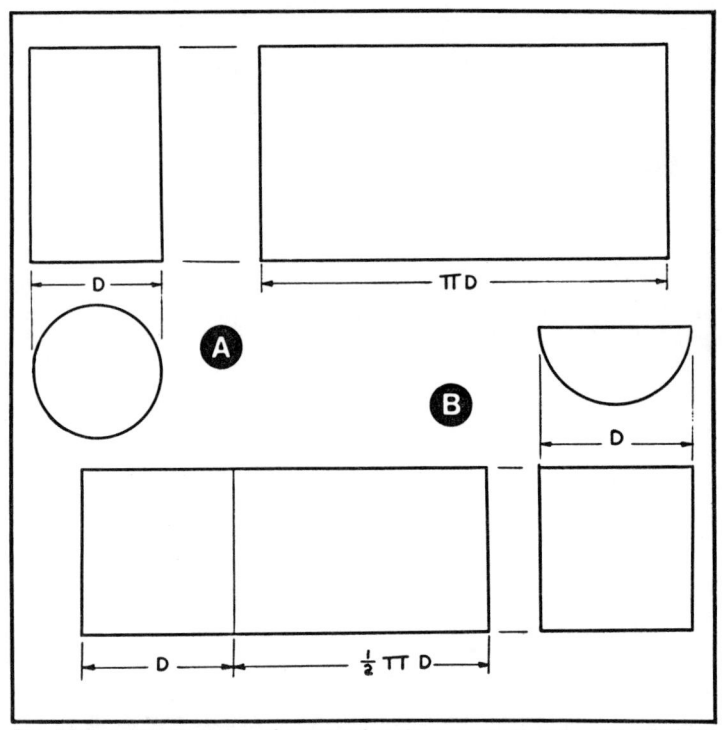

Fig. 12-5. A development of a curved surface must allow for the relationship between the diameter and circumference of a circle.

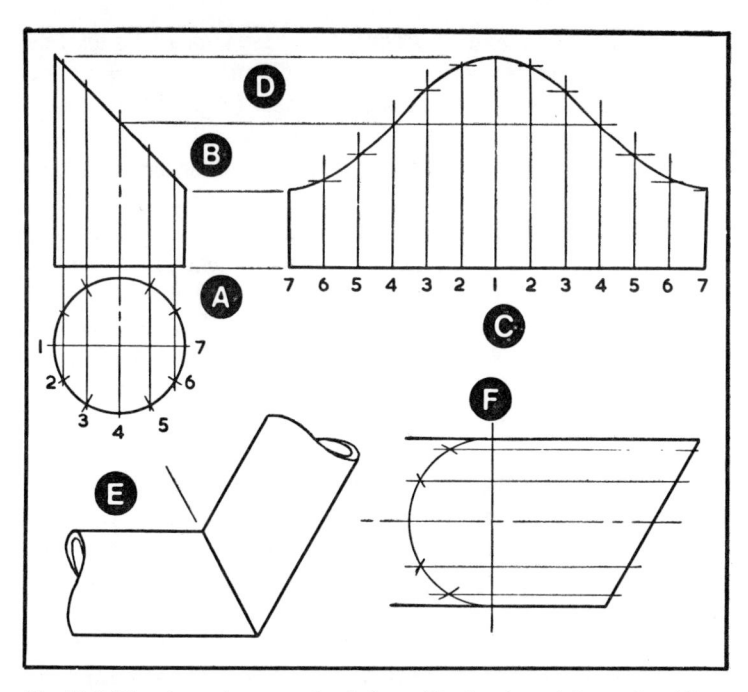

Fig. 12-6. The circumference of a circle and its developed shape should be divided evenly so ordinates can be used to mark heights (A to D). This system is used in developing joints between tubes (E and F).

can be made near this proportion, and then the joint is manipulated to give a bigger or smaller diameter when the curve has to be fitted to another part.

A development of a cylinder has a width equal to π multiplied by the diameter (Fig. 12-5A), which is twice the radius. If it is a D shape, made up of a semicircle and a flat side, you lay out half a full circumference plus a diameter (Fig. 12-5B). In both cases there would have to be a further allowance for a joint.

If there is a diagonal cut across the end of a cylinder, draw a side elevation with the plan view below (Fig. 12-6A). The development is marked out alongside, drawn to its correct limits. Divide the circle into a number of equal parts. Twelve are shown to avoid confusion, but for accurate results you will need more. Project these points up to the elevation (Fig. 12-6B). It will help to number the lines.

Divide the width of the development into the same number of spaces as the plan. Do this geometrically as described earlier, or make experimental steps with dividers until you get an even spacing (Fig. 12-6C). Draw ordinates at these points long enough to take lines projected across. From the elevation, project across to the relative positions on the development (Fig. 12-6D). A few full projection lines are shown. When you have had some practice, you can merely mark the crossings, with the T square or other instrument laid across.

All of these crossings are points on the developed curve. You can get a clue to the way the curve has to go by lightly sketching in freehand through the points, but the actual line is better drawn by bending a spline through the points. For small work, this could be a thin strip of plastic. You can then line in with the aid of French curves, but it is unwise to try to get the first curve with them. There may be a tendency to distort the shape as you try to make too much of a part of the curve fit the French curve.

With thin sheet metal or card and paper, the thickness should not affect the size and shape you get. If you are dealing with thicker metal, work on the neutral axis. That is where the metal bends, and the circumference there may be considerably less than what you will get by working to the outside measurements.

This sort of development is needed when planning ducting or other pipework made from sheet metal. With a right angled corner both pipes are cut at 45°, but for other angles you have to bisect the meeting angle to get the cuts you have to develop (Fig. 12-6E). With long pipes there is no need to draw the full lengths, as it is only the development of an end that matters. Draw a baseline squarely in any position across the pipe elevation and work from that (Fig. 12-6F). With this sort of meeting you may have to make allowance for the joint. There may be nothing extra if the edges are to be welded, but there may have to be flaps or an extra width on one pipe for making the joint.

CONIC DEVELOPMENTS

When the shape to be developed includes tapers, there have to be applications of conic development. The shape may be part of a complete cone, or the conic part may adjoin a cylindrical or square part. If the method of dealing with cones is understood, the technique can be applied in combination with the developing of other parts.

With a square or rectangular cone that has been truncated diagonally (Fig. 12-7A), draw a side elevation above a plan of the base drawn at 45° to horizontal (Fig. 12-7B). The cut in this view is a straight line, and the side view shows the actual lengths of the

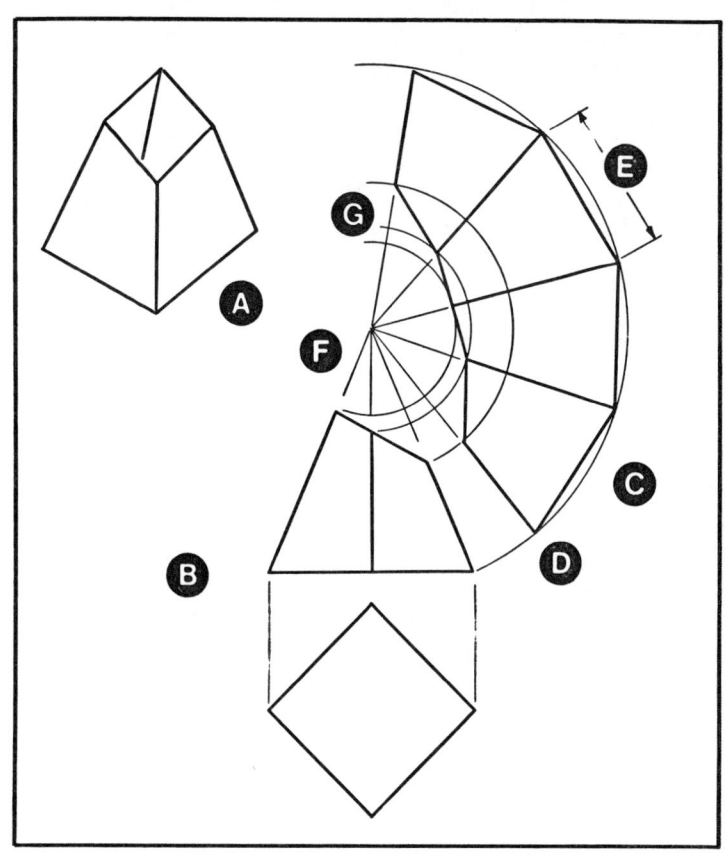

Fig. 12-7. A truncated pyramid is developed about its apex. Then the heights of the cuts on the four corners are marked and joined.

corner in the outline of the elevation. The widths of each side are shown in the plan.

With the apex of the cone as the center, draw curves with a compass through the corners of the base and the parts of the cut (Fig. 12-7C). They have to be long enough to contain the development, and a little practice will show how far to go.

You can use the side of the elevation to start the development, but it will be clearer in your first work to draw a new line from the apex (Fig. 12-7D). Set dividers to the width of the square in the plan, and step off four times around the developed baseline (Fig. 12-7E). Join these points to each other and to the apex (Fig. 12-7F). On the lines representing the corners, mark where the other curved projection lines intersect. Then join them to the shapes of

167

the sides at the truncation (Fig. 12-7G). As with the other shapes, it is usually best to arrange the joint at the shortest corner.

If the cut is parallel with a side, there is a further step to get a true development. In a side view, the edge lines of the elevation represent the distance down the center of a side (Fig. 12-8A). What is needed for the development is the length of a corner. This can be obtained by projecting one side (Fig. 12-8B). In practice this can be drawn over the side elevation more compactly (Fig. 12-8C) when you understand what you are doing. For the first development it

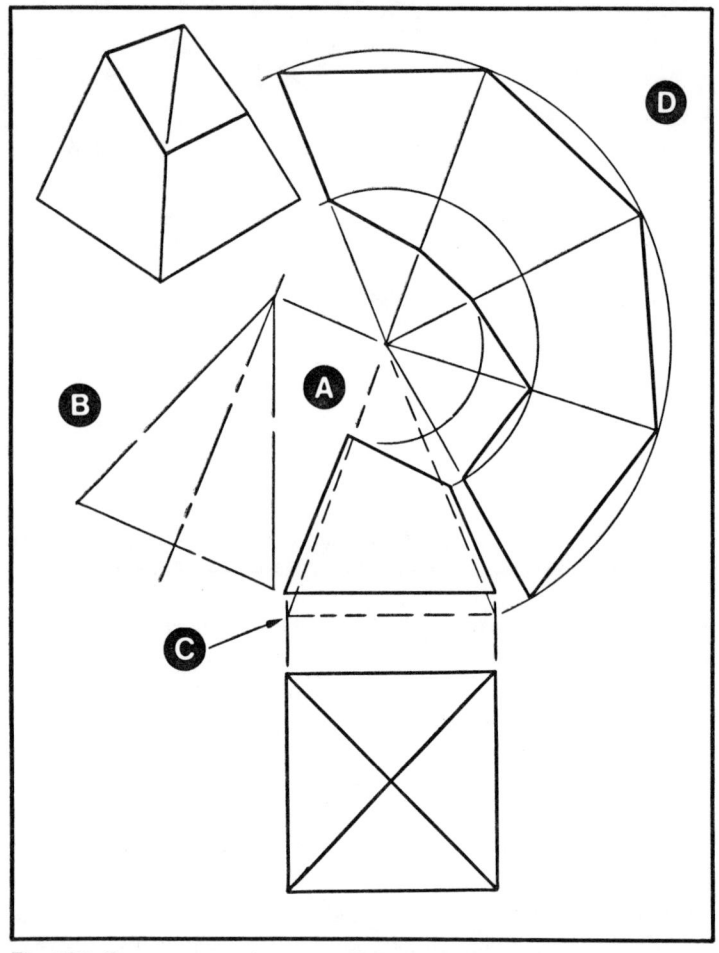

Fig. 12-8. For a truncated pyramid with a cut parallel with a side, the true shape of a side is found first (A, B, and C), then the development is made using this (D).

may be better to draw this separately, including the projected positions of the cut. In the example shown the drawings are superimposed. Draw the curves for the drawing from the projection of a side. Step off the baselines. Mark the cut positions in the same way as the other example (Fig. 12-8D).

ROUND CONES

A round truncated cone development is actually a further step to the stages described in the last chapter for getting the true shape of a cut diagonally across a cone.

Draw a side elevation of the cone with the cut as a straight line and a plan of the base below it (Fig. 12-9A). As the cone is round, the lines represent the length down the slope. There is no need for the step just described for getting the true length of the corner of a square cone. With the apex as a center, use a compass to draw through the corners on the elevation at top and bottom (Fig. 12-9B).

The circumference of the base has to be transferred to the line of the base of the development in order to get the width of the developed shape. One way of getting this is to work from the angle at the apex. Measure this with a protractor, and multiply it by π (Fig. 12-9C). A more common method is to step off similar spaces around plan and development.

Divide the plan circle into a number of equal divisions. Twelve are shown, but it is better to have more. Set dividers to a division on the plan, and step off the same number of divisions around the development (Fig. 12-9D). In effect you are stepping off the corners of a polygon made up of chords across the circumferences. As the chords are being used on curves of different radii, there will be a small error, but in practice this is so slight as to be negligible.

From the points around the curve of the development, draw lines to the apex. It will be advisable to number the points on the plan, elevation, and development in early work. Use the compass to project around from where the truncation cuts the ordinate lines on the side elevation to the matching points on the development (Fig. 12-9E). Draw a curve through these points to get the developed shape of the cut.

OTHER TRUNCATED CONES

Many shapes can be developed by dividing them into parts so the methods described in this chapter can be applied. A problem comes with conic shapes when the taper is slight and the apex

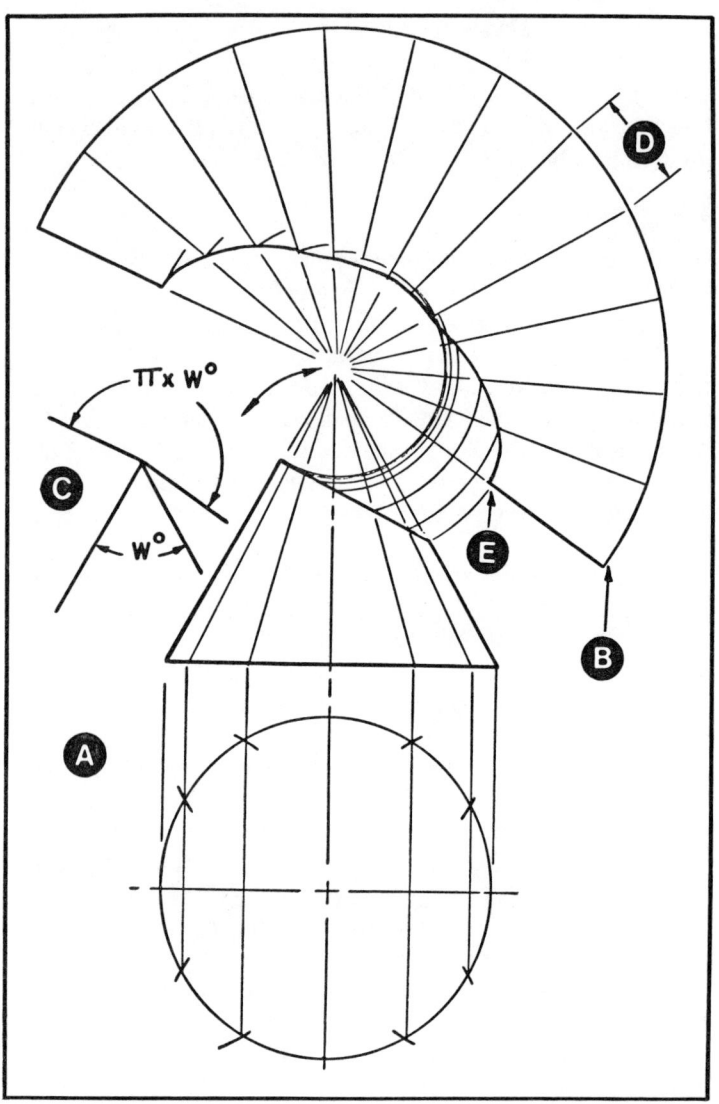

Fig. 12-9. A conical development can be made about the apex by getting the angle geometrically (A, B and C) or by stepping off spacings (D) around the developed base. Parts of the cut are found along the ordinates by projecting around (E).

would be out of reach. It is then necessary to build up the development in a series of panels without reference to the remote apex.

An example is a square cone, cut squarely across the top, but with only a slight taper (Fig. 12-10A). What is needed first is the

170

true shape of one side. Draw a side elevation (Fig. 12-10B). The edge lines of the elevation are the lengths of the center lines of the sides. Project from an edge line. Use the widths at the top and bottom to mark the widths about the projected center line. Join the corners to get the true shape of a side (Fig. 12-10C). Repeat it a total of four times to get the shape of the complete development. You could cut the shape from card and use it as a template to draw around in the further three positions, but it would be more accurate to use geometrical methods.

Use one edge of the shape just drawn as the base for the next shape. Use a compass set to the widths of top and bottom from the corners of this base to strike arcs. Then set the compass to the diagonals of the first shape to draw more arcs from the corners to cross these arcs, and give the positions of the next corners (Fig. 12-10D). Join these points to make the outline of the second panel. Repeat the action twice more, so you get four panels (Fig. 12-10E).

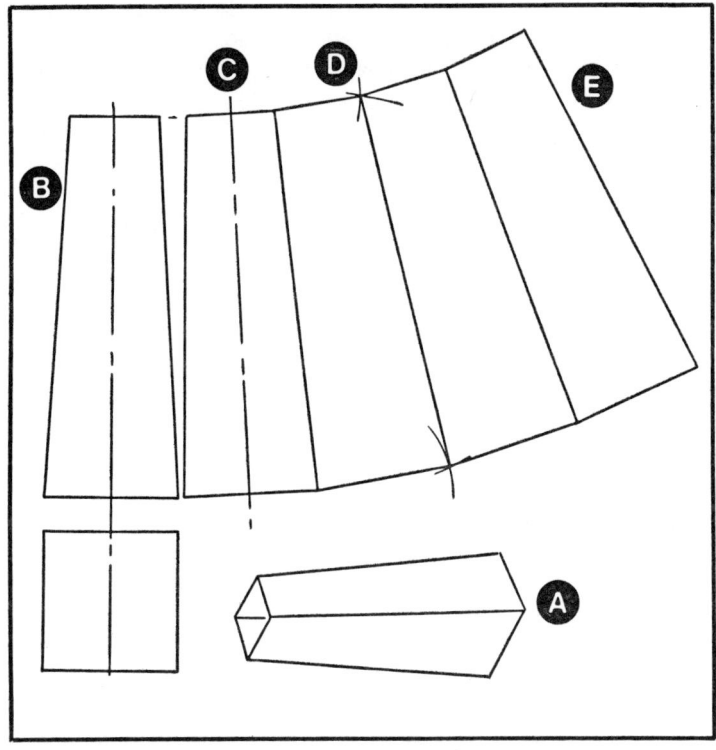

Fig. 12-10. If the apex is too far to be used (A), the shape of one side can be found (B and C), and then repeated alongside (D and E).

If a similar problem arises with a round cone where the apex cannot be reached, it has to be treated like a multi-paneled shape. Curves are drawn through points you obtain.

Draw a side elevation, with circles or semicircles at each end. Divide these circles into the same number of equal divisions. Six may be sufficient (Fig. 12-11A). There is no need to join the marks on the circumference, but the method may be easier to understand if you imagine it as a hexagon enclosed in a circular cone that has to be projected (Fig. 12-11B). Find the true shape of one side and repeat it six times. Then draw curves through the corners. If the hexagons are drawn so flat surfaces are viewed from the side, their lengths in elevation will be the center lines of sides. Project to a center line and mark the widths of the ends, taken from the spacings on the circles, symmetrical about it. Then join the corners (Fig. 12-11C). This is the true shape of a panel if it was a hexagon you were projecting.

Using the method just described for developing a square conical shape, repeat the panel alongside itself for a total of six times (Fig. 12-11D). The corners of this development are actually points on the curves, so if you draw curves through the points, you will have the shape of the development needed (Fig. 12-11E). As can be seen, there is no need to draw many of the lines representing the hexagonal shape, but they are included to illustrate the method and can be omitted when the work is understood. All that the construction has to do is locate the points to draw curves through. Straight lines between the points will not be far off the curved shapes, but get smooth curves by drawing around a spline bent through the points.

OTHER SHAPES

Similar methods can be used when only part of a cone is involved, as in a shape with straight and curved sides (Fig. 12-12A). Draw a plan and two elevations. Treat the total shape as two problems with a division where the curved part starts (Fig. 12-12B). The end elevation gives the true distance down the slope of the sides. The side elevation gives the true lengths along the sides from the division line (Fig. 12-12C). Similarly, the side elevation gives true distance down the center of the end, and the plan view gives the widths of top and bottom. From this information you can draw the shapes of these panels (Fig. 12-12D).

This leaves the curved end, which is actually half of a truncated cone with its apex too far away to be used. Draw semicircles

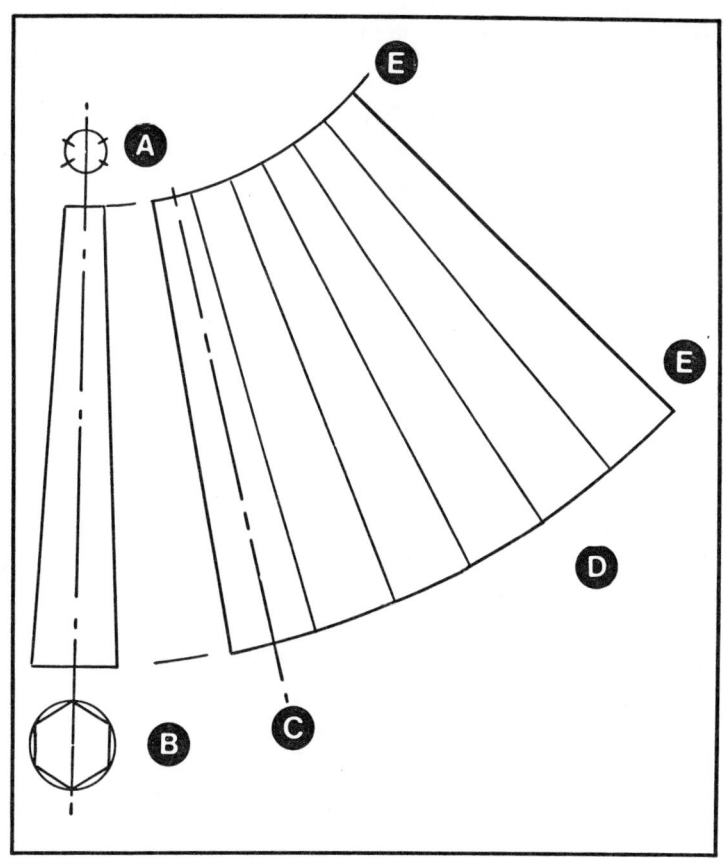

Fig. 12-11. A circular part of a cone where the apex cannot be used may be developed as a polygon, in the same way as the square in Fig. 12-10. Then curves are drawn through the points.

above and below and divide these into similar equally spaced divisions (Fig. 12-12E). From these, find the shape of one panel of an imagined enclosed flat-sided cone. Having found this shape, repeat it a sufficient number of times to complete the half cone. Draw curves through the points (Fig. 12-12F) in the manner just described for a complete round cone.

You now have the shapes of two flat side panels, one flat end panel and that of the curved end. How you assemble them depends on the method of construction. Assume the joint is to come at a square corner, so two sides have to be drawn each side of the conic development. The end panel is added to complete the whole development (Fig. 12-12G), except for any flap or joint allowance.

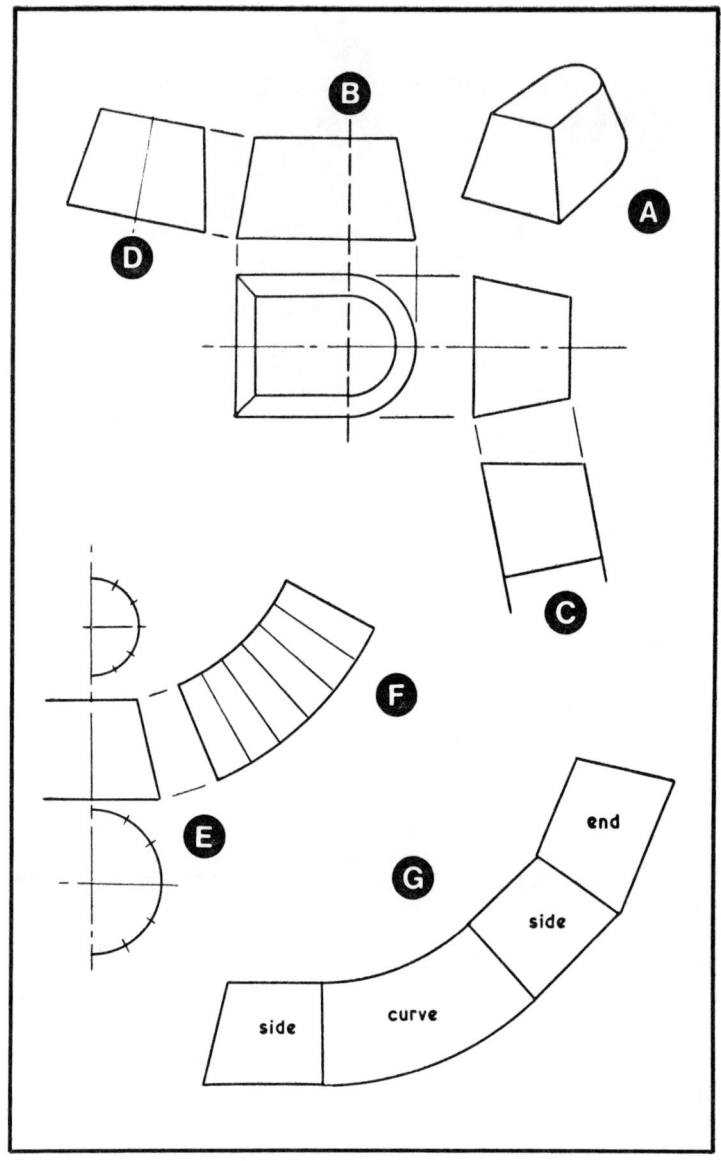

Fig. 12-12. A compound shape that is partially conical has to be developed in parts, which are then put together.

INTERNAL CUTS AND JUNCTIONS

All of the developments given as examples have been external shapes. Similar methods are used if there are cutout parts. There

174

could be any number of holes for bolts or other attachments, and they are obviously round when viewed squarely with the bolt. A hole that is round in a shaped part does not have a round outline when developed. An example is a rod or bolt into a cylinder. The larger the bolt diameter in relation to the cylinder, the greater the distortion when developed.

Draw a plan view of the rod into the cylinder (Fig. 12-13A), and draw a circle to represent the end of the rod. An elevation

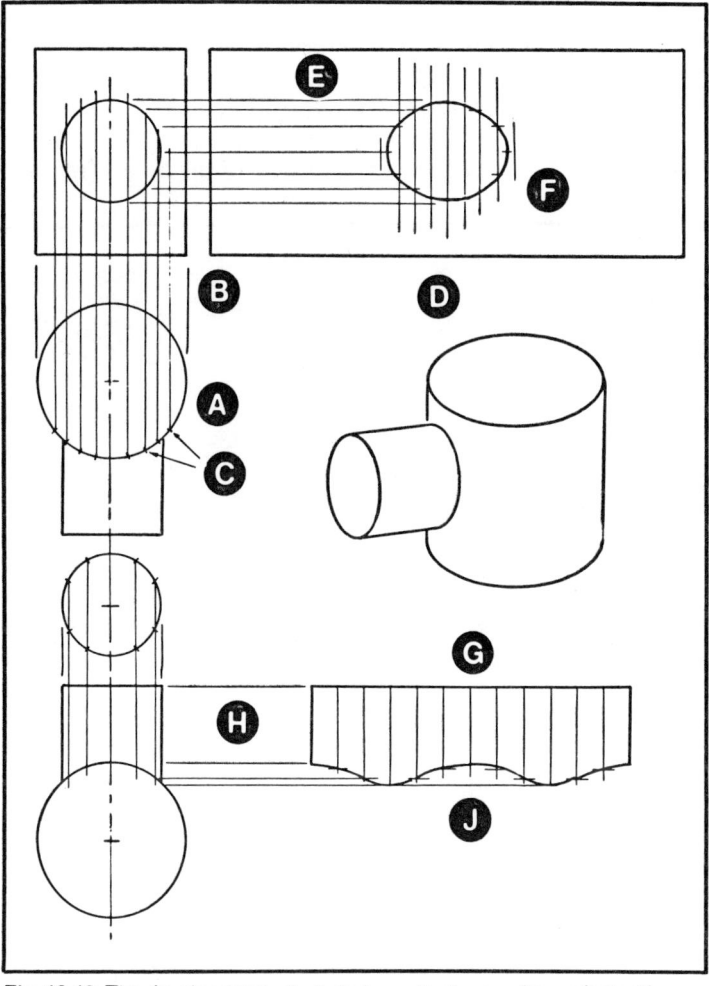

Fig. 12-13. The development of a hole in a cylinder can be projected from a plan and elevation (A to F). The development of a meeting cylinder is separately developed (G to J).

175

shows the rod as a circle on the cylinder (Fig. 12-13B). Make a development of the cylinder outline.

On the plan view draw a number of divisions on the cylinder circumference to include the hole, arranged about the center line of the rod (Fig. 12-13C). Project them up to the circle in the elevation. On the development step off similar divisions about the center line of the hole (Fig. 12-13D). Project across from where the lines from the plan cut the hole to the matching points on the development (Fig. 12-13E). A curve through these points will be an ellipse (Fig. 12-13F).

Suppose the rod or pipe in the same position has to be cut to a matching curve to the main cylinder, so when the two parts come together the meeting edges match, and the developed shape of the surface of the rod of pipe is required. Do not be confused by the construction lines used to get the shape of the hole. Treat the

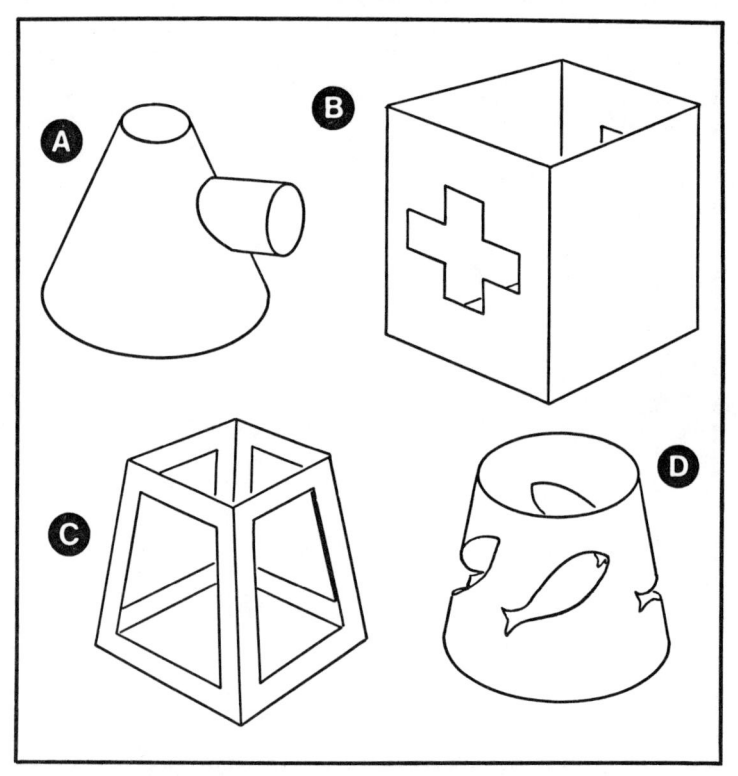

Fig. 12-14. Holes or junctions may have to be developed (A) or may be easy to mark out (B). Spaces in a tapered cone are drawn parallel (C), while decorative cuts can be drawn freehand (D).

joining pipe as a separate problem. What is needed is the shape of the end that meets the cylinder, so divide the circle representing its end view into a number of equal divisions. Start to draw the development, getting the length by multiplying the diameter by π. Divide the length into the same number of divisions as the end circle (Fig. 12-13G).

Project from the points on the circle along the pipe to its junction with the main cylinder, and then across to the matching points on the ordinates on the development (Fig. 12-13H). Draw a curve through the points to get the shape of the pipe's end (Fig. 12-13J).

A pipe into a cone can be treated in a similar way (Fig. 12-14A), except that the development follows the usual curved shape. The ordinates on that and the cone are sloping, instead of parallel as in the cylinder. Other cuts that may make patterns or borders internally are no problem in a straight-line shape, whether parallel (Fig. 12-14B) or tapered (Fig. 12-14C). In the latter case the pattern is drawn squarely on the developed panels or measured evenly inside their outlines.

Irregular shapes on a cylinder may have to be developed in the same way as described for the hole. If the internal cuts are decorative and precision is not very important, they can be drawn on the development (Fig. 12-14D), providing you allow for any reduction in apparent width when the material is rolled or folded. Further information on developments can be found in my book *The Master Handbook of Sheetmetalwork . . . with projects* (TAB book No. 1257).

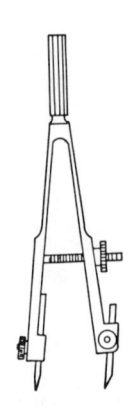

Chapter 13
Fasteners

Many parts which have to be drawn are joined with nuts and bolts or other screwed fasteners. They may have to be riveted, where two or more pieces are held together by the squeezing action of heads on opposite ends of a part going through them. Often the fasteners are drawn quite small, and the draftsman does not attempt to show all their details. Instead, there are certain conventions used that indicate to the reader what is intended. There may be instances where a screw thread has to be shown as it actually appears. Unless this is essential to the understanding of the drawing or the action of a part, though, it is acceptable to use one of the conventional methods of indicating threads.

You should have an understanding of screw threads and the methods of riveting, particularly if your drawings are intended to be used in an engineering shop. Otherwise, you may be indicating a fastening practice which is not practicable.

SCREW THREADS

How the principle of the screw originated is not known. Certainly Archimedes made use of it more than 2000 years ago. A thread follows a *helix* or *helical* curve around a rod. It is sometimes wrongly called a *spiral*, which is actually flat, with the curve increasing in size as it is drawn around a point. If a wedge-shaped piece of paper is wrapped around a rod, its path is a helix (Fig. 13-1A). The actual path in a side elevation can be projected by

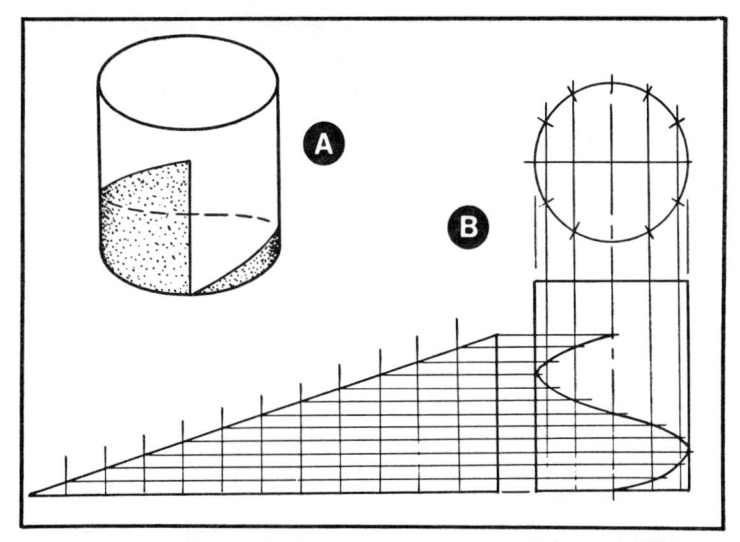

Fig. 13-1. A wedge wrapped around a cylinder follows the same helical path as a screw thread.

locating points with a method similar to that used for developments in Chapter 12. Draw the wedge-shaped development and project to ordinates (Fig. 13-1B).

Certain terms used in relation to screw threads should be understood (Fig. 13-2A). The pitch of an ordinary thread will be the same as one complete revolution of the helix. All threads used on a drawing are assumed to be of this single type unless indicated, but there are multiple threads. The single thread has one ridge around the rod. A nut running on the thread will advance a distance equal to one pitch with each revolution. If there are two ridges starting at opposite sides of the rod, there will be two helixes side by side. A nut would then advance a distance equal to two pitches with each revolution. The amount of advance is called the *lead* and is equal to two pitches (Fig. 13-2B). Less common are triple or three-start threads, where the lead is three times the pitch. Other terms used in relation to these threads are the same as for single threads.

The outer diameter of the thread is the *major diameter*. The bottom of the thread is the *minor diameter*. At the average depth of thread is the *pitch diameter*. The major diameter is that of the diameter of the rod on which the thread is cut and, theoretically, the depth the thread goes in the cut. In practice there must be some clearance. Similarly, the minor diameter is theoretically the size of the hole through the nut in which the internal thread has to be cut, but it is actually slightly larger to allow for clearance. These differences may be ignored in drawing, but if actual sizes have to be quoted there are tables which give recommended drill and other sizes for particular threads.

Early threads were cut so one external thread fitted one internal thread. With industrial development there had to be some standardization so parts could be interchanged. Bolts that were quantity-produced had to fit nuts that were also made in quantity and not necessarily by the same manufacturer. Unfortunately, there was no national standard and certainly no international one, so there were many standards brought into use. All general purpose threads have a V section. A sharp V (Fig. 13-3A). is difficult to cut, and the top is liable to break. It is better to curve or flatten both top and bottom, as in the *American National Unified* thread (Fig. 13-3B). The *British Standard Whitworth* is similar, but has an angle of 55° (Fig. 13-3C). These two have been the most common threads in use in the English-speaking world. With the increasing adoption of metric threads, though, ISO metric threads are being used (Fig. 13-3D). Tops and bottoms of the threads are flat. The profile is not very different from the American National Unified thread, although the measurements are metric. ISO is the abbreviation indicating a standard adopted by the International Organization for Standardization, and it is applied to many things besides threads. These international standards are coming increasingly into worldwide use.

If there is much endwise thrust, there is more strength in a square thread (Fig. 13-3E). There are practical difficulties in making and engaging two parts, so it is more usual to slope the sides of the thread. *Acme* (Fig. 13-3F) is one version. B & S worm is similar but deeper. A knuckle thread has semicircular tops and bottoms to the profile (Fig. 13-3G) and is not as strong as the previous two threads, although it is convenient where split nuts and other parts have to be engaged. If the thrust is greater in one direction, as in a vise screw, the thread can be angled to suit. A common form is the *buttress* thread (Fig. 13-3H).

Within the various standards there are standard pitches, which may be abbreviated to t.p.i., meaning threads per inch. There are coarse and fine thread versions of most standard threads.

All common screwed parts have right-hand threads. That means you turn a screw clockwise to enter it into an internal thread. With a wrench on a nut you pull around clockwise. Unless

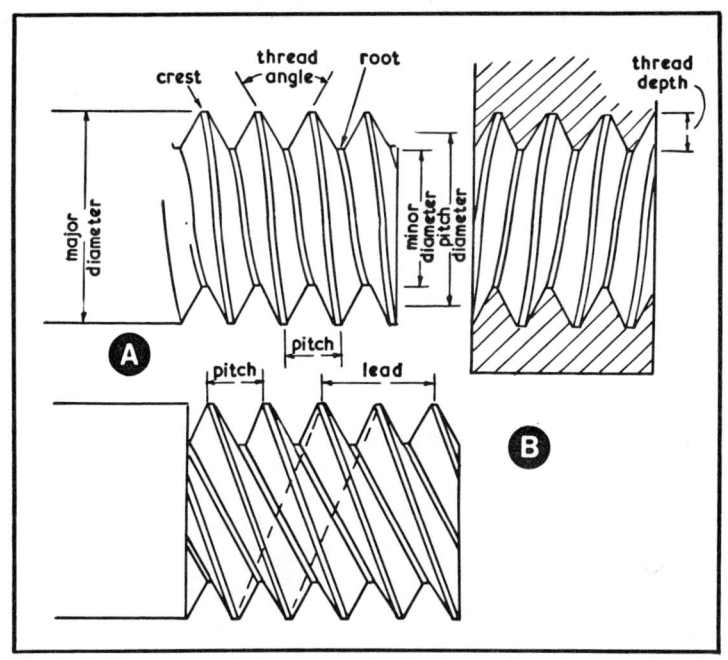

Fig. 13-2. The names of the main parts of a screw thread.

there is any instruction to the contrary, all screw parts are assumed to be right-handed. A left-handed thread is used where two screw parts have to work in or out of a central member when it is turned, as in a *turnbuckle* (Fig. 13-3J).

REPRESENTING THREADS

Drawing a screw thread in a truly helical form is time consuming, as it involves geometric construction. The first simplification is to adopt straight lines to represent the threads. If a fairly close approximation to the true appearance is needed, threads are drawn as sharp Vs, with the crest at one side opposite the root at the other side. If these outlines are joined with straight lines, the appearance is approximately correct and looks well if the root lines are heavier than the crest lines (Fig. 13-4A). For a section through an internally threaded hole, the threads on the far side of the hole will slope the other way (Fig. 13-4B). If the threaded hole is to be shown hidden, only the profiles of the threads are shown (Fig. 13-4C). Greater dotted detail would only lead to confusion. In end view of the rod or hole, the depth of the thread is indicated by a solid line about three-quarters of a complete circle (Fig. 13-4D).

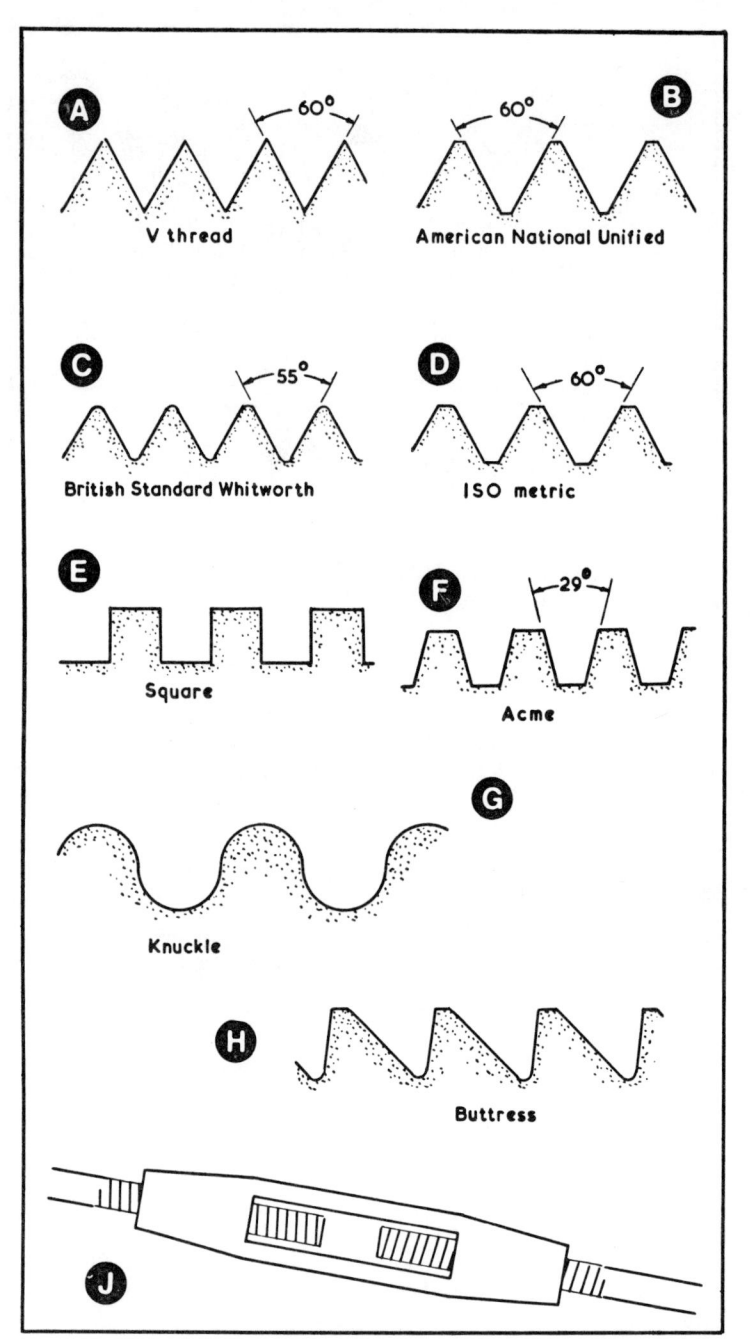

Fig. 13-3. Some thread forms. The use of right and left-handed thread (J).

In another representation of a thread, there are alternate lines at the slope of the thread, but without the profiles at their ends. The crest line goes to the outside, but the thicker root line stops at what would be the root size in profile (Fig. 13-4E). The angle of the lines is found by drawing a slope with an advance of half the pitch (Fig. 13-4F).

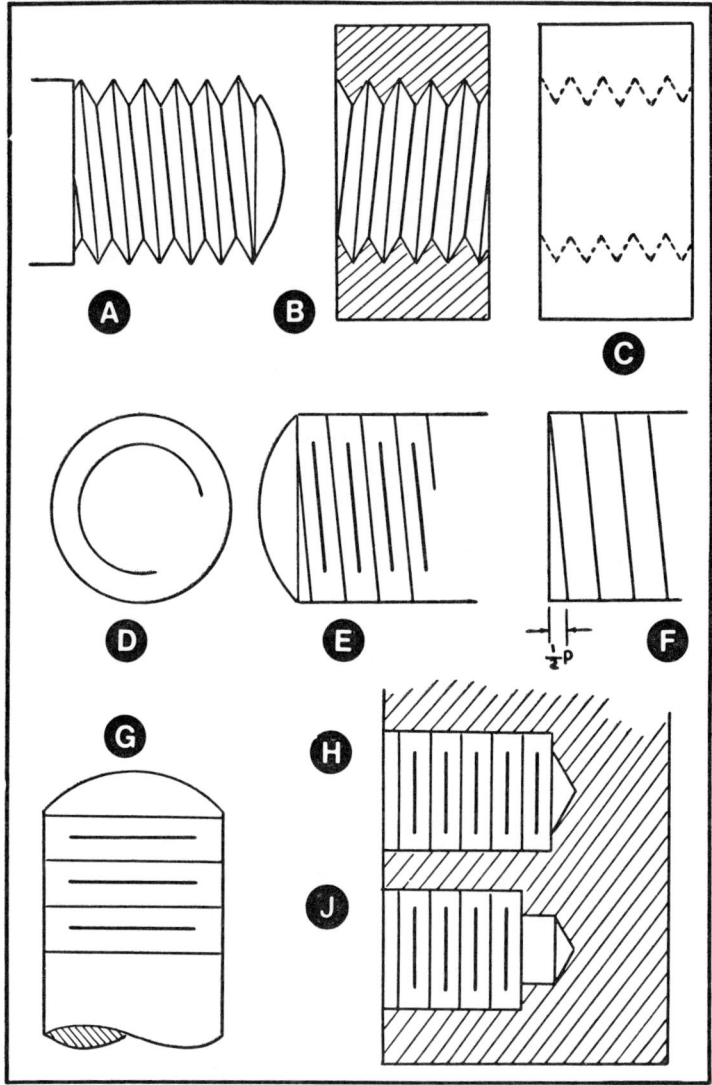

Fig. 13-4. Methods of representing screw threads on drawings.

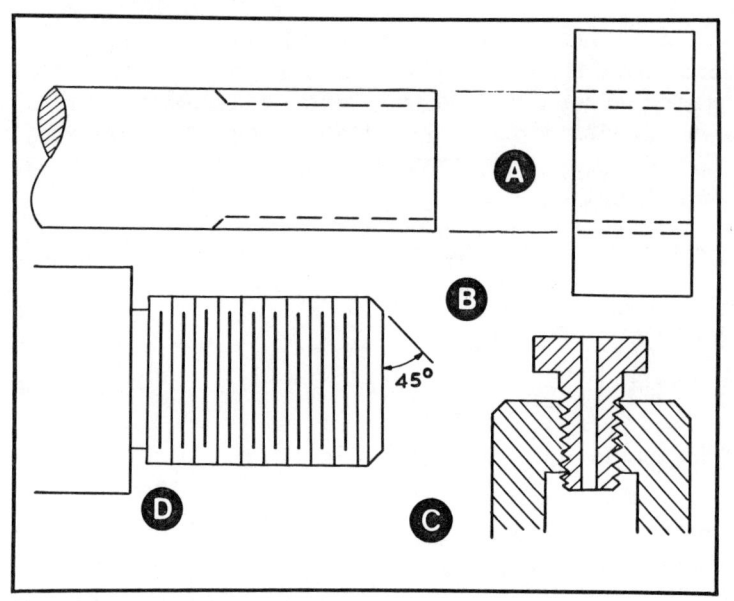

Fig. 13-5. Typical ways of indicating threads, including in section.

That method gives a good indication that the part is threaded, but it is necessary to maintain a steady angle in drawing. It is more usual to draw the lines squarely across (Fig. 13-4G). This means that the threads look the same whether left-hand or right-hand and external or internal.

If an internal thread does not go right through (a *blind* hole), frequently a tapered point is shown, as it would be left by a drill (Fig. 13-4H), and the root diameter hole is shown carried further, if that is permissible. Screwing to the bottom of a hole is difficult (Fig. 13-4J).

An even simpler way to represent threads is used in some drawing offices. Parallel lines are drawn at the depth of the thread, whether external or internal (Fig. 13-5A). A note alongside may then indicate the type of thread to be used. Bolts and screwed rods usually have their ends beveled at 45° for about the depth of the thread (Fig. 13-5B). This facilitates entering a screw in a nut and helps in the manufacturing process. It should be indicated by the draftsman whatever method is used to show threads.

If a section has to be shown with screwed parts assembled, section all parts (Fig. 13-5C). This is clearer than trying to show the rod part not sectioned, with threads across, although a plain spindle in similar circumstances might be left whole. The screw profiles are then shown approximately to scale and sharp angled. Profiles exact to scale might be too tedious to draw if the scale is small.

If a part is threaded close to a shoulder or the head of a screw, there is a practical difficulty in screwing fully. Cut a narrow groove to the depth of the thread close to the shoulder (Fig. 13-5D).

NUTS AND BOLTS

A bolt has a head and is screwed only partway from the end. If the thread is taken fully to the head, it is more correctly called a screw. If it is necessary to distinguish it from a wood screw, it may be called a machine screw or a metal thread screw. Heads of screws and bolts, as well as their nuts, may take many forms. Hexagonal ones are fine for engineering purposes, while square ones are found in nuts and bolts for general use. Heads may be round or countersunk to take a screwdriver, like wood screws. A coach bolt has a shallow rounded head with a square neck below it to pull into wood and prevent turning.

There are many standard nut, bolt, and screw forms. Details are provided by the manufacturers and should be followed if these parts are to be drawn exactly. There are certain conventions used in drafting that show what is intended, but may not be absolutely exact to what will be used.

A bolt head, whether hexagonal or square, has a chamfered top (Fig. 13-6A). This removes sharp corners and makes the fitting of a wrench easier. The underside is usually flat across, but there may be a washer face of the same diameter as the distance across the flats (Fig. 13-6B). The bolt end is chamfered, and the length of thread is indicated (Fig. 13-6C). A nut looks like a bolt head and is usually only chamfered on the outer surface (Fig. 13-6D). It may be chamfered on the other surface, particularly if it is intended for locking by tightening against another nut.

In a conventional drawing of a nut and bolt, the head thickness is three-quarters of the bolt diameter (Fig. 13-6E). A nut is slightly thicker at seven-eighths of the diameter (Fig. 13-6F). For general engineering bolt heads, the distance across the flats is 1½ times the bolt diameter (Fig. 13-6G). For heavy-duty general purpose bolt heads, this may be increased by ⅛ inch for all sizes.

To draw a bolt and nut, make parallel lines for the bolt and a circle for the end view of the head. On the circle draw a hexagon with one flat towards what will be the elevation (Fig. 13-7A). Draw on the thickness of the head and the nut, and then project the faces

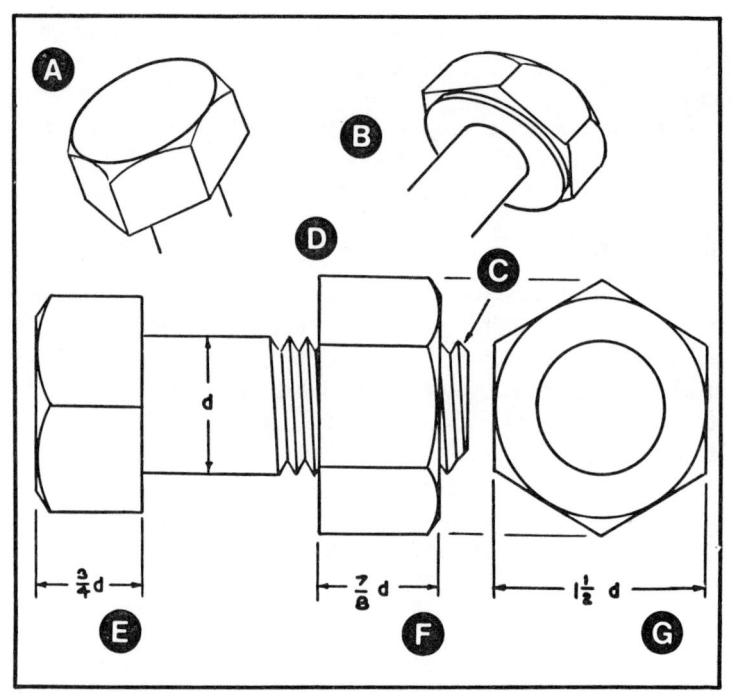

Fig. 13-6. Proportions of nuts and bolts as usually drawn.

of the hexagon to them (Fig. 13-7B). The circle and the hexagon around it show the chamfered corners in plan, but to draw them in elevation, the wide face is drawn with a radius of 1½d and the others a radius of ⅜d, where d is the bolt diameter. To complete the view of both head and nut, draw chamfers at 30° each side (Fig. 13-7C). Bevel the end of the bolt, and draw a conventional representation of screw thread. Put a washer face under the head, if that is required (Fig. 13-7D).

It is common to draw heads and nuts as viewed in that way, but they could be drawn with a corner to the front (Fig. 13-6E). Square heads and nuts are usually drawn with a corner to the front. The distance across the flats is the same as for a hexagonal shape, so the same size wrench can be used. A similar circle is drawn, but with a square around it (Fig. 13-8A). Project this to the head and nut on the elevation, and suitable curves can be drawn at a radius of ⅞d (Fig. 13-8B). If a head or nut is to be shown as viewed from one side, there is just one surface showing (Fig. 13-8C). The chamfer on square heads and nuts is drawn at 45°, whatever the view (Fig. 13-8D).

There are many special locknuts, available, and a drawing will have to be made to suit. To show the conventional method of locking with a second nut, a thinner nut is drawn above the main nut and in line with it (Fig. 13-9A). Another traditional method of locking uses a *castellated nut* with a cotter and split pin through it. The nut may be of normal depth and slotted (Fig. 13-9B), or there may be an extension above the hexagonal part with the slots. Another method of locking near an edge or into a small hole is with a *tab washer,* bent up a nut face and down over the edge or into a

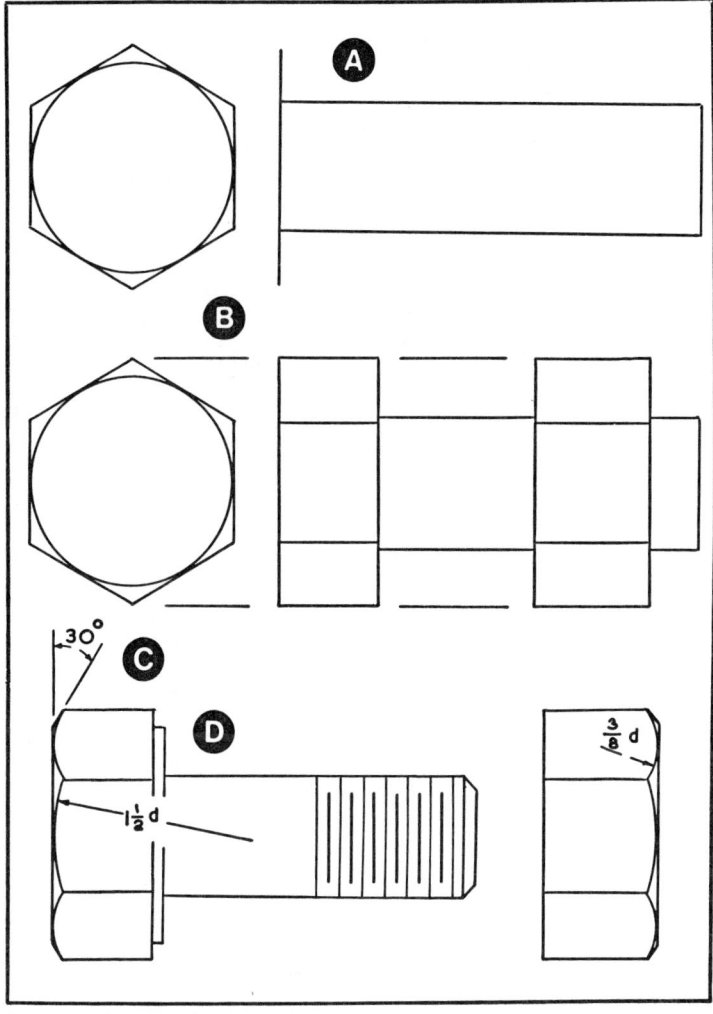

Fig. 13-7. Steps in drawing a bolt and nut.

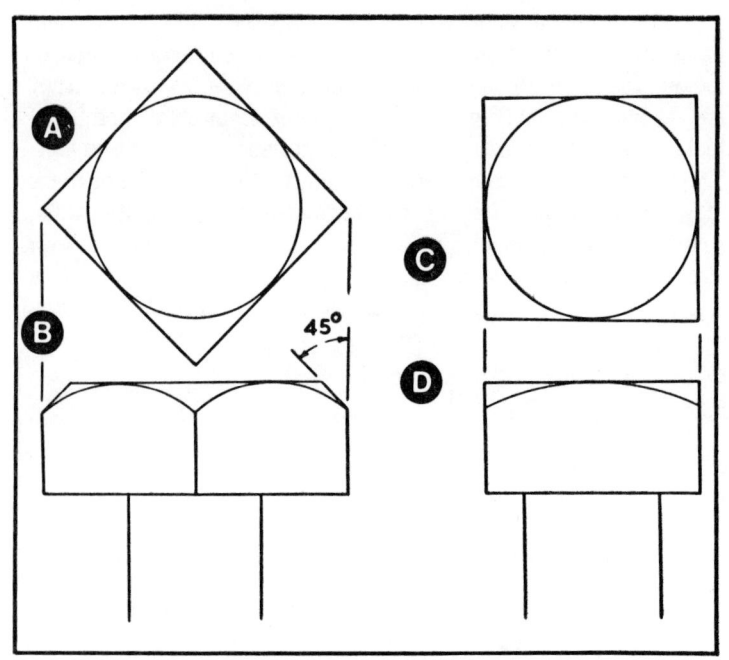

Fig. 13-8. Drawing details of a square-headed bolt.

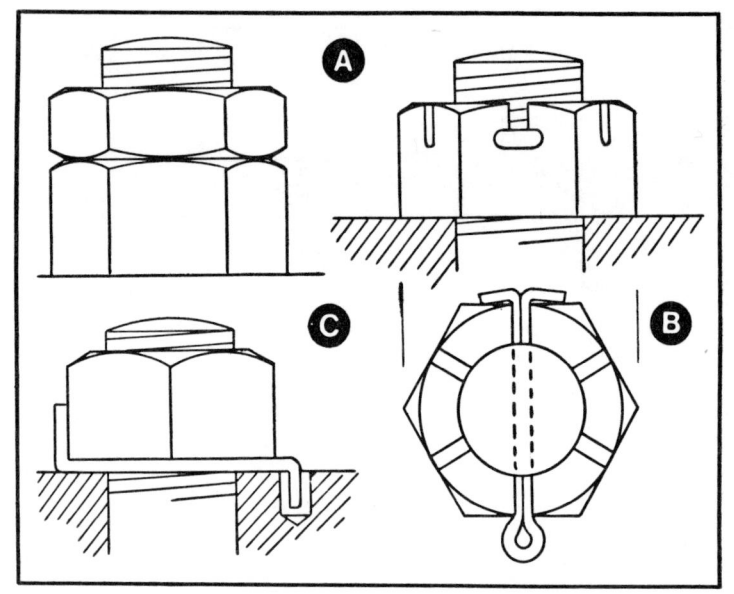

Fig. 13-9. A locknut (A), a castellated nut (B), and the use of a tab washer (C).

hole (Fig. 13-9C). Some locknuts incorporate a friction arrangement to resist vibrating loose. Epoxy adhesive is also used for locking.

Smaller machine screws may have a variety of heads, mostly to take a screwdriver (Fig. 13-10A). The head may have a socket to take a hexagonal wrench (Fig. 13-10B). If the screw is without a head, it is a setscrew or a grub screw. It may have a screwdriver slot (Fig. 13-10C), but now more commonly it has a hexagonal socket (Fig. 13-10D).

If a wood screw had to be shown in detail, its thread is a rather similar helical problem to a metal thread, but with the complication of a taper to a point. It is more usual to indicate the thread form with a few freehand lines (Fig. 13-10E), unless there is a need for greater accuracy.

RIVETS

Rivets are usually supplied with one head already formed. The other is made after the parts are assembled, either by hand ham-

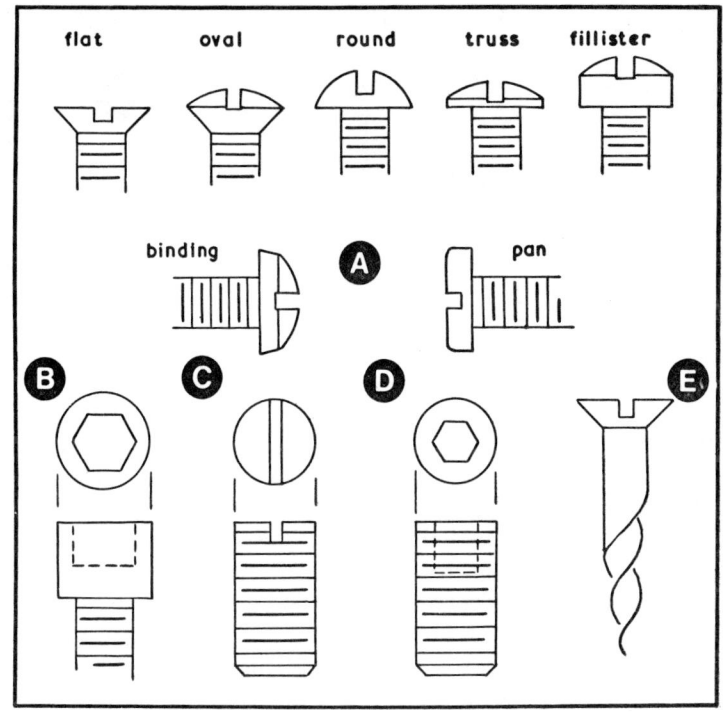

Fig. 13-10. Typical machine screw heads and a wood screw.

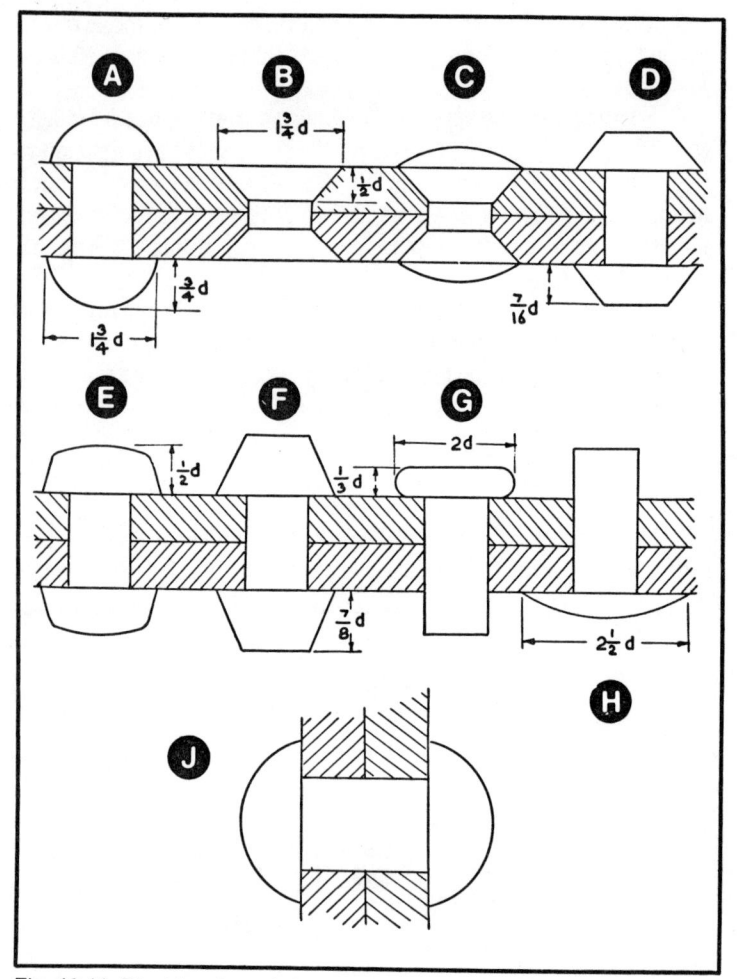

Fig. 13-11. Rivet heads and the proportions they are drawn.

mering or with a power tool, finishing with a die or snap to shape the head. Larger rivets are fitted hot, so they tighten further as they shrink when cooling.

The common *round* or *button head* size is drawn in simple proportions of the diameter (Fig. 13-11A). A countersunk head is drawn at 50° from a depth of half the diameter, and this gives an overall size of about 1¾d (Fig. 13-11B). A *raised,* or *oval head* is like a countersunk one with a curved top (Fig. 13-11C). The name *pan head* is given to a tapered one (Fig. 13-11D) and a more rounded one (Fig. 13-11E). A deeper version is a *cone head*

(Fig. 13-11F). Less common are *flat* (Fig. 13-11G) and *truss* or *wagon box heads* (Fig. 13-11H).

If a section is shown with a rivet through it, do not section the rivet. Hatch the adjoining parts at different angles (Fig. 13-11J).

For plate joints, the rivets may be in a single line along a lap joint (Fig. 13-12A). There may have to be a double line, with the rivet spacing staggered (Fig. 13-12B). If one side has to be flush, there may be a butt strap or fish plate on the other side (Fig. 13-12C).

A problem comes when thin plates have to be joined, and the rivet must be countersunk on one side. So much would have to be taken out by countersinking that the plate would be weakened. Instead, the plates are punched to the angle of the countersink, so the head goes in without any metal having been removed. Then the head at the other side is closed over the conical part (Fig. 13-12D).

It is unusual to put any lines which might be regarded as purely decorative on a mechanical drawing, but with round head rivets

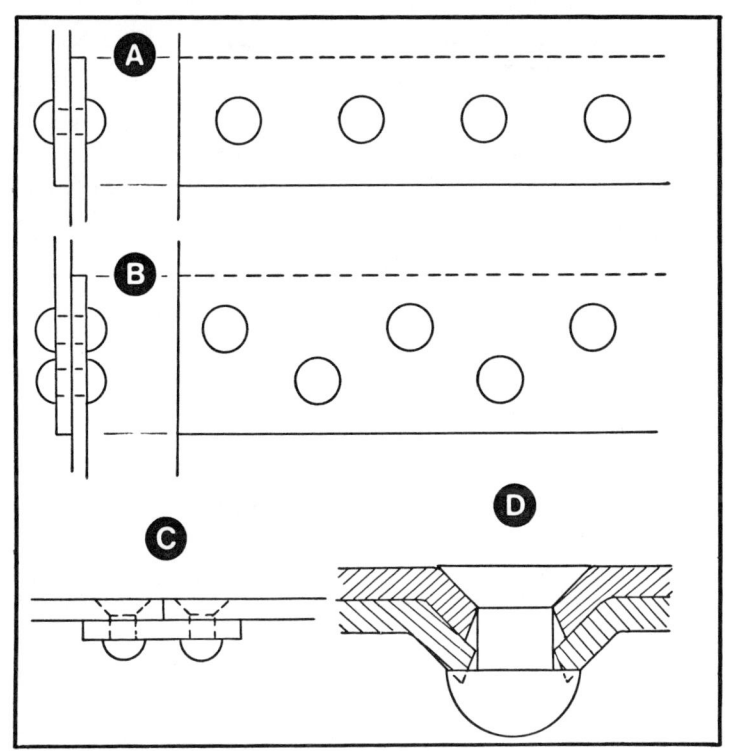

Fig. 13-12. Riveted joints (A, B, and C). How to countersink thin plates (D).

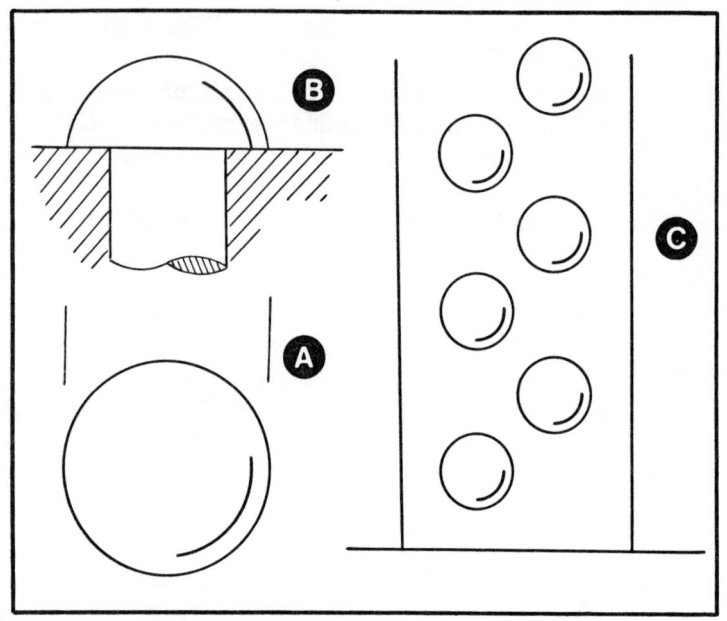

Fig. 13-13. Method of shading rivet heads.

there may be shading added to give an indication of roundness. In plan view the shading is done with a compass near the edge and going for about one-quarter of the circumference (Fig. 13-13A). In a side view the shading is done always at the right in a similar way (Fig. 13-13B). It is always assumed that the light causing the shadow comes from the left. With a row of rivets, the effect is to draw attention to the fact that rivets have been drawn and not just holes (Fig. 13-13C).

OTHER FASTENERS

A draftsman concerned with furniture making or other forms of woodworking may have to draw nails, wood screws, and dowels, as well as many cut joints. In the woodworking industry many details are left to the workers in the shop, who are used to the standard methods of construction. The draftsman then has to indicate main sizes without needing to give much detail. If wood screws have to be shown, some guidance has already been given (Fig. 13-10E). A side view is usually drawn so the slot is seen from its end (Fig. 13-14A), although if a row of heads are being shown in plan view the slot is more often drawn at 45° (Fig. 13-14B). If Phillips heads are to be used, put a note saying so and leave the heads blank. A dowel in a wood section is best shown shorter than the hole and with its ends beveled (Fig. 13-14C). There is no conventional way of showing nails. To a small scale, they are simple lines.

For fabrics, leather, and upholstery there may have to be a variety of fasteners and grips shown. Either they have to be described and their location indicated, or a drawing must be made of the actual items. There are no conventional ways of indicating them.

KEYS

If a pulley, crank, or other part with a hole through it has to be secured to a shaft so it can only turn with it, it may be possible to drill for pins or other fastenings. Also, the two things may be arranged to mate with square or other angular parts, or the two may be welded together. The more traditional way of securing to a shaft uses a key in a keyway. In most cases it is possible to remove the parts later if ever required. A plain key is a simple rectangular strip (Fig. 13-15A). A gib key has a raised end so it can be more easily removed (Fig. 13-15B).

In most cases there are matching slots in the pulley and the shaft (Fig. 13-15C), with half the key in each. Where there is a lesser load, it may be sufficient to have a *flat key* (Fig. 13-15D). An even simpler arrangement is a *saddle key* (Fig. 13-15E) which holds by friction, but which may have a tapered keyway in the pulley for

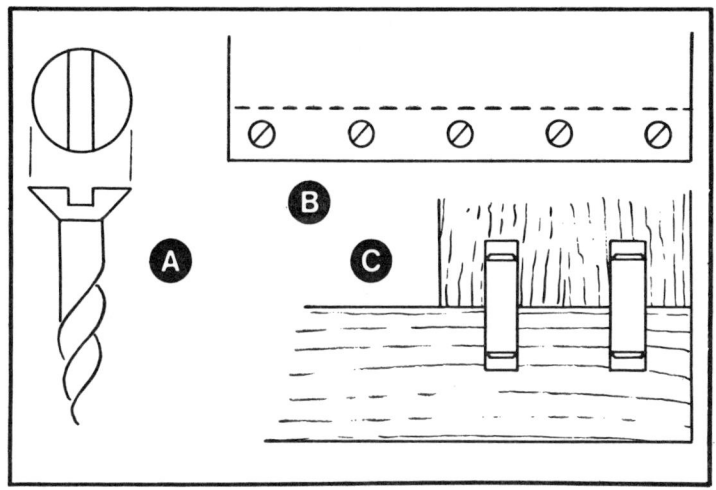

Fig. 13-14. Wood screws (A and B) and dowel sections (C).

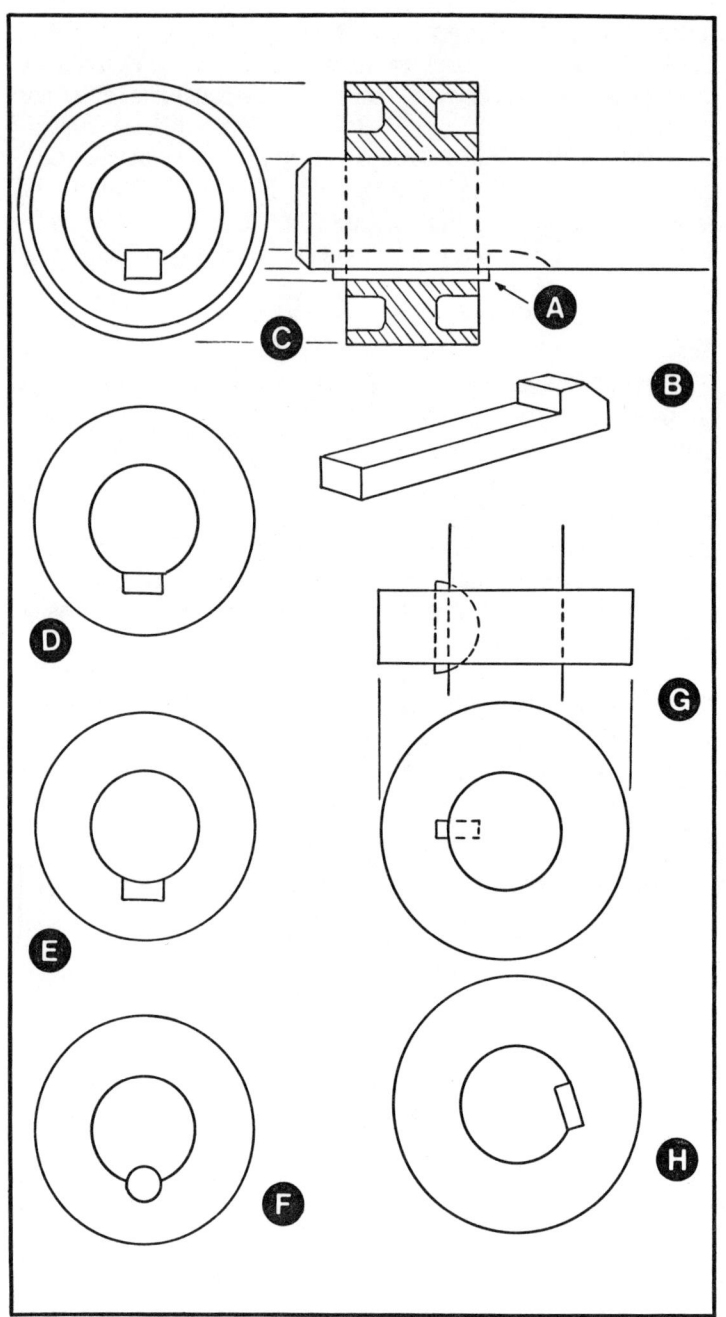

Fig. 13-15. Keys and keyways.

tightening. The key could be a piece of round rod (Fig. 13-15F). A different and much used key for machine tools is the *Woodruff key*, which is slightly less than semicircular and fits into a curved slot in the shaft (Fig. 13-15G). A *Lewis key* fits angled slots so it can provide the maximum strength when the load comes one way (Fig. 13-15H).

WELDING

Although joining by welding does not use fasteners, this is a related process. A drawing of parts that have to be welded together does not show the actual welds, but what is to be done is indicated by symbols. The parts are shown in the intended relation to each other, but symbols alongside indicate what welding is to be done.

There is a very large range of symbols, and little purpose would be served by showing only a selection of them. The whole range, together with a long list of letters used to indicate the type of weld, will be found on a chart published by the *American Welding Society*. Any draftsman concerned with providing drawings and instructions for welding should have one of these charts available for reference.

The main symbols consist of angled arrows to which are added side marks to show what type of weld is required and its relation to

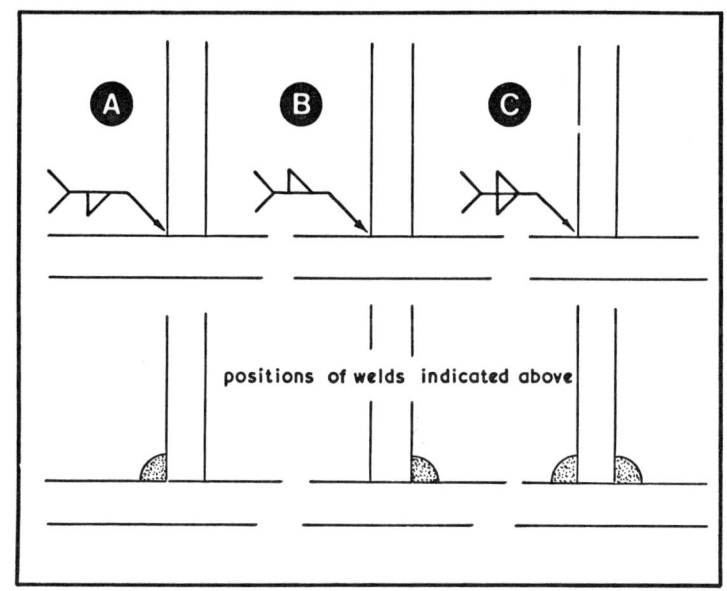

Fig. 13-16. Typical symbols and the welds they indicate.

the indicating arrow. Suppose one plate meets another at a **T** junction, which is a common place for welding. A triangle on the same side as the cranked arrow means that there is to be a fillet weld on the arrow side of the joint (Fig. 13-16A). If the arrow is put on the side of the line opposite to the cranked arrowhead, a fillet weld is intended to be at the side further from the arrow (Fig. 13-16B). A triangle at both sides of the arrow line means that there is to be a fillet weld at both sides of the joint (Fig. 13-16C). Among other marks that may be put on the arrow are indications of plug, spot, seam, and many other welds.

Chapter 14
Inking

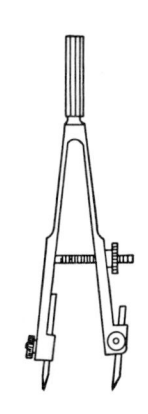

It would be unwise for an original drawing to be taken into a shop and used by the workers making something from it. It would soon become dirty or torn. If deteriorated to the point that it was useless, there would be no record left for future use. Where the draftsman is also the person working from the drawing in the shop, the original drawing might be used as the only copy. Otherwise, it is normal for the master drawing to never leave the drawing office. Instead, there are copies made from it. These are used on the bench, in the field, or anywhere a drawing is required. This means that the draftsman has to prepare the drawing in a way suitable for reproduction.

There are a large number of reproduction methods in use. Not all reproduce the drawing the same size, although that is usually what is required. For record purposes the drawing can be copied on to microfilm and stored compactly, but that method of reproduction is not the concern of the draftsman. For most methods of copying, it is necessary for the drawing to be on something with a degree of transparency. Ideally, the material would be perfectly clear, but good copies can be made when the degree of clarity is rather less.

TRACING MATERIALS

There are tracing papers which can be worked on in the same way as ordinary paper, but it is possible to see through them. For general work, these are most economical and can be obtained in various thicknesses—the thicker paper being stronger. At one time if a drawing was expected to stand up to the making of a large number of copies, it was made on tracing cloth, which was basically linen cloth treated to finish as a paperlike material, usually pale blue and with one matte and one glossy side. Although this cloth may still be obtainable, it has been largely superseded by plastic film, which may be had in several thicknesses and with one or both sides matte. It is also possible to get thin drawing paper, sometimes called *detail paper,* which can be used for early drawings. Temporary prints from this paper may be needed for consultation on the design or other purposes, but which are not intended to be permanent. This paper has enough transparency for making prints, but it is comparatively fragile.

An important requirement of the paper or film is stability; it should not expand or contract. Some papers will, although the movement may be only slight. Modern tracing cloth is stable, but the most stable material today is polyester plastic film. Besides its dimensional stability, it is unaffected by moisture and does not show its age. Some other materials discolor as they get older, and this may affect the quality of reproduction. The word *vellum* is sometimes used to describe some tracing paper, but that name does not give it any special quality.

Most methods of reproduction involve light passing through the material to affect a sensitive reproduction material below. The lines drawn have to be as opaque as possible to prevent light passing through them. The light then affects the background, but the lines remain unaffected.

Ordinary pencils used for drafting and general work do not make a very opaque line. A print from ordinary pencil lines may be just readable, but there will not be much difference between the background and the lines in the print. It is the contrast between the lines and the background that provides clarity. There are pencils made specially for use on tracing film which make a more opaque line, with a better quality of reproduction. They are not much use on paper or cloth.

INDIAN INK

Although pencil is convenient and quicker to use, the best quality tracing for reproduction by any method is made with *Indian ink.* For mechanical drawing the Indian ink should be black, not so much because of its color, but because this is the most lightproof type and will produce the best line when printed. Make sure that the ink obtained is intended for mechanical drawing. It will be free-flowing and quick drying. It will also adhere well to the surface without chipping or smearing after it has dried. The ink may be described as waterproof or nonwaterproof. The nonwaterproof inks are more suitable for artists' use and you should specify waterproof technical drawing, mechanical drawing, or Indian ink.

Many draftsmen find that the ordinary ink is suitable for all purposes, but there are some special ones. One type is specially made for use on tracing film, with improved adhesion, but this type is slower drying and may remain slightly sticky. If the film has a good matte surface, ordinary Indian ink should suit. Much ink drawing is now done with technical fountain pens. Their makers provide special ink for them, so it is advisable to get the ink to suit the pen. Use of a wrong ink may cause clogging, so the pen ceases to function and has to be cleaned.

Indian ink is available in various quantities, but it is convenient to have a small bottle, even if it is topped up from a larger container. There are 1-ounce and 2-ounce bottles fitted with a device in the cork for picking up a small amount of ink to deposit in the pen or other instrument. This may be just a quill from a feather, or a plastic version of it pushed into the stopper. An alternative is a plastic or glass tube into which ink can be sucked by pressing the soft rubber stopper. Indian ink is very messy if spilled, so it is advisable to fit the bottle into a stand (Fig. 14-1), or at least put it on blotting paper.

If a surface gets greasy, it may not take ink evenly. A line drawn will be broken or vary in width. Sufficient grease to cause this may come from a hand sliding over the paper or film. Talc or chalk can eliminate this trouble, but it must be used sparingly. One arrangement has the talc powder wrapped in a cloth and the top tied. If this is lightly bounced on the surface, a small amount of powder passes through the cloth and can be wiped away by lightly using another clean cloth. Only light use is needed to remove grease.

RULING PENS

For all line work, whether straight or curved, the general purpose tool is a *ruling pen*, which may be in any of the forms otherwise used with a pencil, such as a straight handle or as a compass. Any of the work done by a pencil can be done with a pen. Most sets of instruments allow for a ruling pen to be substituted for the pencil.

Fig. 14-1. To reduce the risk of spilling, a bottle of ink can be fitted into a stand with a broad base.

A ruling pen consists of two blades, with a screw adjustment to move them in relation to each other, with a general appearance something like a pair of tweezers. One blade, into which the screw pulls, is fairly stiff, while the other is reduced for part of its length to make it flexible (Fig. 14-2A).

The nibs or points are rounded. For long lines, they are better fairly wide (Fig. 14-2B), as they are easy to keep straight against a straightedge and hold more ink. For thin and curved lines, it is better for the points to be narrower (Fig. 14-2C). The blades should be as thin as possible where they touch the surface. Ideally they have no thickness there, but such knife edges would cut the paper or film. Obviously, both blades should be the same length. Round the ends slightly in thickness by rubbing on an oilstone. Some plastic film tends to wear pen points, so check your pen ends periodically. Rub them down if they have started thickening.

Ink is put between the two blades with the device in the stopper of the bottle. Never dip a ruling pen in the ink. There should never be ink on the outsides of the blades. If you get any there, wipe it on a cloth or blotting paper before starting to draw. The thickness of the line is controlled by the adjusting screw. Even if you adjust so there is the absolute minimum space between the points, you will not get as narrow a line as you expected if the ends are thick. They should be maintained as thin as possible without the risk of cutting.

As the ink is quick drying, there is the possibility of it drying in the pen. Unless a pen is to be used again almost immediately, wipe between the blades with a piece of cloth before the ink has a chance to harden. There are solvents for ink, but really hard ink has to be soaked a long time to free it. If hard ink has to be scraped, use a blunt knife blade so the inner surfaces are not scratched. A pen can be obtained with the stiff blade arranged to pivot (Fig. 14-2D), so it is easier to clean inside.

Experience will show how much ink to put in a pen. For a thick, long line you will have to put in more than for a short fine line, but too much ink may fall out in use. Usually, not much more than ¼ inch from the points is the most that is advisable. A blot caused by spilled ink on the drawing is difficult to erase completely, so never fill a pen held over the drawing. Do this over scrap paper. Ink can be just as much a nuisance on the floor or table. If ink is spilled there, soak up what you can with cloth or blotting paper. Then wipe with a damp cloth before the ink starts drying. If the spilled ink is on your drawing, use the erasing methods described later in the chapter.

Ruling pens are made in many other ways, but for most purposes the types described will do all that is necessary. There are pens with widened blades so as to hold more ink for long lines. There are twin pens that can be adjusted to draw parallel lines at different spacings such as are required for roads on a map. Ends of pens may be cranked or with gaps, so it is easier to see where the pen actually is drawing.

TECHNICAL FOUNTAIN PENS

Technical fountain pens carry the ink in the body of the pen and feed it to a fine tube that serves as a nib. The diameter of the tube controls the thickness of line that can be drawn with it. This means that there has to be a separate pen or a changeable point for each thickness of line that is to be drawn. Although the pens are very convenient, sufficient pens to draw all the widths of lines normally needed are costly compared with the ruling pen, which adjusts to any thickness of line. For curves there have to be special compasses or adapters to hold the fountain pens, unless the ruling pen point is used in place of a pencil in a normal compass and adjusted to match the fountain pen lines.

Some technical fountain pens fill with a piston action while dipped in a bottle of ink. Others have a cartridge to take out and fill. As with other pens there is the problem of ink drying in the pen, but

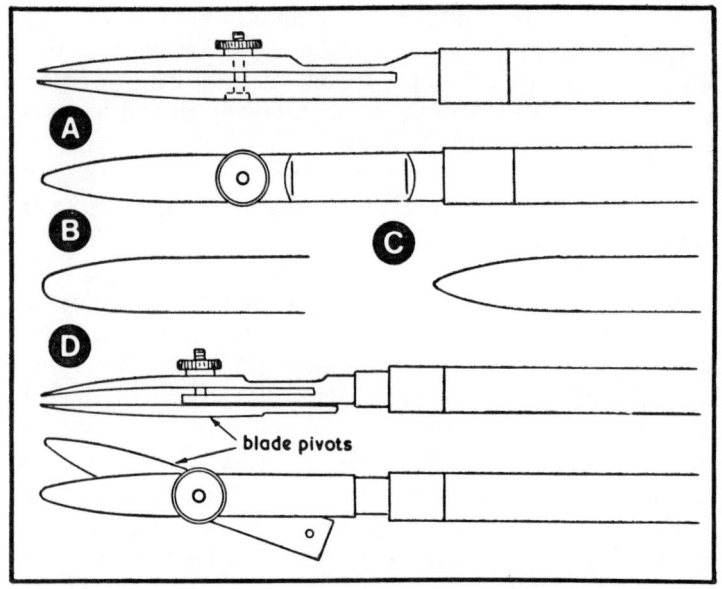

Fig. 14-2. The sprung parts of a ruling pen (A) may have broad (B) or narrow points (C). Some pens open for cleaning (D).

the usual arrangement for cleaning is a weighted wire inside the tube. This can be operated by shaking the pen up and down. There may be an airtight cap to put on the pen when out of use. For some pens, there is a special sealed container to take several pens. Technical pens are satisfactory and convenient, but it is important to read the makers' instructions so as to get trouble-free use from them.

Much lettering on ink drawing is done with guides that use technical pens or special pens which work in a similar way. If lettering is to be done freehand, the pen should be of the ordinary nib type as commonly used for writing before the days of fountain and ball pens. There are fine nibs made for mapping and artists' use, and one of these types should be chosen to suit your technique. Not everyone puts the same pressure nor holds at the same angle, so it is advisable to experiment. Then keep to the one type that suits you. Nibs do not have a very long life, so buy several along with a holder that you find comfortable.

DRAWING WITH INK

There are different ways that an ink drawing can be made. The lines can first be made lightly in pencil on the tracing paper or film.

Then they are inked over, and any surplus pencil marks are erased. Make a pencil drawing in the ordinary way on another piece of paper. Then put the tracing paper over it, and trace the lines in ink. The first drawing need not be produced in the finished state it would have to be if nothing else is to be drawn over it, so long as you put on all the information needed when you start tracing. The tracing will then become the master drawing and be kept for record purposes, while the first drawing that went below it can usually be discarded.

A little practice will show how to hold and use the pen for the best results. Slope a ruling pen slightly towards the direction it is moving (Fig. 14-3). Keep the gap through it parallel with the way it is going. Normally have the stiff blade against the straightedge or other guide. Do not press any harder than is necessary for the pen to work.

When looked at in the direction it is going, the pen should be upright (Fig. 14-4A), so both points are making contact with the surface. The inner point will then be slightly clear of the edge (Fig. 14-4B), with no risk of ink creeping under the edge if the outside of the blade is dry. If the pen is tilted towards the edge, ink may be directed into the angle and under the edge (Fig. 14-4C). With too much tilt that way, the different contact of the points will cause uneven lines, even if ink does not go under the straightedge and

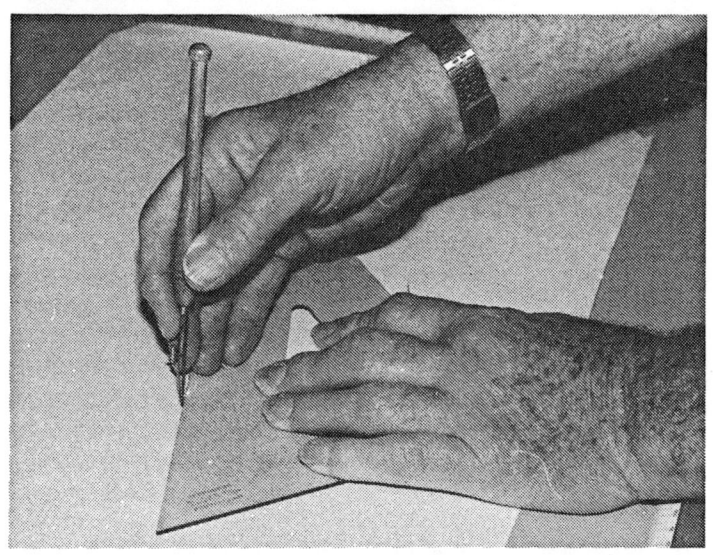

Fig. 14-3. A ruling pen may tilt slightly towards the direction it is to move, but should be upright when viewed from the end of the straightedge.

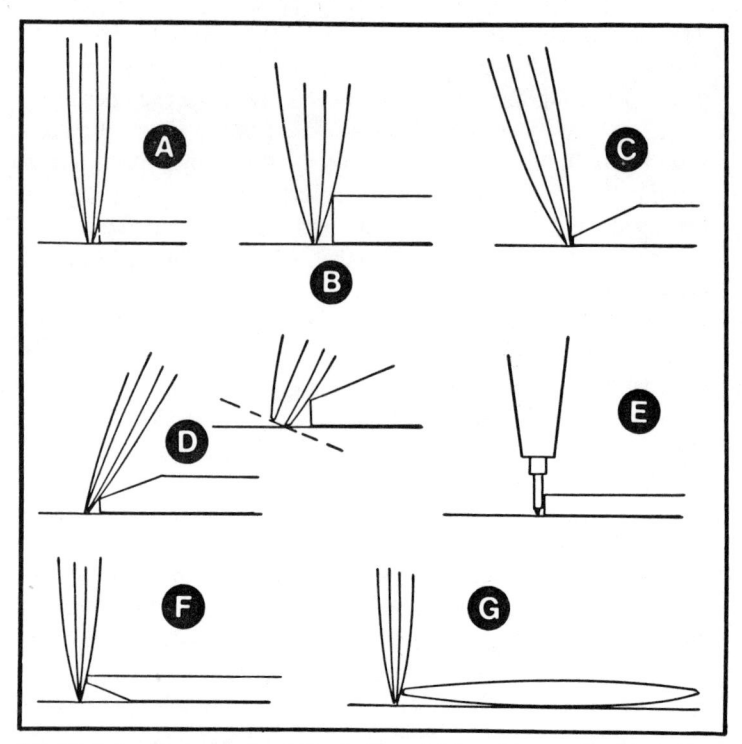

Fig. 14-4. A ruling pen should be held upright and be used so there is no risk of ink going under the straightedge.

cause smearing. Similar trouble will come if the pen is tilted much the other way (Fig. 14-4D).

A technical fountain pen must be used upright (Fig. 14-4E) to keep the end of the tube in even contact with the surface. When using a French curve, there may be difficulty in keeping a ruling pen square with the edge of a tight curve, although it is possible. The upright attitude of a technical pen allows it to follow an intricate shape easier.

For precision drawing, a pen should be used along a straightedge close to the paper. With care in use, there should be no risk of ink going under the edge, but where less accuracy is acceptable a straightedge can be turned over (Fig. 14-4F) so the guiding edge is clear of the paper. A scale or other straightedge with an elliptical section gives clearance without lifting as much (Fig. 14-4G).

When drawing curves with a compass, the pen point should be upright in relation to the surface. Most compasses have the leg hinged, so this can be arranged. Do not try to draw a circle with the pen blade sloping. Some small bow compasses have no adjustment of angle, but their range of movement is slight. The amount the pen may tilt is acceptable. With larger compasses, angle the arm to suit (Fig. 14-5). This is even more important if it is a technical fountain pen mounted in the compass. When turning the compass, be careful to keep the instrument upright.

Although Indian ink dries fairly quickly, it does need a short time to harden, particularly with thicker lines. It is possible to become impatient and try to pass your hand or a triangle over a line before it is dry, with consequent smearing. You may be able to put the damage right with erasing, but it is a nuisance and may be noticeable. Never try to blot Indian ink lines. That would spoil their quality, either roughing their edges or making them less opaque for printing. They must dry naturally.

Ink lines may be in many widths compared with those possible with pencil, so you can show particular parts with lines of different thickness if you wish. Dimension lines will usually be the thinnest, although in some drawings there may be thinner projection lines. You may need to check what method of reproduction is to be used, as some will not show very fine lines. The relative widths of lines should then be adjusted accordingly.

SEQUENCE OF INKING

In general, trace fine lines first. They will dry almost instantaneously, so there is little fear of smearing. Try to go all over the

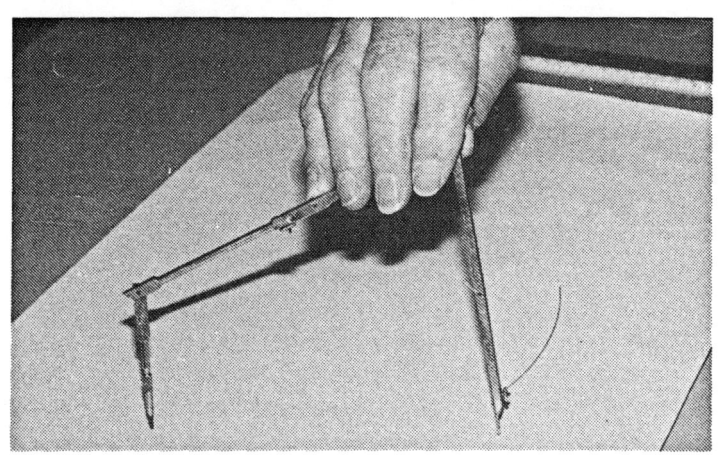

Fig. 14-5. Care is needed to hold over the central arm when drawing a curve with a compass using an extension.

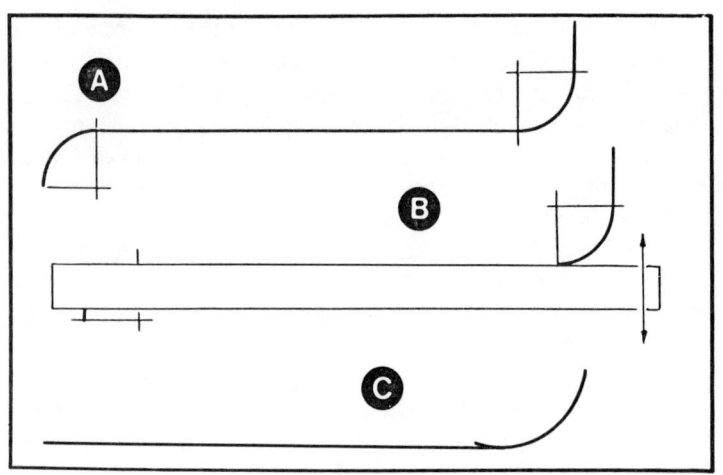

Fig. 14-6. Draw curves first when inking and adjust straight lines to match.

drawing dealing with lines of one thickness before moving to the next thicker ones. Exceptions are curves. On most drawings it is best to do all compass work first. This is particularly important if a part of a circle joins a straight line (Fig. 14-6A). The reason is that you have no way of varying a compass curve once you have located its center. It is possible to move a straightedge slightly to line up with a curve (Fig. 14-6B), but not to vary the curve if the straight line is drawn first. It helps to also locate the meeting points on the pencil drawing, so you know where to stop with the compass. It is better to go too far than to stop short, when there would be a kink in the line. If the curve goes too far (Fig. 14-6C), you can correct by erasing.

As you progress through thicker lines, do all of one thickness in one direction first. Start at the top and work down, doing all horizontal lines. Make sure they are dry; then do all vertical lines. Let them dry, and fill in diagonal and other lines. If you have to do any hatching of sections, leave that until you have the outlines drawn so you know the limits for the lines.

If there are irregular curves to be drawn around a French curve, draw these lines at the same time as others of the same thickness. Make the curves before any straight lines are to meet. Ink in arrowheads on dimension lines. Usually these can be done freehand with a fine nib. Be careful that they come to fine points. It is easy to get them too thick by going too much each side of the line.

Lettering, including figures in dimension lines, can usually be done last. There is sometimes a problem when guides are used in knowing how much space will be taken up. In that case it may be better to do lettering at an earlier stage. Figures for dimensions could be put in position before drawing the dimension lines on each side of them.

Lettering can be done freehand using pencil guide lines, either on the tracing or on the paper beneath. There are many types of guides for lettering with varying degrees of complication. A simple form is a plastic guide, supported clear of the surface, through which a technical pen can be fitted (Fig. 14-7). A special pen may be provided, having a reservoir to be filled with the ordinary device from a bottle. Because of the form of some letters in a stencil such as B, the guide has to be moved to two positions in some cases. With the guide against the T square, it can be slid along to get the letters where you want them. A little practice shows how to space them evenly. The guides are available in several sizes and with a variety of symbols, as well as upper and lower case letters and figures.

Another lettering guide uses a template with the letters cut out. Instead of the pen being put through it a scriber traces the letter while a technical pen does the drawing above the guide.

ERASING

Try to keep all ink only where it is wanted. There will almost certainly be occasions when a line has gone too far, hatching has gone over the border, or you have smeared a still wet line. Remember that what you are doing is removing something opaque so what is left is transparent, for the sake of printing. It is no use removing the ink and leaving a surface that obscures light in printing. Usually, attending to the fault improves appearance as well, but transparency is the first consideration.

An ordinary eraser intended for pencil will not have any effect on ink. There are erasers that include fine particles of grit which will remove ink. These are similar to some typewriter erasers. Another type has a bunch of filaments of fiber glass projecting from a holder. It will erase ink. As the fiber glass wears, the holder can be adjusted to extend more filaments. There are eraser materials to use in a powered holder. All of these erasers remove ink and part of the surface of the material in a general area. They cannot be used for fine and close corrections, such as one line going a short distance too far over another. Trying to erase that would affect the line that is correct as well.

For close erasing, as well as quite long lengths of lines, it is preferable to scrape. The blade should be finely sharpened and

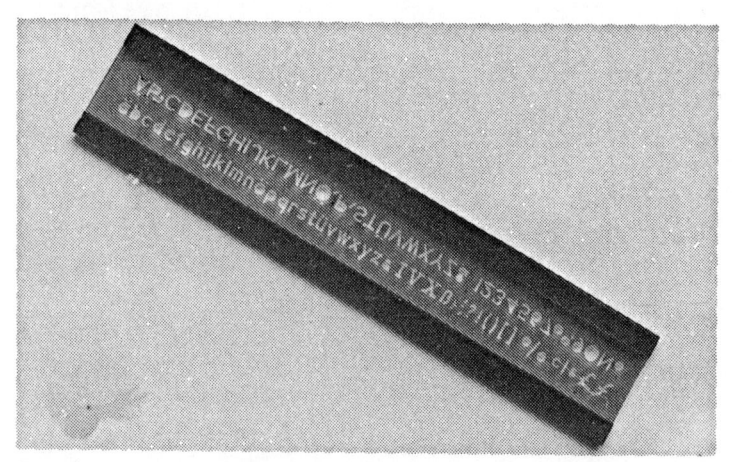

Fig. 14-7. A stencil for lettering using a technical fountain pen.

pointed. The corner of a razor blade might be used if it is wrapped or otherwise given a grip. There are finely pointed knives sold for the purpose by artists' suppliers. They may also sell things that look like nibs to fit in a handle, but which have cutting ends. Scrape sideways with the cutting edge sloping very slightly from the surface. Although it is possible to scrape just the offending piece of ink, it is almost impossible to avoid scraping a layer off the surface of the material. That does not matter unless it is very thin material and there is a risk of going right through.

An erased surface does not take ink in quite the same way as the surrounding unerased surface. If you draw a line across a part you have rubbed or scraped, the area across the affected piece is likely to finish wider and with a ragged edge. It may be better to draw any correcting lines before you erase what was wrong. By careful scraping, it is possible to erase along both sides of a correct line without damaging it.

If you have to draw over an erased surface, first follow with an ordinary soft pencil eraser over it. Then burnish it by rubbing with something hard and smooth. This should seal the surface so there is little risk of the line becoming ragged. It is worthwhile using a scrap piece of tracing paper or film to draw a few lines that you can use for experimental erasing.

REPRODUCTION

A draftsman is not usually involved in the reproduction of drawings, except that he must make sure he draws in a way that suits the technique used. The oldest method of reproduction is called blueprinting. The print shows white lines on a blue background. The sensitive paper is put under the tracing and exposed to a light. Usually this involves feeding through a machine so the two are closely in contact as they go around a light source. A suitable paper can be obtained for exposure to daylight. For smaller drawings, the tracing and sensitive paper may then be put in a frame and exposed to daylight until the background has been affected. After exposure, the blueprint needs a brief soak in water to fix it. A further chemical treatment can be used to intensify the blue, followed by another wash. The print has to be hung to dry before it is ready for use. Blueprints are still used, but having to wait for the paper to dry and the need for space if many prints have to be dried are obvious disadvantages. The name blueprinting is sometimes used incorrectly to describe collectively all methods of reproduction.

More printing is done by the diazo process, which may be called dyeline or white print. The print has a dark line on a white or near white background. Lines may be black, blue, or red. Besides paper, the sensitive surface can be on plastic film, so it is possible to print a second tracing from the first. This may be done when a large number of paper prints have to be made, and the second tracing can be used to make them. The original tracing is kept as a master, from which more copy tracings can be printed as required.

In the diazo process the sensitive paper or film is put behind the tracing and exposed to a light in the same way as for blueprinting. It is then fed through ammonia vapor, usually in the same machine, while the tracing emerges without going into the vapor. The print comes dry from the process and is ready to be used immediately. As it does not get wet, there is no risk of distortion, which may occur when paper is soaked.

Another diazo process differs in the way the image is fixed on the paper. Instead of passing through ammonia vapor, the paper passes over rollers which moisten it with a developer. This does not saturate the paper, but allows it to emerge only slightly damp, so it dries in a short time. There is no smell with this process, which may be described as semidry. In the process using ammonia, most machines have to be vented to the outside atmosphere. There is a slight ammonia smell coming from the paper, so some forced ventilation to the room is advisable.

There are other processes including electrostatic and thermographic, but for same-size prints the methods just described are usual. There are photographic ways of enlarging or reducing. Microfilms reproduce and store drawings on very small films which can be viewed in an enlarged form on a scanning device.

Technology is making the copying, storing, and reproduction of drawings ongoing processes. A draftsman should be aware of developments as far as they affect the way he makes his drawings. Although many methods can use pencil drawings, it is likely that there will always be a need for good ink drawings. A draftsman who claims to be master of his craft should be able to produce finished drawings in ink.

Chapter 15
Three-Dimensional Drawings

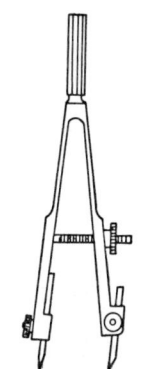

A drawing made by orthographic projection shows two or more elevations and one or more plan views, with sections and detail drawings, so a craftsman or other user of the drawing has all the information he needs. Each part of the drawing shows two dimensions, such as length and breadth, with all the detail measurements in those directions. For the thickness or depth the reader has to refer to another view, with possibly a further reference to other views for details.

This is the only way that an object with much detail can be shown so all the parts are correct to size and in the right proportion to each other, with angles given as they actually are and circles drawn truly round. For nearly all communication between a draftsman and the user of his drawing, this is the way a drawing should be made. A regular user of such drawings is able to look at the drawing and visualize what is illustrated, and go ahead making the object, extracting information from the various views without difficulty. His interpretation is the result of experience. In many cases he will have received training in drafting himself, if only to the limited extent needed to understand how the drawing is laid out and the various views are produced. For technical work, there is no satisfactory substitute for this conventional way of making drawings.

There are many other people in industry who want to get information from a drawing. For anyone with little or no technical knowledge or training, the orthographic drawing may be a complete mystery or only partially understandable. There may be some appreciation of what the object looks like when viewed squarely towards one face, but some of the other information may not be understood. If the reader visualizes anything complete, it may not be the shape the drawing really portrays.

PICTORIAL VIEWS

For such users of drawings, there has to be a portrayal of the thing in a more pictorial form. The shapes in all directions are shown in a way close to what they would be as seen by looking at the actual object. In the case of simple objects, this could be dimensioned. There would be no need for other drawings. In other cases the three-dimensional view would supplement the orthographic projections, which would give the sizes to be worked to. The added view helps a less experienced reader understand what the drawing is all about.

Even when it is assumed the reader has sufficient technical knowledge to follow the normal drawing, a three-dimensional view may help him and the draftsman to visualize the final concept. This is particularly so if the whole thing is complicated. The run of the hydraulic pipes in an airplane may be correctly shown in elevation and plan, but their positions and paths may not be easily understood if there are many other details on the drawing. A more pictorial view, with the pipes picked out in color, shows them instantly and helps a draftsman check that he has the scheme laid out correctly.

An architect usually draws a house or other building in a pictorial way, sometimes with trees and people included to show relative sizes. It may be a technically correct three-dimensional view, or it may be more of an artistic impression, but it is still what the designer visualizes as the appearance of the finished work. The builder may get his information from many other drawings, but this one helps him to more easily understand what he is producing. In that case the draftsman has to be something of an artist, but there are other things which can be illustrated by normal drafting methods, even when the draftsman is not artistic.

INSET VIEWS

A simple thing, which can have all the information needed shown by two elevations and a plan, can be set out so there is an area of paper left above or below the end elevation. This may carry the title, scale, and other information. It is also a convenient place to put a pictorial view, if that would help the reader understand what the other views represent. For instance, a wooden box could have all the necessary sizes shown on the two elevations and plan, but a pictorial view with the lid partly open gives an instant picture of what the box will look like (Fig. 15-1).

In this case the picture is not to any scale. The vertical lines are drawn upright, but the other lines lead off towards a vanishing point, as they would when viewing the actual box. Such a picture is probably best drawn freehand to a shape you think seems correct. Then it is lined in with the help of a straightedge.

This is also a useful way of showing engineering parts to a reader who is only semitechnical. There might be more than one part drawn as they would come together but separate in an *exploded* view, possibly with screws shown ready to be inserted (Fig. 15-2). This sort of illustration is being increasingly used to instruct people with little technical knowledge who are engaged on assembly work. The drawing might be artistically made without regard to actual sizes, or it might be made in one of the ways described later in this chapter.

A problem comes when what has to be drawn is quite complicated. A considerable amount of time might be taken up in producing the three-dimensional drawing in addition to the orthographic views. Another artist-draftsman might produce a pictorial representation of the item. Then the original draftsman puts in the technical details which have to be emphasized. Despite the amount of work involved, such a final drawing may be justified, particularly if it is needed for a large number of users, such as would happen with an airplane, car, or other complicated piece of engineering when those engaged in servicing need a clear drawing for reference.

OBLIQUE DRAWING

A simple way of providing a pictorial representation is to draw one view as seen squarely from the front. Then show the other way by drawing appropriate diagonal lines from it to indicate thickness (Fig. 15-3A).

An oblique drawing is formed on three axes, which should be drawn first. Two are horizontal and vertical lines, obviously square to each other and of any length that will enclose the drawing. A diagonal line goes from their meeting point. The angle of the diagonal line is not important and may have to be chosen, so it does not conflict or obscure a cut or slope that you want to show. If there

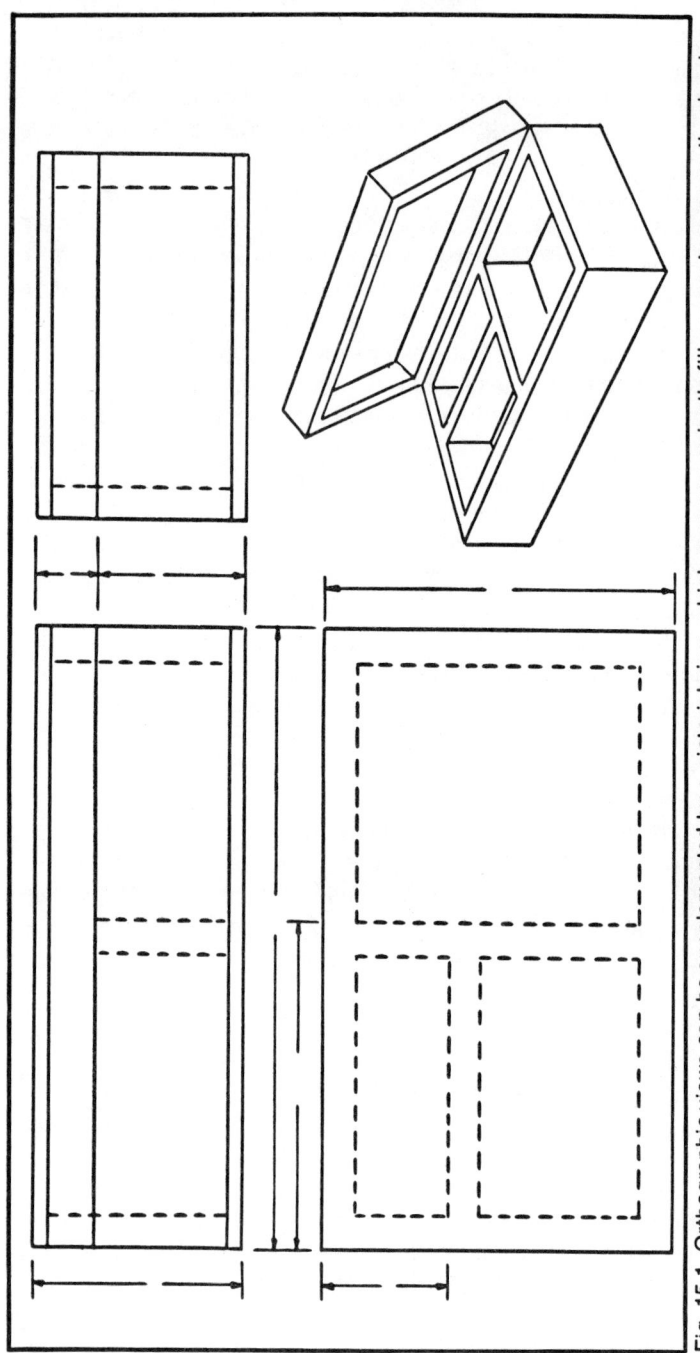

Fig. 15-1. Orthographic views can be supplemented by a pictorial view, which may conveniently fill a vacant space on the drawing.

214

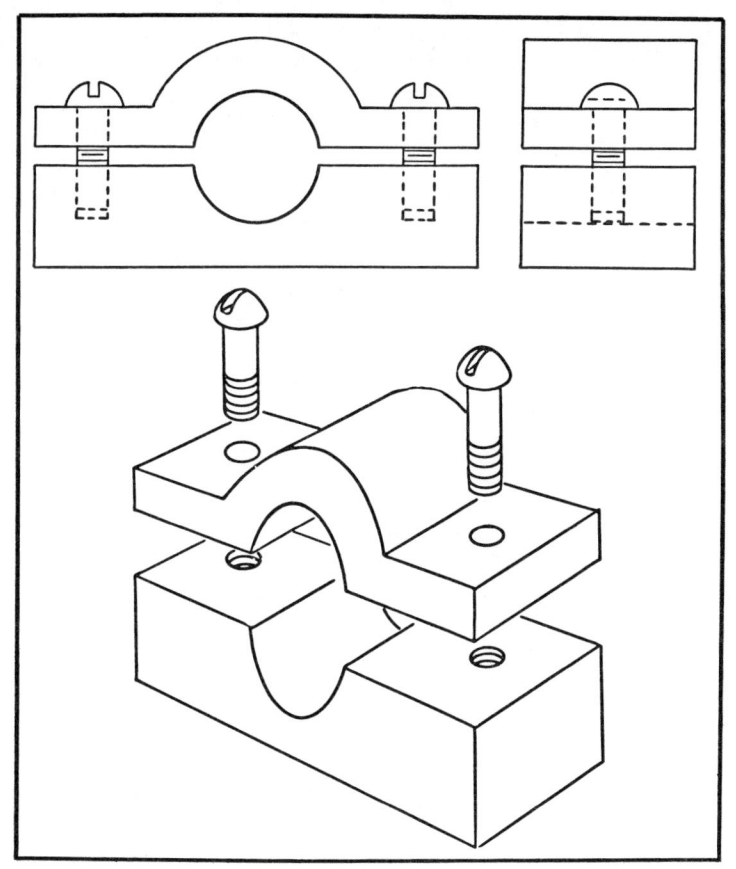

Fig. 15-2. An exploded isometric drawing may clarify details not obvious in orthographic views.

are no other reasons for choosing an angle, it is convenient to make the diagonal axis at 45° or 30° to horizontal.

Suppose you want to draw a solid piece of metal with a hole through it. With a 30° projection the hole shows on the front (Fig. 15-3B), but at 45° you can show a part of it at the back as well (Fig. 15-3C). This shows the reader immediately that the hole goes right through and is a better picture in any case.

The direction of the angular axis does not necessarily have to go upwards to the right, although that is more common. It could go upwards to the left (Fig. 15-3D), if that fits your space better or some detail is better illustrated that way. It could go downwards, particularly if what you are drawing will normally be viewed upwards (Fig. 15-3E).

When deciding how to arrange an oblique drawing, there are two main considerations. If it is something much longer one way than the others, put the long dimension parallel to the picture plane (Fig. 15-4A) unless there is a good reason for illustrating it in some other way. Normally the long dimension would be horizontal, unless putting it vertical matches the way it would be in use. If it is something with curves, angles, and other complications much more in one direction than the other, let that surface come to the front (Fig. 14-4B). You can then draw all those details true to shape in just the same way that you would on an orthographic view.

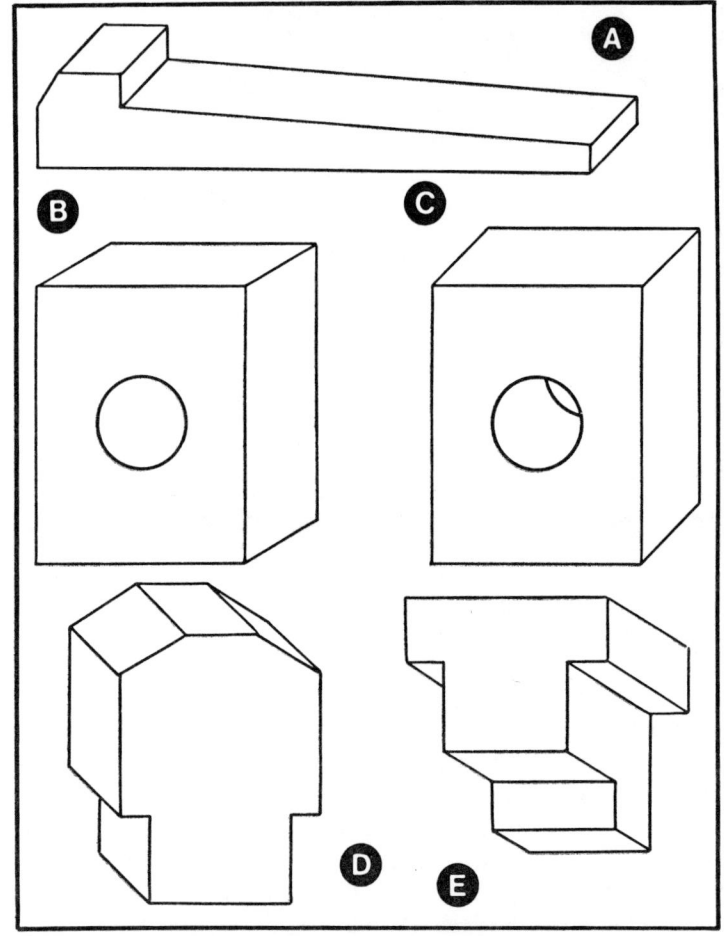

Fig. 15-3. Oblique projections give a good idea of shape. They do not always have to be drawn at 45° and can be as viewed from below.

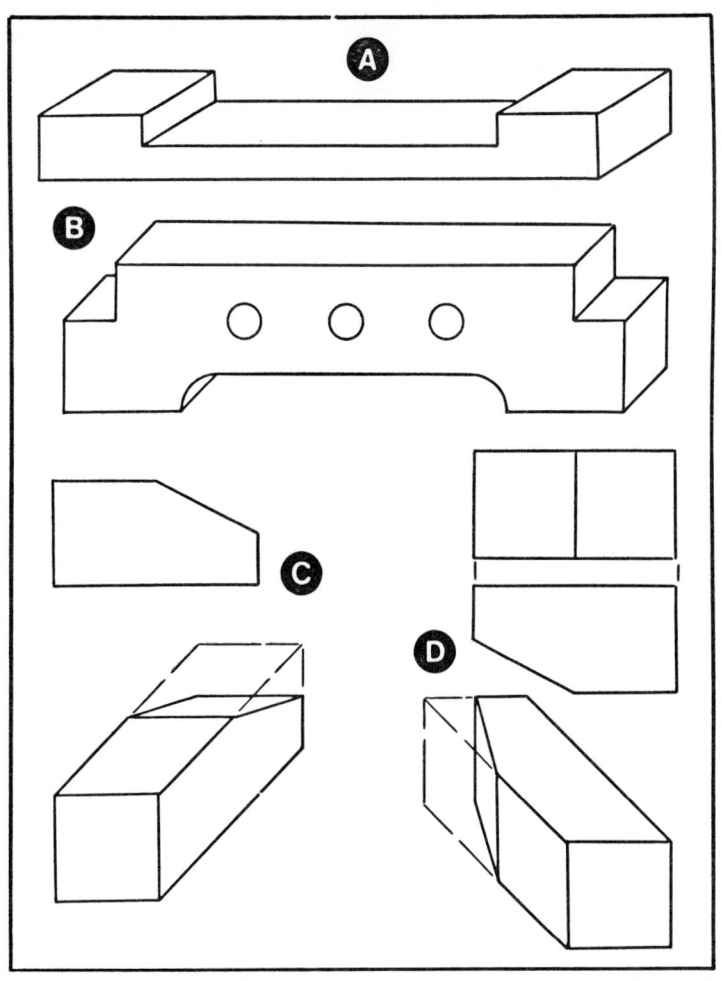

Fig. 15-4. Long or curved surfaces should usually be at the front. Diagonal surfaces will not be true to scale when projected.

All dimensions on the oblique drawing which come parallel with one of the three axes you started with are true to size or scale. A problem comes when there is a slope to be illustrated in the diagonal plane. In that case it will not be shown to its true length along the slope. What will still be accurate will be the locations of its ends on any lines parallel with the axes. To draw a slope on a diagonal view, measure along vertical and diagonal lines and join them (Fig. 15-14C). For a slope the other way, measure along horizontal and diagonal lines (Fig. 15-14D). These are cases where

217

you may need to choose an angle for the diagonal axis that will allow you to show a width in the sloping part.

CIRCLES DRAWN OBLIQUELY

There is a further problem in drawing if you cannot include all the circles in the front view, but there has to be a representation of a circle in a diagonal view. The circle is then projected as an ellipse, but you have to get it to the correct size.

The method can be illustrated by considering similar-sized circles on all three visible faces of a cube (Fig. 15-5A). Draw the

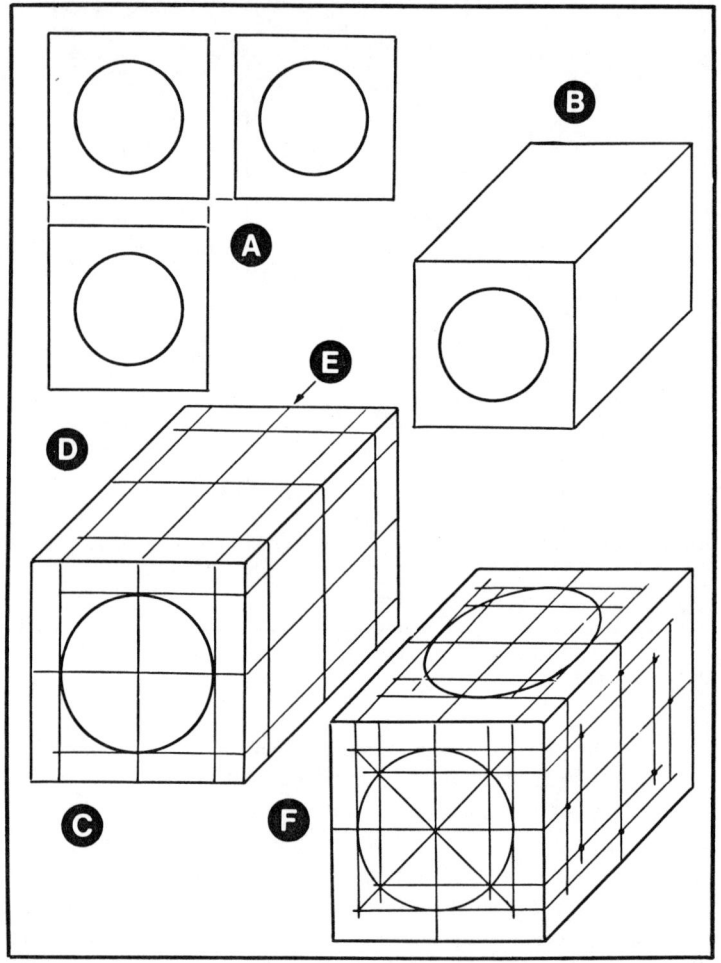

Fig. 15-5. Although a circle is true on the front view, it becomes an ellipse when it is projected on to the other surfaces of an oblique projection.

218

full circle in position on the front face (Fig. 15-5B). Draw a square with its sides as tangents to the circle, and continue them to the edges and on to the other faces with lines parallel to the axes (Fig. 15-5C).

The shapes on the other face mark the limits of the ellipses that have to be drawn (Fig. 15-5D). How they are drawn depends on the accuracy needed. It is possible to draw the ellipse axes and use the method described earlier to produce an approximate ellipse with compasses. If an ellipse template is available, it may be possible to find an opening that matches. This will give a better result. If it is a large ellipse that has to be drawn accurately, points on it can be found geometrically. A line is drawn through them.

Another way of getting points on the ellipse is to project ordinates from the circle. The points where the ellipses have to touch the squares can be found by projecting center lines (Fig. 15-5E). You can get the positions of other points on the circle midway between the center lines by using diagonals to the front square (Fig. 15-5F). By projecting these you get four more points on each ellipse, and those positions should be enough to enable you to draw an acceptable ellipse in the diagonal views. Do not make the mistake of drawing diagonals in the projections of the square.

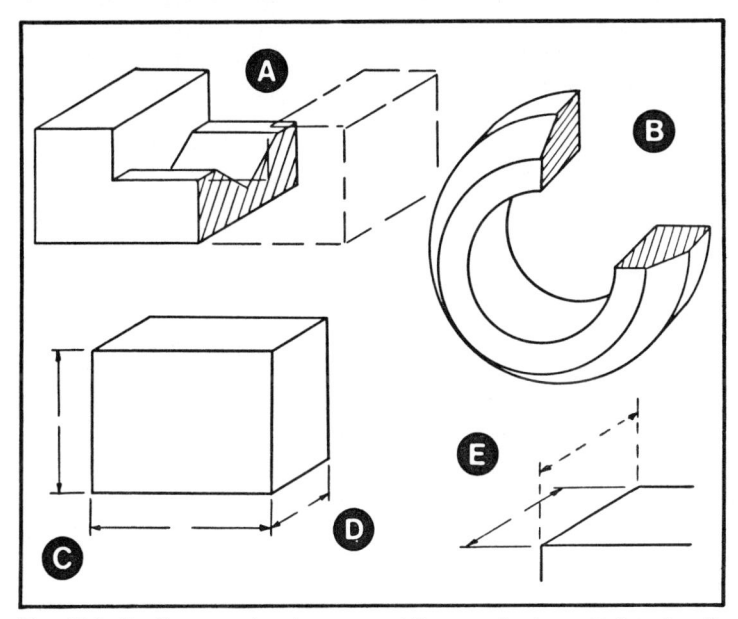

Fig. 15-6. Sections can be shown on oblique projections. Relate the dimension lines to the directions of the surfaces.

219

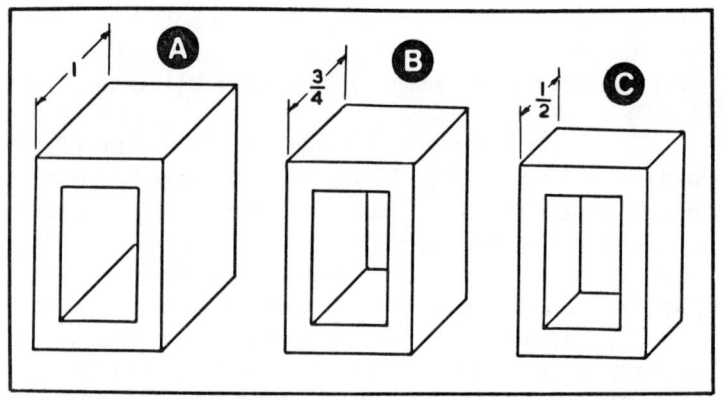

Fig. 15-7. If the diagonal lines are full-size, the effect is to make the object look thicker, so the scale that way may be reduced.

As they are not parallel to axes, distances along them would not be correct.

Usually an oblique drawing shows a complete object. It is possible to show a section, either across the complete drawing (Fig. 15-6A) or of only part of it (Fig. 15-6B), if that would help the reader understand the picture better. Use the usual section hatching or other markings on the cut surfaces.

Dimension lines can be put on an oblique drawing. For the front view they are arranged in the usual way (Fig. 15-6C), and that is another reason for having the more complicated view to the front. Dimensions the other way are arranged parallel with axes and between lines projected parallel with axes (Fig. 15-6D). Usually the projections for dimensioning can pick up dimension lines two ways, but if they are independent, either way of projecting is satisfactory (Fig. 15-6E).

An oblique drawing with the measurements in the diagonal direction full-size or to the same scale as the front view looks thicker than the object actually is (Fig. 15-7A). This may not matter, and that method of drawing is often used. If the picture is to be more realistically drawn, however, there can be a smaller scale for the diagonal direction. Reducing it to three-quarters of the front scale improves appearance (Fig. 15-7B), but this can be brought down to half (Fig. 15-7C). That is usually more convenient as it means the same scale can be used, providing any measurement diagonally has to be doubled to get the true size. This sort of reduction is often used by furniture draftsmen. The method of projecting half-size diagonally is sometimes called *cabinet* oblique drawing. The term *cavalier* is used for full-size diagonal projection, but there is not an obvious reason for the choice of that name.

ISOMETRIC VIEWS

A cornerwise view of an object is often loosely called *isometric*, but not all of these views are correctly described in that way. The general name is *axonometric*. Even isometric has two versions. A view that may be assumed to be looking down diagonally towards a corner (Fig. 15-8A), with the three main axes at 120° to each other, is isometric. A similar view, but from a longer angle, drops the angular lines to 105° to vertical. This is a dimetric view (Fig. 15-8B). If the viewpoint is slightly to one side, but at the same height as in the isometric view, the shape can be achieved by tilting the axes (Fig. 15-8C). This is called *trimetric*.

In all of these views the dimensions in all directions are true, and all lines are parallel with the axes. This means that isometric or other views differ from a pictorial view, in which the diagonal lines converge as they go away from the viewer towards a vanishing point.

There are isometric drawings and isometric projections, although many people regard the terms as interchangeable. In an isometric drawing all dimensions are actual or scale size (Fig. 15-8D). This is what any draftsman will normally draw. It tells the reader what the shape is and gives him lengths in the directions of the axes which he can measure and use. As with full-size oblique projection, the effect is to make the object look bigger than it is.

In an isometric projection, not only are the diagonal lines apparently receding away from the viewer, but the vertical ones are assumed to do so as well. In other words the whole cube is tilted towards the viewer (Fig. 15-8E), instead of being regarded as standing upright. The effect of this is to make the vertical corners appear shorter. If a cube is tilted towards the viewer until the diagonal back to front is horizontal, all sides will look shorter than their actual measurements. The actual sizes could be calculated by trigonometry, but in this cube the apparent length in the isometric projection is about four-fifths of full-size.

By drawing in this proportion, the cube is given a better relation to actual size (Fig. 15-8F) than if full-size measurements are used (Fig. 15-8G). A draftsman uses isometric drawing in practice rather than isometric projection nearly every time. Actual sizes are usually of more use to the reader than a reduced size that may look more like the overall size, but does not have lengths in the direction of the axes that can be measured easily.

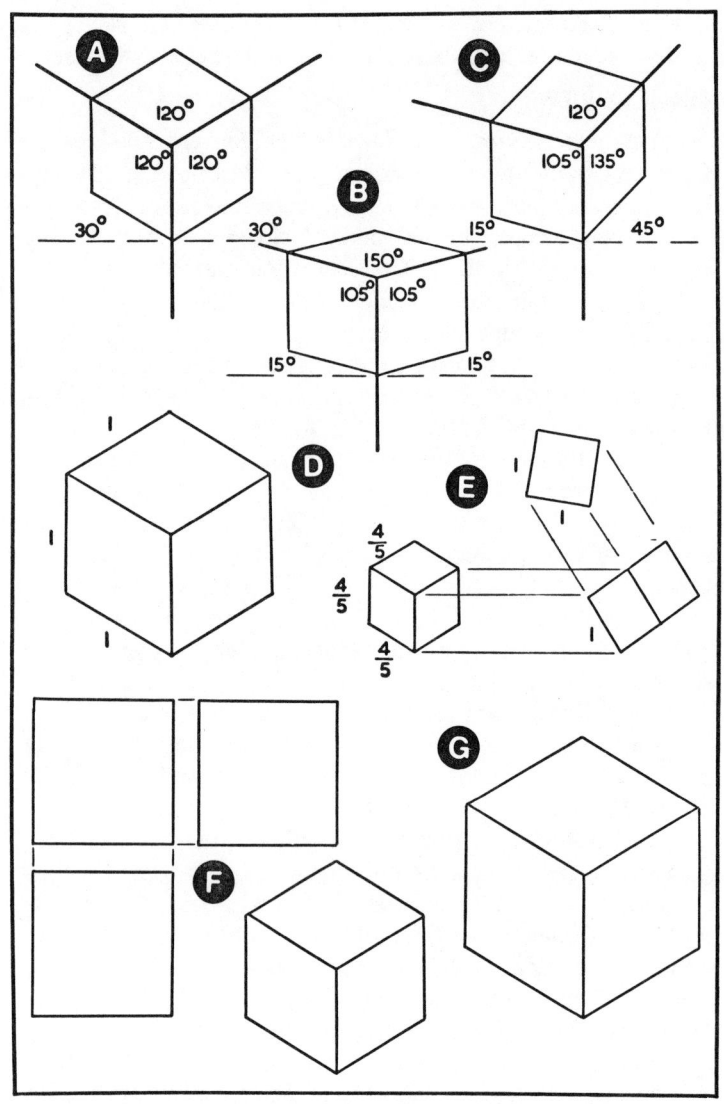

Fig. 15-8. The standard isometric angles (A) may be flattened (B) or the view moved to one side (C). A scale isometric view looks oversize (D) and may be reduced (E).

Dimetric and trimetric projections are rarely used, but if so the methods used are similar to those for isometric drawing, except for the different angles of the initial axes. Only isometric drawing is described here.

ISOMETRIC AXES

Start by drawing three axes at 120° to each other. With one vertical, that makes the others 30° to horizontal (Fig. 15-9A), so the other lines to be made are easily kept parallel by using the 30° triangle. Nearly all isometric drawings are made on these axes so they finish as viewed from above the front corner (Fig. 15-9B). If it

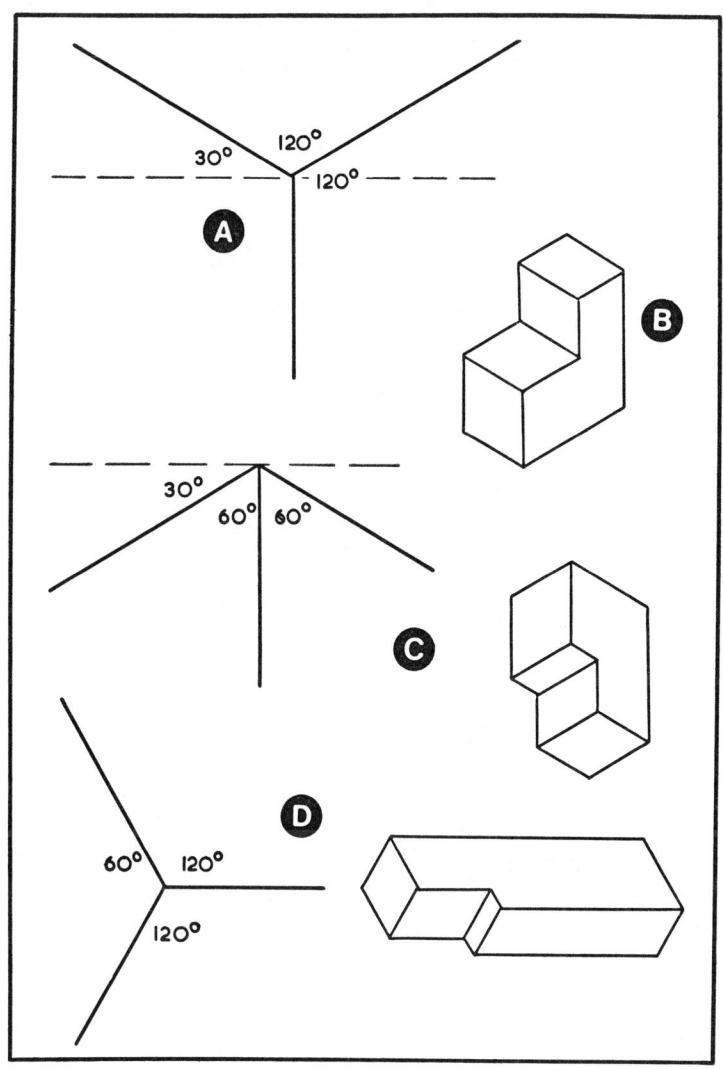

Fig. 15-9. The standard isometric view is from above (A and B), but it could be from below (C) or from one side (D).

223

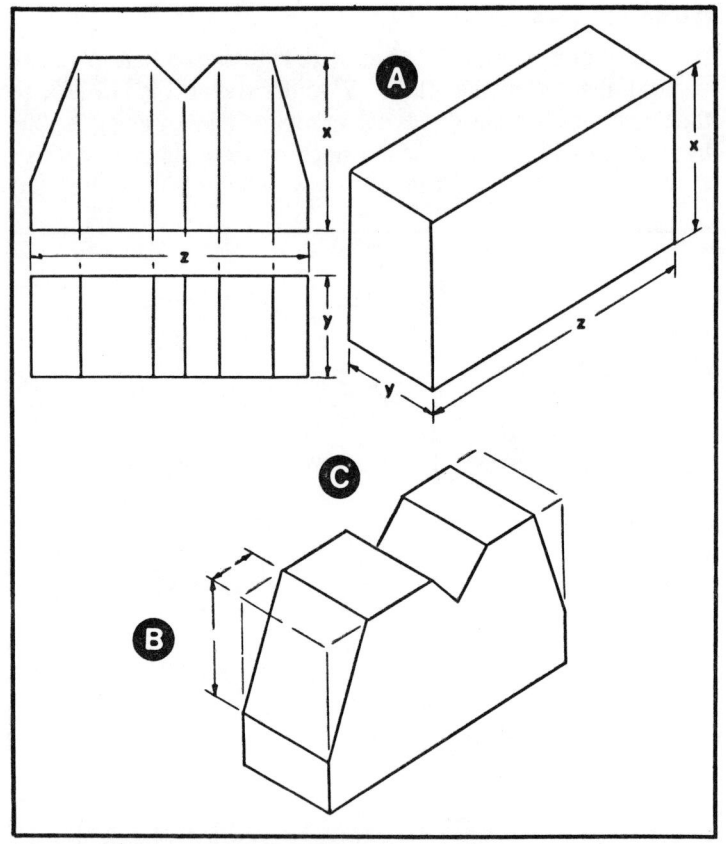

Fig. 15-10. Diagonals on an isometric view are drawn after the outline (A) has been drawn by measuring along the axes (B).

is something that would normally be viewed from below, the arrangement of axes could be reversed (Fig. 15-9C). Less often the axes could be arranged for a view from a side, with one axis horizontal instead of vertical (Fig. 15-9D).

When making the isometric drawing, all lines that are parallel to an axis are drawn their true length (Fig. 15-10A). As with oblique projection, any line that is not parallel with an axis cannot have its true length shown. The positions of its meeting or cross lines parallel with axes will be correctly located. Get these positions along the lines, and join them for the slopes (Fig. 15-10B). There is not the opportunity of varying a view by altering the angle, as in oblique projection. Sloping lines will have to be drawn as they come, possibly showing a face or having it hidden (Fig. 15-10C).

If the thing being drawn is irregular in outline, it will usually help to first draw an enclosing skeleton outline of a rectangular block that would touch the article at its extreme points. This allows you to locate positions at the limits of lines which have to be drawn, so you can join them to get a good representation of the shape in isometric view. For instance, suppose a part of the object is an irregular truncated cone (Fig. 15-11A). You can draw the outline of a top to the full size of the base of the cone at the level of the truncation. From the plan, you can draw the shape of the truncated top at this level (Fig. 15-11B). With this positioned, draw in the diagonals to complete the view (Fig. 15-11C). It could be that some views are almost entirely made up of lines which are not parallel with axes. In that case their ends have to be located on lines that are parallel with axes, so they can be linked and the pattern built up.

ISOMETRIC CURVES

As no surface is square to the front in isometric projection, any circles have to be drawn as ellipses. As there is no circle on that drawing to project from, an elevation or other flat view has to be used. Draw sufficient ordinates across an enclosing square to get points on the curve (Fig. 15-12A). Repeat these on the isometric surface with lines parallel to axes. Measure or use dividers along these lines to locate the positions of matching points that a curve can be drawn through (Fig. 15-12B).

If a curve is not part of a circle, a similar method can be used. Draw a side view with the curve to its true shape (Fig. 15-12C). Draw enough ordinates across the curve, either horizontally or vertically, depending on which way will give more satisfactory crossings of the curve. Repeat these ordinates at a similar spacing on the isometric view (Fig. 15-12D). Check the heights and distances of crossing points of the curve along these lines, and plot the same distances on the matching lines on the isometric view (Fig. 15-12E). Draw a curve through the points you have marked, and project to the other side. If you are doubtful about a particular part of the curve, put in one or more ordinates to cover it. Ordinates do not have to be equally spaced, so long as the spacing on the isometric view matches that on the side view.

In making an isometric drawing, it will probably be advisable to draw a plan and two elevations, if they do not already exist (Fig. 15-13A). They will provide your references and will be essential if you have to include curves or irregular shapes.

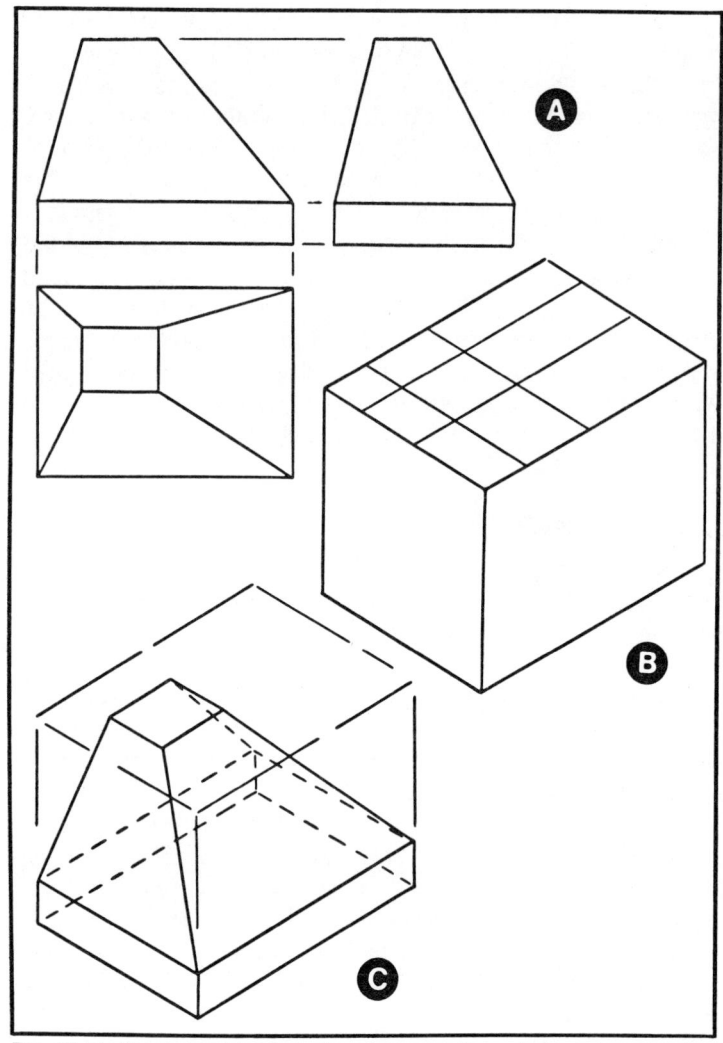

Fig. 15-11. The isometric view of an irregular figure is found by locating points on a containing rectangular shape and joining them.

Start by drawing the three axes lightly (Fig. 15-13B), and follow with main outlines parallel with the axes (Fig. 15-13C). If there are extending parts that include lines which are not parallel with the axes, draw parallel lines that will touch and include them, so you can locate key points (Fig. 15-13D). At this stage, treat all the lines you make as construction lines. Do not draw them heavily until you know that a line is correct and within its limits. It is a

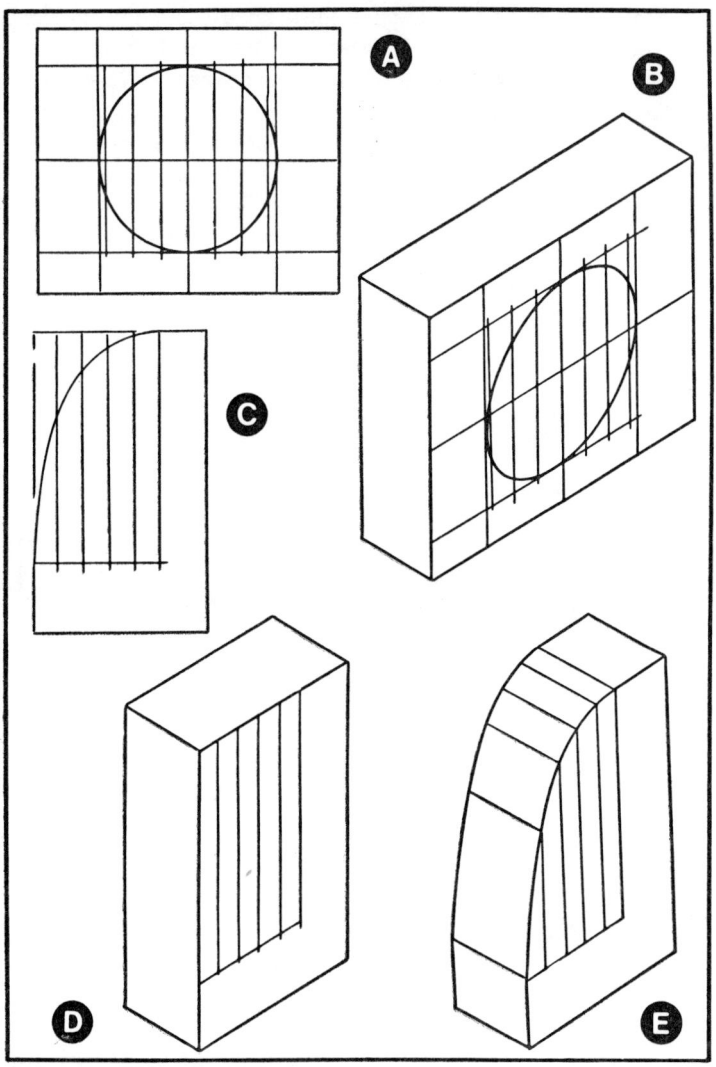

Fig. 15-12. A circle has to be projected as an ellipse (A and B) by using ordinates. An irregular curve can be projected in a similar way (C, D and E).

nuisance if you discover later that you have to erase part of a line you have already treated as final.

If there are curves, whether openings or outlines, get their shapes and draw them (Fig. 15-13E). Link them with the straight lines, erase unnecessary lines, and line in all other parts to complete the drawing (Fig. 15-13F).

If you wish to show a section on the isometric drawing, put the usual section lining on all cut surfaces (Fig. 15-14A). If there are to be dimension lines, keep them parallel with the axes and between lines projected parallel with axes (Fig. 15-14B). You can align the figures with the faces, assuming they are to be read below or from the left (Fig. 15-14C). It is probably easier for the reader if the figures are upright or *unidirectional* (Fig. 15-14D).

If the item to be drawn is long or its final position is horizontal, it may be better to work about a horizontal axis instead of a vertical

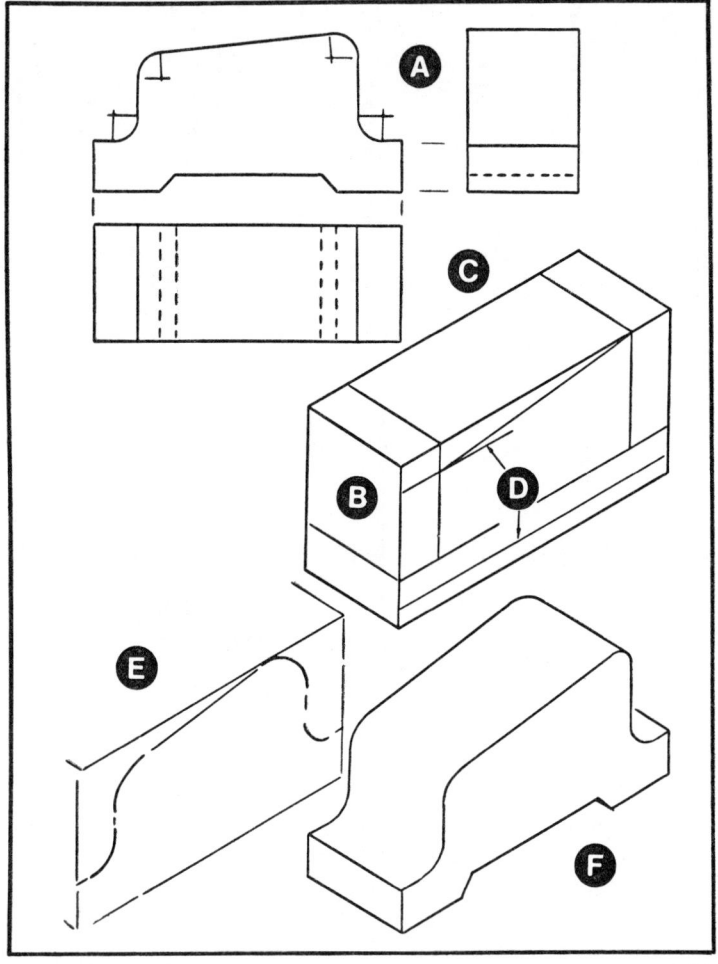

Fig. 15-13. A shape with slopes and curves is first projected with straight lines, and then the curved parts are added.

one (Fig. 15-15A). Prepare the three axes (Fig. 15-15B). Construction is no different from the other way, providing you remember that the horizontal axis is the equivalent of the vertical one on the more usual upright projection.

Similarly, the isometric view may be better made with the finished view apparently overhead. In that case the main axes are drawn sloping down from a vertical one (Fig. 15-15C) instead of upwards. Construction is the same as the usual vertical projection, except that it is upside down (Fig. 15-15D). The same result would have been obtained by reversing the paper.

CHOICE OF VIEWS

The choice of views to be drawn will be dictated by the needs of the probable users of the drawings. Orthographic projections are usually all that are needed in manufacturing industry where the users are familiar with the system. Even then it may be necessary to provide additional illustrations for the benefit of a customer or even administrative people within the industry, who may not un-

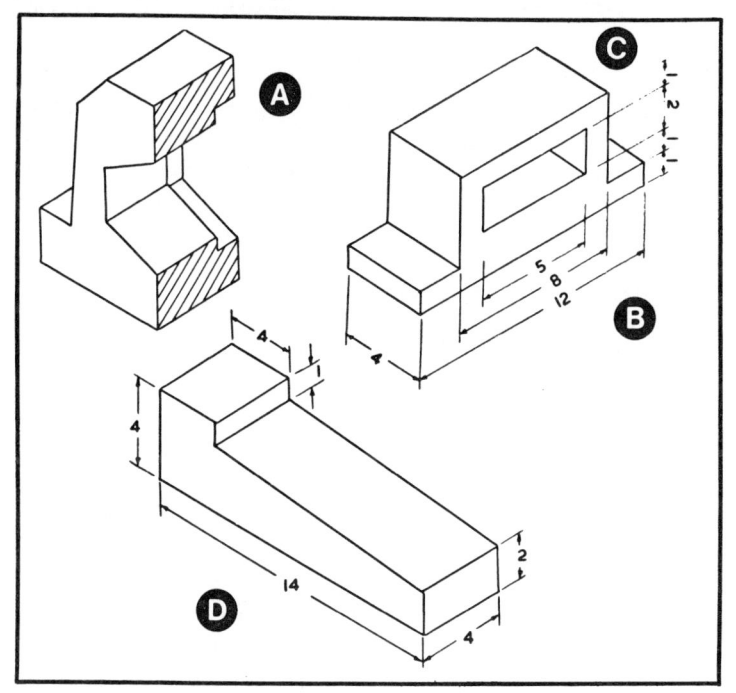

Fig. 15-14. There can be sections on isometric views. Dimensions may be all as read from the bottom or from the bottom and the left side.

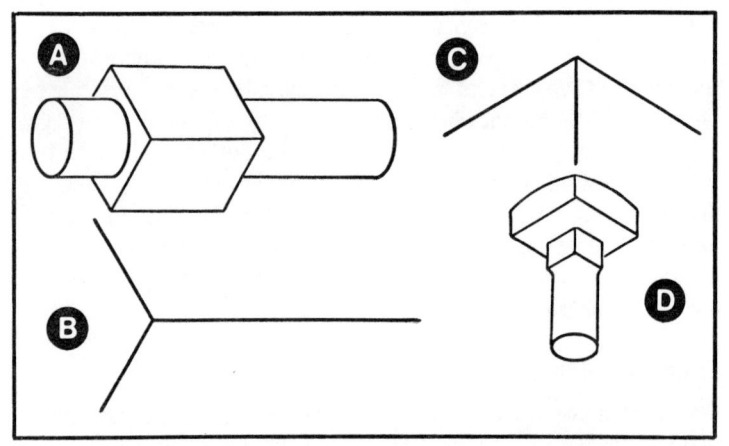

Fig. 15-15. An isometric view from the side (A and B) or from below (C and D) is based on a rearrangement of the axes.

derstand normal working drawings. Certainly, if there is any doubt about the ability of users to understand a drawing, there ought to be some sort of a pictorial view, usually to complement the more usual views.

Someone else may decide whether to use oblique or isometric views. Isometric views are more like the views expected of the finished products. If the drawing is to be worked from and measured, an oblique view has advantages, both for the person using it and for ease of the draftsman making it. If choice is left to the draftsman, an oblique drawing is more often the one selected.

In general drafting practice, the primary aim should be to make good drawings by orthographic projection, using whatever elevations, plans, sections, and other views are necessary to convey the information required. The different views should give all the sizes and other details required. The reader should be able to find all the information he needs somewhere on the drawing. That mechanical drawing ought to be regarded as complete. It is only where there is a special need for a pictorial view that isometric or oblique views should be included. A competent draftsman working for the benefit of a competent craftsman should not have to draw pictorial views, unless there is something very unusual that might be better shown that way. It is only when skill at reading drawings is not up to the required standard that the additional work should be needed. The ability of a draftsman to provide these views when required, however, is an important part of his range of accomplishments.

Chapter 16
Laying Out Drawings

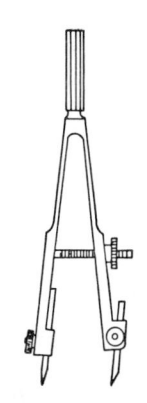

One problem in addition to those associated with the actual techniques of mechanical drawing is the planning of drawings in relation to how they fit on the paper. When something has to be drawn so a reader can understand it and get all the information he needs from it, you have to decide what views you will use, what other sections, detail drawings, and other breakdowns of the construction are needed, and then decide how you will lay them out. You may wish to put everything on one piece of paper, or the thing may be so complex that a large number of separate drawings are needed. In a very large drawing project there may be several draftsmen involved. Someone has to be in charge of overall planning, while the others take care of parts of the operation or deal with certain details. One draftsman may handle general construction, while another deals solely with something like electrical work or details of assembly.

These considerations come first, so what you visualize will form a complete coverage of what is needed to graphically provide all the information. After deciding what is to go on a particular drawing, there is the practical problem of laying it out. If standard sizes of paper are used, you need to choose a scale that will allow the views drawn to conveniently fill the area in a way that will be of most use to the reader. If paper and border sizes are not standardized, it is easier to adapt the outline to suit the drawing at the scale you choose. Even then you have to consider practicalities, both of the drawing and of the user. If there are several sheets covering the project, it would be unwise to have one tiny piece of paper and another very large one if the parts illustrated have to be related to each other. The scales are then widely different.

NUMBER OF DRAWINGS

Many projects can have a general drawing, which gives overall dimensions, with three or more orthographic views. The reader then has an ideal of what he is to make or visualize, with the important sizes, and an understanding of how the assembled object will appear.

If it is something that is not easily visualized from an orthographic drawing, there may be a need for a pictorial view. This could be by oblique, orthographic, or isometric projection, if it is expected that the user will want to take measurements from it. In cases where the draftsman and user are accustomed to working together, or the user of the drawing is known to be familiar with normal drafting techniques, the standard orthographic projections will convey all that is wanted. Otherwise, there is a case for a three-dimensional view as well. If that view is not to be measured, the draftsman may prefer to give the picture more as it would appear by starting with isometric or oblique axes, but altering the angles of other lines so they are directed towards a vanishing point. Those above eye level slope down more than isometric lines do, while those below eye level have an increase in their upward slope.

Besides the general drawing, there will be other drawings showing details of construction and assembly. If it is a fairly simple project, the details should be included on the general drawing. If there are many details to be shown, it is better to have a separate detail drawing, or more than one if it is that sort of project. Make sure details are cross-referenced to parts of the general drawing or covered by accompanying notes, so there is no doubt about the meaning of any drawing.

In some cases there may be alternatives or additions. With a simple thing like a second acceptable way of dealing with a corner, it may be possible to include it as another view on a general drawing. If there are extra things that could be included or alternatives to adapt the object to another purpose, it is better to keep these things to extra supplementary drawings.

MATERIAL SCHEDULES AND LISTS

Many drawings need to be accompanied by lists of materials showing the various parts giving their widths, lengths, and thicknesses where appropriate or other details such as rod or tube, as well as the type of material and any other relevant data. If there are only a few parts, the list can come on the drawing. Even if it is a lengthy materials list, it is the practice in some offices to let it take up part of the drawing space. Otherwise, a long material schedule may be better on a separate typewritten sheet. The advantage of putting it on the drawing is that it cannot be lost. The disadvantage is the time taken to hand letter the list in a ruled grid that takes up considerable space on the drawing.

Where a materials list is appropriate, a draftsman should outline its contents provisionally quite early in the drawing project, so he keeps to sizes. It also allows him to dispose of many dimensions on the drawings, particularly some of the smaller ones. If the reader learns from the schedule that all shelves, for instance, are ¾ inch thick, he will not be looking for the dimension on every shelf drawn. The draftsman thus has no need to draw what could be confusing and tedious small dimension lines and figures.

PROPORTIONS

If the border size has to be a standard one, consider what views and details you want to include within it. Do a little arithmetic so you know what space each view will take, but allow for dimension lines and any notes. Avoid bringing parts of the drawing too close to the border, for the sake of appearance. You may need to cut scrap paper to about the size of views, so you can experiment with positions on the drawing paper.

If the drawing is to consist of only three views, choose a scale that makes the best use of the space. It would be rather silly to find that the drawing only occupies one corner (Fig. 16-1A). Choose a scale that fits the drawing neatly within the border, but do not adopt an unusual scale for the sake of getting the fit of the drawing just right. For shop use, it is better to use the scales that divide by two such as one-half, one-fourth, one-eighth, one-sixteenth, and so on. These are 6 inches to 1 foot, 3 inches to 1 foot, 1½ inches to 1 foot and ¾ inch to 1 foot, all of which can be measured with the usual shop rule marked to at least 1/16 inch divisions. If it is a metric scale drawing, or your drawing is very small in relation to the whole thing, it may be preferable to use a scale that is in tenths. In any case, consider the user adopting a scale.

With a reasonable choice of scale, the drawing is proportioned to make the best use of the paper size (Fig. 16-1B). If there are only the usual two elevations and a plan view, you will be left with an

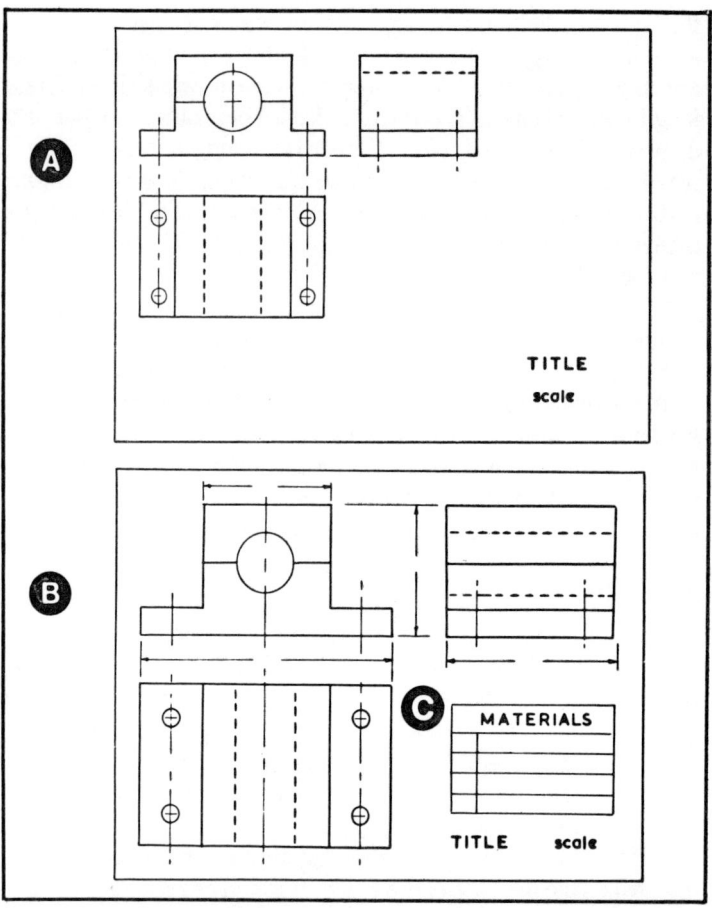

Fig. 16-1. If a drawing will only fill part of the space, its scale is wrong (A). With a better scale, the drawing has a more balanced appearance (B).

unused corner. Maybe you could put a pictorial view there (Fig. 15-1). You have to include a title and other information, as well as a scale. If a material schedule has to go on the drawing, that could also be fitted into the space (Fig. 16-1C).

It is likely that other information will have to be included, such as the name of the draftsman and his employer, possibly the name of the customer, a reference number of the drawing, and details of what part it plays in a series of drawings. In some offices the paper is already printed with a block into which you enter details. A grid to take the material schedule may be printed or have to be drawn above this (Fig. 16-2A). If you are working on your own account, you may draw the title, schedule, and other information in any way you wish. If you are establishing a standard practice, it is unwise to be too elaborate (Fig. 16-2B). Keeping to the layout with later drawings may prove to be tedious when the initial enthusiasm has

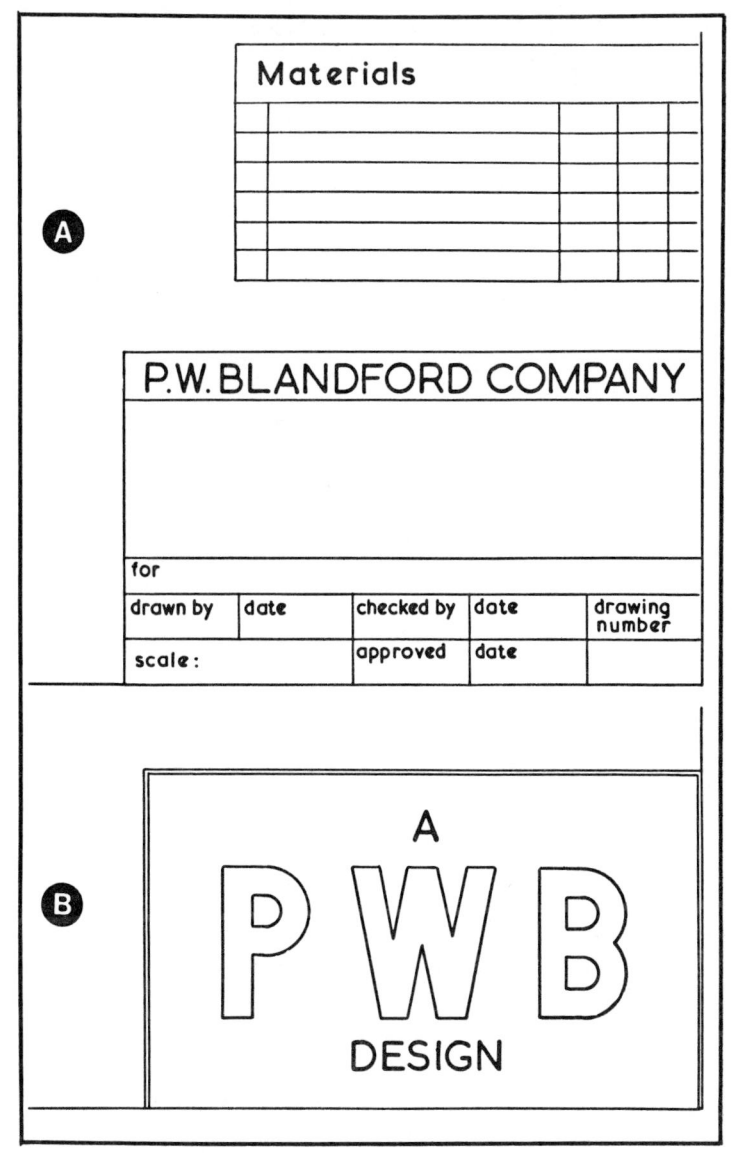

Fig. 16-2. Information can follow a standard pattern (A). A title design should not be too complicated (B).

Fig. 16-3. A design is originated by a few preliminary sketches showing possible ways of achieving the desired result.

worn off. If many drawings are to be produced, it is advisable to give each its unique number, so referring back is easy. Any drawing filing system can thus be arranged efficiently.

BASIC DESIGNING

Suppose a bookcase has to be produced. You need to know the sizes of books to be accommodated and their numbers. This information tells you the spacing to allow between shelves and their total length. You need to know how large the bookcase is, and whether it is to hang or stand on the floor. Balancing the two requirements, you have to put together enough shelves at the required spacing within the bounds of size suggested. It has to be a compromise. You may first make freehand sketches (Fig. 16-3), so you can discuss the design with others. If one design is agreed, you may make a drawing with the minimum of lines, but to scale, to show the proportions. If this seems satisfactory, you can proceed with a general drawing.

In the case of this book the drawings have to be proportioned to suit the page size. It is convenient to make a general drawing on one page and use a detail drawing as another page. In practice the whole project could be dealt with on one wider piece of paper. Sufficient information can be provided by two elevations. Although a plan view could be included, it would not show anything that could not also be read from these two views.

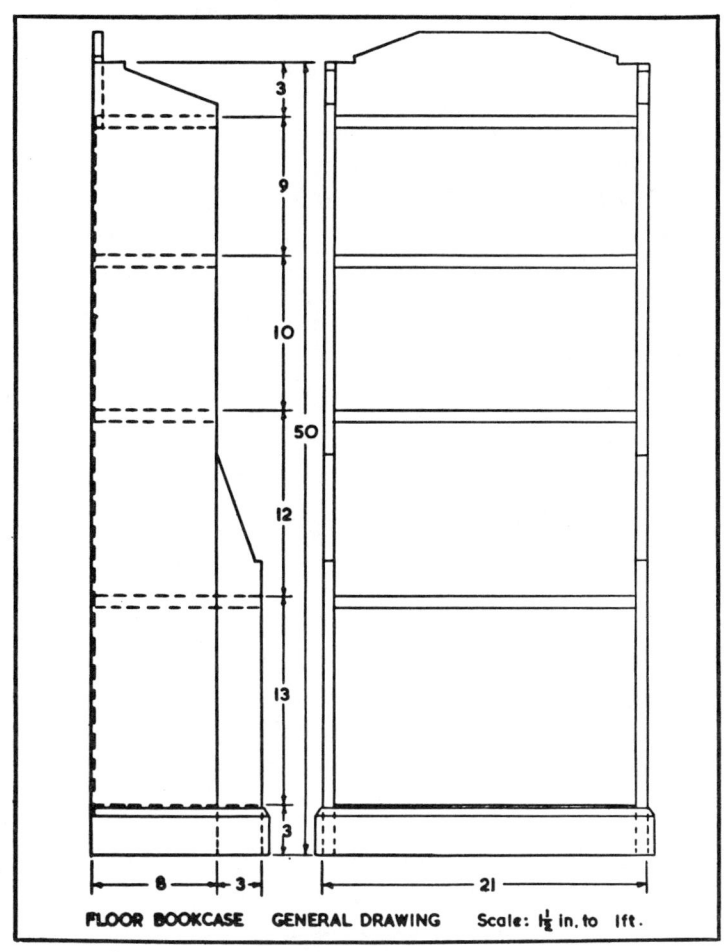

Fig. 16-4. Two views of the design provide all the main dimensions.

The scale is chosen so the two elevations and their dimension lines only leave a small amount of space within the border (Fig. 16-4). The dimensions given are the main ones, but they locate key points and sizes. Other sizes will come from the material schedule (Table 16-1) or from the detail drawing.

This is a project where anyone with a little experience in furniture making could visualize the bookcase by looking at the general drawing. Suppose it is necessary to have a pictorial view, either to clarify a point or to make the design understandable to others. There could be an oblique drawing made, looked at squarely from the front, with the sides sloping away, either to cavalier or cabinet scale. There could be a standard isometric drawing, which the reader could measure. All measurements are given on the orthographic drawings, however, so scaling a pictorial view is unnecessary. In that case it is better to adapt an isometric drawing to a more pictorial three-quarter front view (Fig. 16-5A). If the illustration is intended to clarify appearance for the sake of those working in the shop, that is sufficient. If it is to give an intended user or customer an idea of appearance, there could be some graining and a few books added (Fig. 16-5B).

Instruction has to be provided on the constructional methods to be used, and these things all come on the detail drawing. In the case of this fairly simple piece of furniture, all the details may come on one drawing.

Table 16-1. A List of Materials Complements
the General Drawing by Giving the Sizes of All Parts.

MATERIALS

number	part	width	length	thick
2	sides	8	51	5/8
2	side extensions	3	26	5/8
3	shelves	7 3/4	21	5/8
2	shelves	10 3/4	21	5/8
1	top back	6	21	5/8
1	plinth	2 7/8	23	1/2
2	plinths	2 7/8	12	1/2
1	back	21	48	1/4 ply

Fig. 16-5. A pictorial drawing (A) helps shop workers to visualize the project, but a more elaborate drawing (B) may be better for showing to others who are interested.

The sides have to be widened to accommodate larger books. The pieces needed to make up the width are fully dimensioned (Fig. 16-6). The shelves fit into dadoes, and an example is more clearly shown in an exploded isometric view than if orthographic projection is used. This also applies to the top corner, where the rabbet for the top can be illustrated as well as sizes in the vicinity.

239

As the plywood back is thinner than the top part above it, a section shows how the top goes behind the top shelf, and the plywood back fits into a rabbet.

At the bottom a *plinth* goes around and overlaps the bottom shelf. It is not intended to reach the shelf height. There are two possible treatments shown in section—one for hand beveling and the other to be machined.

How details are related to the general drawing depends on the project. In this case there are short notes or captions which indicate what the detail is all about, but it may be better to have key letters by the appropriate part on the general drawing. Then mark the detail with the same letter. You could mark each: detail at A, section B, or assembly at C as appropriate. The important consideration is that the reader should have no doubt about what the drawing is telling him.

CHAIR DRAWING

Another example is a doll's chair, which is a reproduction of a traditional type, without any exact prototype, and is mainly an exercise for a woodturner. Drawings have to provide all the information he needs and relate it to other parts. This sort of project is not precision engineering, and a certain amount of detail work can be left to the craftsman. If one chair finishes ½ inch higher than another, it would not matter. Both chairs should stand square and symmetrical.

All of the information might be put on one piece of paper. To get this in the book, it is arranged across two pages (Fig. 16-7). There have to be several different scales. All views that are dimensioned are orthographic views, while pictorial drawings are provided where that system would better show how the parts are fitted. Where details are not obvious, they have been given the same letter as the related parts on the two elevations.

This is another project where a full plan view would not be much use. Instead there are two elevations, with dimensions common to both of them. Notice that the leg spread is given to the centers of the legs as they are round. Notice also the use of phantom lines to show that the table swings upwards, and the solid lines to show the table in the other view have been omitted to show more detail of the back. A draftsman should show what is necessary to make the drawing understandable and not slavishly include every projected detail from one view to another, if that might become confusing.

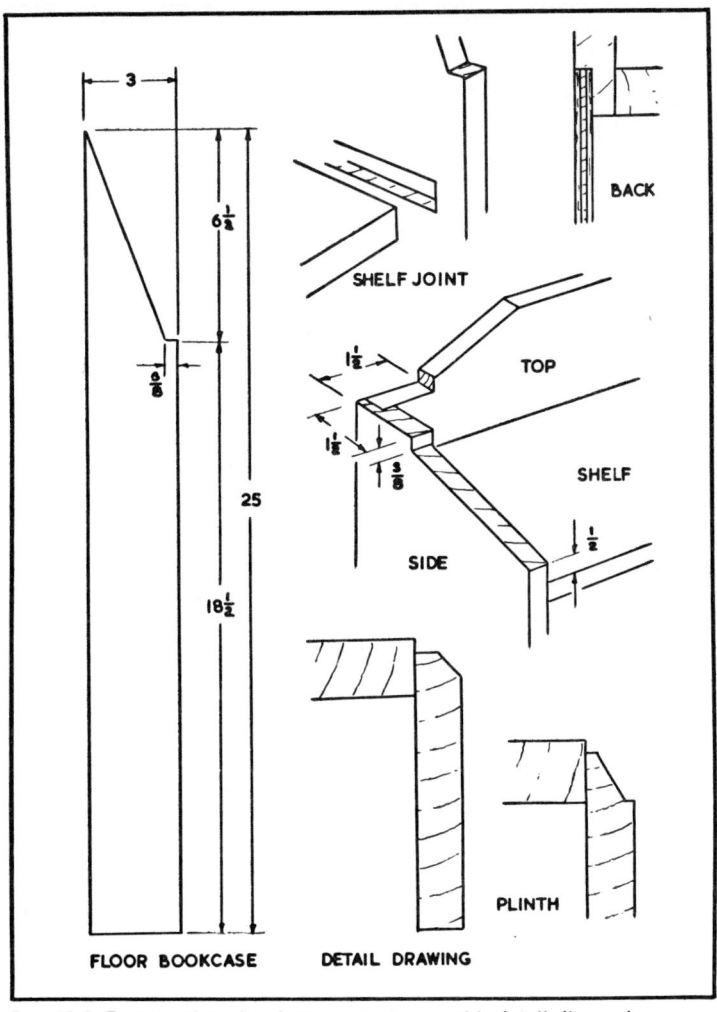

Fig. 16-6. Details of the bookcase are shown with detail dimensions.

The important thing in plan view is the seat, which is fully dimensioned. The arms are shown in relation to it to indicate that they are parallel and level with the inner edges of the upright parts of the back. Below it is a secondary elevation, put there to show that the back is to slope, so its top finishes level with the rear edge of the seat. The curved leader line takes the eye to a detail of the joint of the side of the back into the seat, indicating how the mortise is cut to be clear of the top of the leg. Another leader line from the arm shows its joint to the back. The joints are not dimensioned, as

241

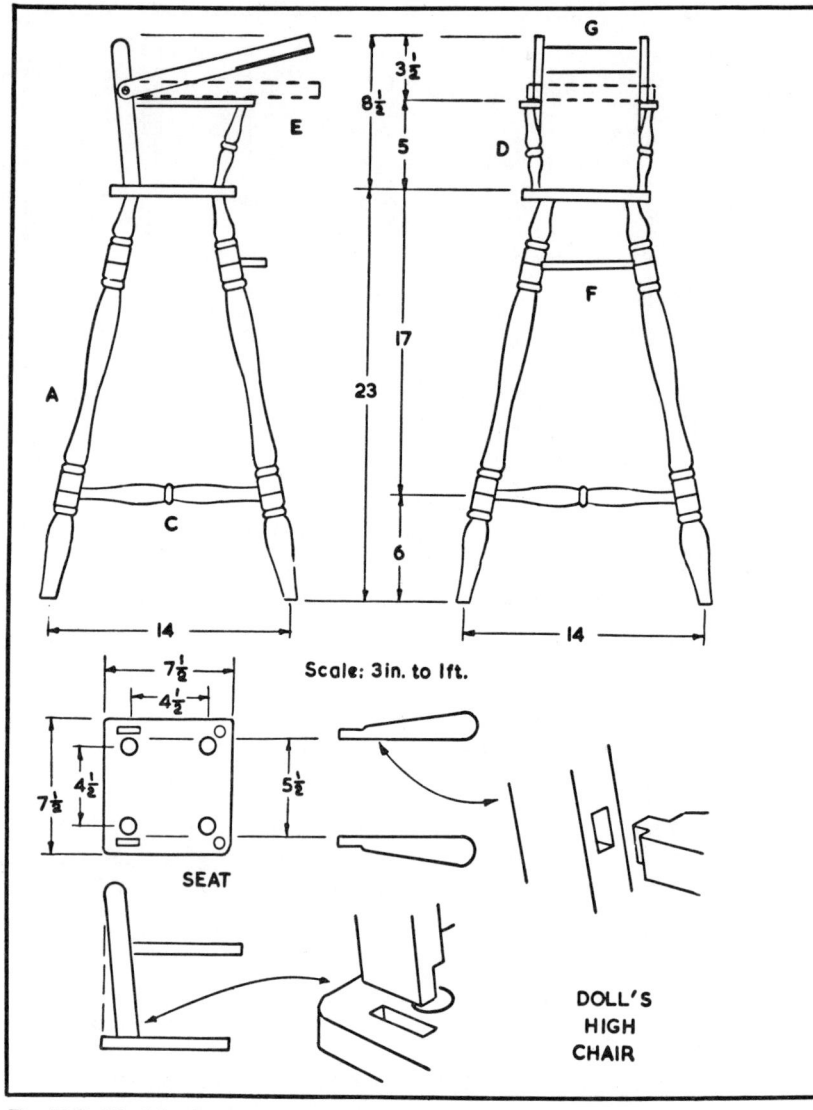

Fig. 16-7. All of the details of a doll's high chair are shown on one drawing, with letter references to indicate where the separate parts apply.

most woodworking craftsmen understand how to proportion the parts. If the drawing was intended for a beginner or for use in a school shop, there may have to be more detail and dimensioning.

A turner requires key dimensions. He will turn beads and free-flowing curves that satisfy his eye, providing he has an indica-

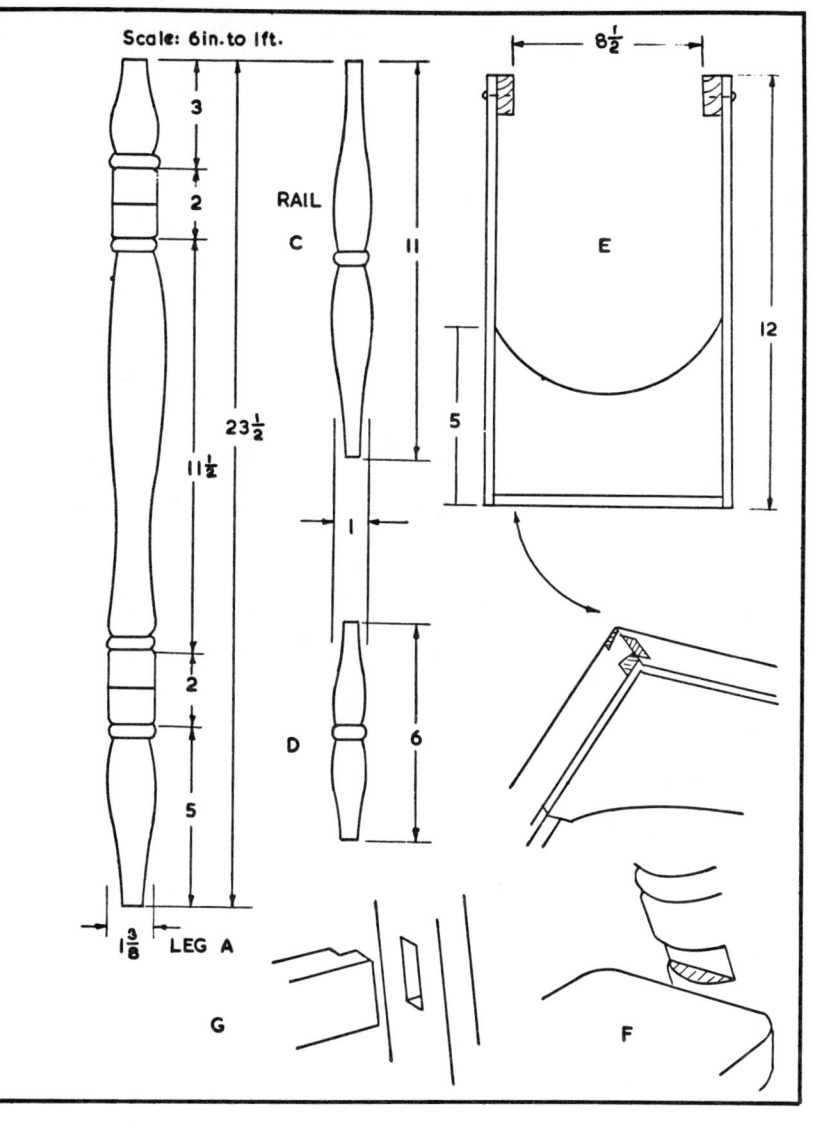

tion of what is required. If four legs and four rails or two arm supports match each other, it does not matter if in another chair there are slight differences in turning details. A draftsman should not try to impose exact curves on a turner, when they are purely decorative, as the turner will have a better idea of a pleasing shape since the wood is cut under his tools. Parts that form joints or have to fit holes are a different matter, and the turner expects to find those details on the drawing. The drawings of turned parts, shown here, provide all that a turner needs to know. For the sizes at the ends, he will use a piece of scrap wood drilled with the same bit that will be used in construction.

The joint for the rail across the back is shown pictorially. The hinged table sizes are better shown orthographically, but the detail at the corner is clearer drawn pictorially as seen from underneath. The footrest and its notch are shown by an exploded view.

This doll's high chair is typical of a woodworking project in which the draftsman is able to assume that the craftsman using the drawing understands normal woodworking joints and other practices. There is no need to indicate every small detail; nor is it necessary to show the amount of rounding to be given to edges or how to check assembly. This liaison and understanding between draftsman and craftsman is common in furniture making and most other woodworking industries.

ENGINEERING DRAWING

In metalworking it is possible to work to much finer limits than in woodwork. Consequently, the man who handles the metal expects to get more precise information from a drawing. Very little detailing is usually left to the man working the metal, even to the extent of the amount of chamfer or rounding, that would not be specified for wood.

Some engineering drawings are quite large and have considerable detail. If a drawing about 4 feet across was reduced to the size of this book, most of the detail on it would be lost. An example has to be one that does not require much reduction. For clarity, lettering has to be larger than it would be on a drawing.

The drawing is of a flanged bush to fit on a shaft (Fig. 16-8). All of the sizes are decimal, as the machinist will be using a micrometer or other instrument capable of working to three places of decimals.

As with earlier examples there will be a preliminary sketch, or there may be no doubt about requirements if the part has to mate with something else. As the part is symmetrical about a vertical center line on the circular view, it would be possible to only draw a half with the center line at its edge, but the full circle has been chosen. The two views project from each other and could be regarded as two elevations or an elevation and a plan, depending on which way the drawing is viewed. A second view would have the same outline as the one it is projected from. There could be a full

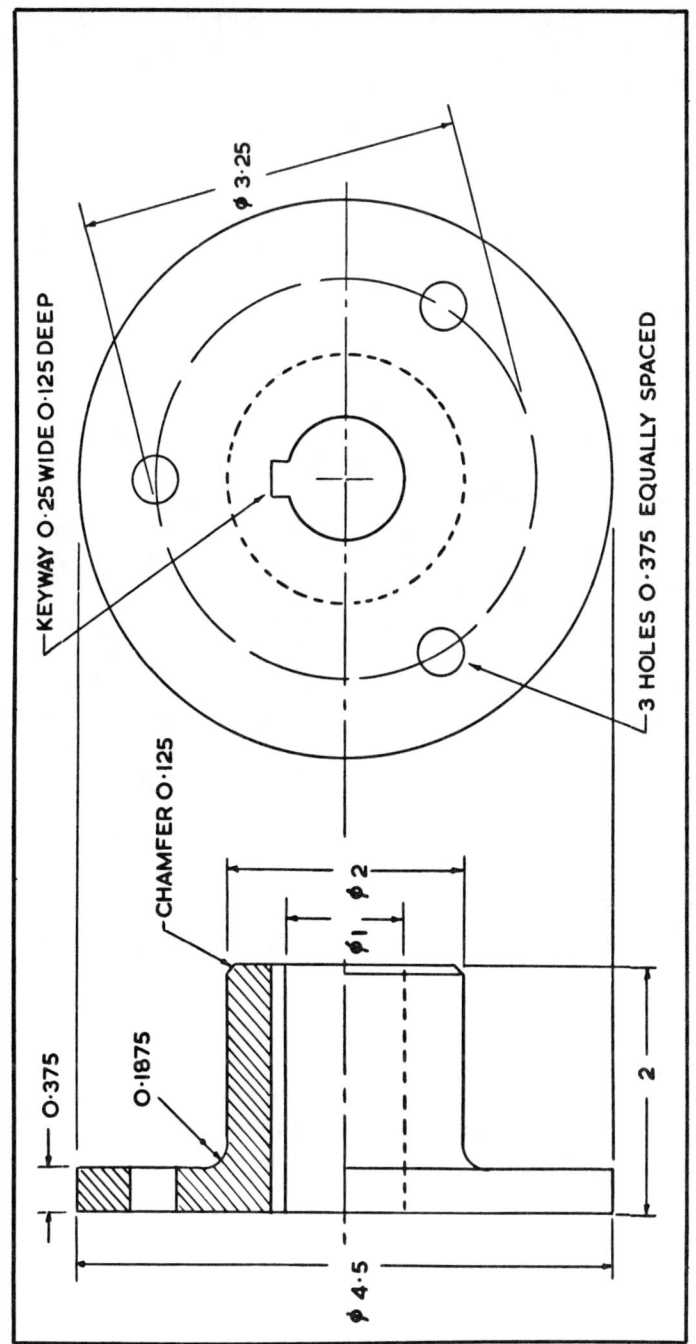

Fig. 16-8. A drawing of an engineering part uses decimals and shows all dimensions, including minor ones.

section as an extra view, but in this case half of one view is sectioned and the other half is shown solid. In this way the keyway and hole can be shown in the section, while the other side shows just the external view. There would be no particular advantage in having section and complete views separately.

The scale has to be chosen to make the best use of the paper. The two views could be close, if that would fit them in better, or the spacing could be greater, with longer projection lines. Taking the views too far apart would not look right. Dimensions common to both views could come between them or be put outside. Dimension lines to side views are neater than to a circle, although that is the obvious way to give the pitch circle for the holes.

Where dimension lines are long enough to contain the figures, that is where they are placed. In this case they are all made to be read the right way up of the drawing. It would be acceptable to have those in vertical dimension lines to be read from the left. Notice that where the thickness of the flange does not leave much room for the figure, it is put outside using reversed arrowheads. This is also done with the radius of the fillet.

Although the main parts are in a proportion that allows direct dimensioning, the keyway is very small so that dimension lines on it would be microscopic compared with the rest of the drawing. One way of showing this would be to draw just the keyway part to a large scale, probably alongside the main drawing, with a leader or note showing what has been done. In this instance, as it is something with just a width and depth, there is a leader to a note giving these sizes.

In engineering the symbol π is often used to indicate diameter, and it is shown in the appropriate place on this drawing. Dimension lines do not always have to come in the same position. It is unwise to make leaders longer than necessary. By arranging these things around the main part of the drawing, it is possible to get a drawing that looks balanced. As well as providing all the necessary information, a draftsman, proud of his work, should try to get a reasonably even coverage of his paper. With some things the shape does not always lend itself to this treatment, but a balanced general appearance should be the aim.

Chapter 17
Engineering Drawings

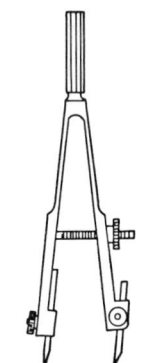

Draftsmanship has been most closely tied with *mechanical engineering* ever since the Industrial Revolution. Although there are many other applications of draftsmanship, it is the needs of machinery building and using that have made the greatest calls on draftsmen. The whole technique of drafting has been associated more with mechanical engineering than with any other branches of engineering or with such things as furniture making, building, and display work. These trades have some affinity with mechanical engineering, but have developed their own techniques and practices. It is drafting for mechanical engineering that still forms the basis of all drafting. The fundamentals described in most of this book might be considered aimed at the mechanical engineering draftsman, but other draftsmen, whatever their ultimate occupations, must be aware of and use these techniques. All draftsmen are primarily mechanical engineering draftsmen to some degree.

There are many divisions among the people who call themselves engineers. Engineering today is an occupation that is becoming increasingly specialized, and draftsmen may have to limit their work accordingly. The wise draftsman makes sure he is aware of more than just the branch of engineering for which he is catering. The craftsman using the drawing may know just the narrow subject he is dealing with, but the draftsman who is providing him with information should also know about related branches of engineering and other associated activities that might affect what he is designing or drawing. Without this wider knowledge, he might start the production of something which someone working on an associated product might find unsatisfactory. It might have been better suited to the whole product if something, possibly quite minor, had been a different shape or a different size.

Besides the many divisions within mechanical engineering, there are specialized branches which tend to stand apart such as *electrical engineering, hydraulic engineering,* and *aircraft* or *aerospace engineering.* Some engineering has links with other industries, and it may sometimes be difficult to draw a dividing line. Structural engineering is associated with *civil engineering,* which in turn has more links with the building industry. The draftsman is interpreting the work of an architect, who does not have much in common with a mechanical engineer. A draftsman concerned with hydraulics such as hot water systems or plumbing is really dealing with the building trade, although the mechanical part of the work is engineering. Yet he could be dealing with the hydraulic system of an airplane, which is purely engineering.

Structural drafting is dealt with in the next chapter. The work of a draftsman concerned with most other branches of engineering is covered here.

DIMENSIONS

Sizes in America and most of the English-speaking world have been in feet and inches. The larger and more unusual measurements that have grown by tradition and were based on man's stride and how much an ox could plow in a given time, or similar unprecise things, might be regarded as obsolete and only of historical interest. They are not absolutely irrelevant. Anyone having to deal with things like old documents may still have to interpret such measures as *chains* and *furlongs.*

There will be the eventual adoption of the metric system as the sole method of measuring. The period of transition, though, can be expected to last for a long time. Other countries, who are closer geographically or have other ties to France, where the metric system was invented, are further ahead with conversion to metrication than the United States. Draftsmen concerned with international implications must be ready to use one system or the other, or even produce drawings in which the old and new systems are combined, possibly because of the need to include a standard item made to metric sizes in a construction measured in inches.

For most engineering drawing, dimensions up to 72 inches are most conveniently stated in inches. Over that, feet and inches are used. Yards are not used in engineering. If all the measurements on a drawing are in inches, there is no need to indicate that they are at every place. The distance can be just a figure (Fig. 17-1A). In most cases inches can be assumed. If there is any possible doubt, put a general note on the drawing: "All dimensions in inches." It is common to use the symbol for inches rather than print the word in full or abbreviated (Fig. 17-1B). If the size is in feet as well as inches, the symbols to indicate this should always be used (Fig. 17-1C).

Whether to use common fractions or decimals depends on the drawing or the practice of the office. With most mechanical engineering products made to quite close tolerances, measuring instruments that read to at least one-thousandth of an inch are usual in the shop. This is three places of decimals or further, and that sort of accuracy cannot be conveniently represented by common fractions. It is better to quote fractional measurements as decimals. It is always better to quote three places of decimals, even if some places are zero. For instance, ⅝ inch is .625, but ½ inch is .5, and it would be written .500. Because the decimal point is not very obvious and could be overlooked or even not appear in a poor reproduction, it is better to put a figure before the decimal point, to draw attention to it, even if it is zero for a size less than 1. So the above example would be written as shown (Fig. 17-1D).

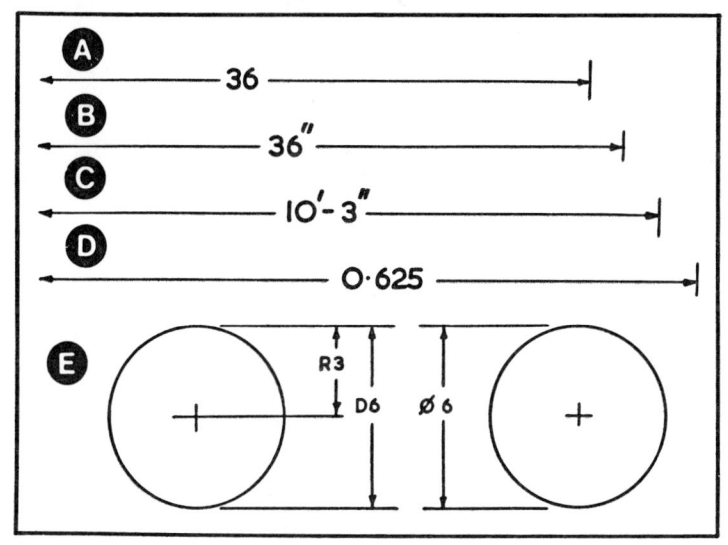

Fig. 17-1. Make sure it is clear what dimensions are and indicate their relation to circles.

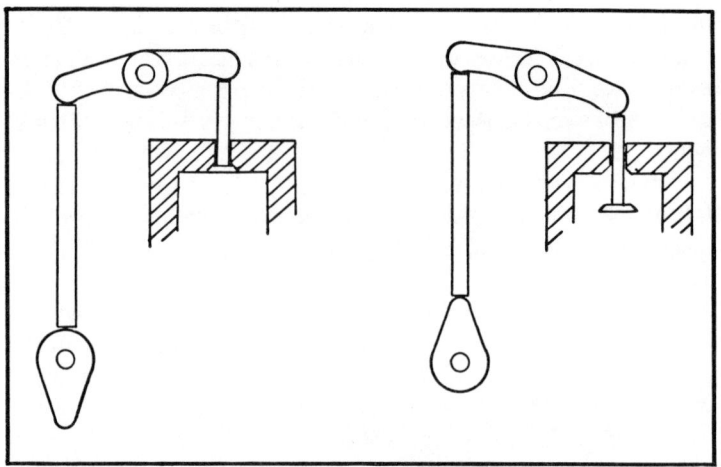

Fig. 17-2. A cam rotates to convert its rotation to the linear movement, in this case to move a valve.

It is very unusual on a mechanical engineering drawing to let any dimension line come within the solid line bounds of the object being drawn. If the lines have to come within the drawing, make them obviously there for the purpose and positioned in a way that would not let anyone confuse them with construction or other lines. Never use a center line as a dimension line. If any dimension line refers to the size of a circle, attention may be drawn to the fact by putting D before or after a diameter measurement, or R if it is a radius. With the metric system comes the use of a crossed circle to indicate diameter (Fig. 17-1E).

CAMS

A *cam* converts rotary motion to some other sort of motion, usually a straightforward up and down action. Cams in great variety occur in many kinds of machines. A draftsman should understand what is required and how to lay out a cam to get the required response. An everyday example of a cam is found in the operation of the valves of an automobile engine (Fig. 17-2).

As in the automobile cam, a simple example is the need for something to be lifted once in each revolution. The amount of rise has to be known. In a simple instance, the end of the rod rests on the cam. For about half a revolution, it follows a circle concentric with the shaft and does not lift (Fig. 17-3A). At one side the cam projects by the amount the rod is to lift (Fig. 17-3B). As this part comes to the contact point with the rod, it is lifted and allowed to return as it passes (Fig. 17-3C). The rod may be kept in contact by its weight if it is vertical, but more often there is a spring to hold it against the cam. A point would wear. Either the end of the rod is flat, or it may have a roller (Fig. 17-3D) to follow the profile of the cam.

The movement does not have to be directly in line. The lever arm could be cranked and worked on a pivot to alter the direction of movement (Fig. 17-3E). If the arms are different lengths, the amount of movement at the operating point can be varied. For instance, if the roller arm is twice the length of the other, the cam will have to lift the roller twice as much as the intended movement at the other end.

A cam can be arranged to lift more than once in a revolution, to pause in the action at some point (dwell), or to vary the movement as required. In many machines a large number of cams cause many motions to be made at various places at the appropriate times, so complex actions are performed. This sort of work is seen in machinery used in things like bottle filling and sealing or the wrapping of food products.

A cam of the type being considered may be called a *plate cam*. Suppose the follower is to rise to a certain point, stay there briefly,

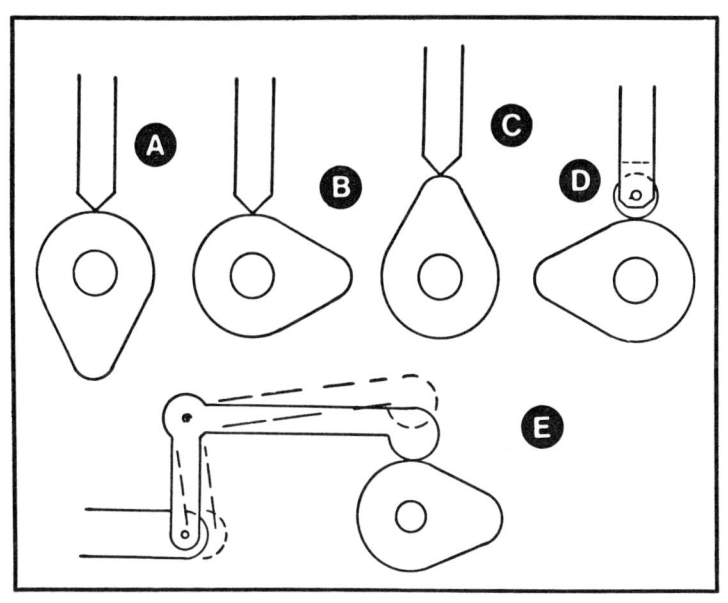

Fig. 17-3. A cam may be pointed or have a roller, and a cranked piece may change the direction of motion.

and then rise more to a point where it is to stay the same length of time before dropping to the original position. You need to know the sequence and length of time for each part in relation to the revolution of the cam. With this information draw a cam displacement diagram, based on a straight line, divided according to the proportion of a revolution each action is to take (Fig. 17-4A). The length of the line does not have to match the circumference of the cam, but it must be divided in relation to a full revolution of 360°.

Draw the shaft and the hub of the cam. Consider the direction of rotation, and draw radiating lines at the appropriate angles to match the angular spacing indicated on the diagram (Fig. 17-4B). Divide the portions that involve movement into a number of spaces that can be used to find positions on the slopes. For simplicity, these are shown arranged at 30° intervals, but for greater accuracy more divisions would be desirable in practice (Fig. 17-4C).

A pointed follower would interpret the movement exactly. For a roller there could be different areas of contact as the profile changes, giving slight differences in timing. It is necessary to allow

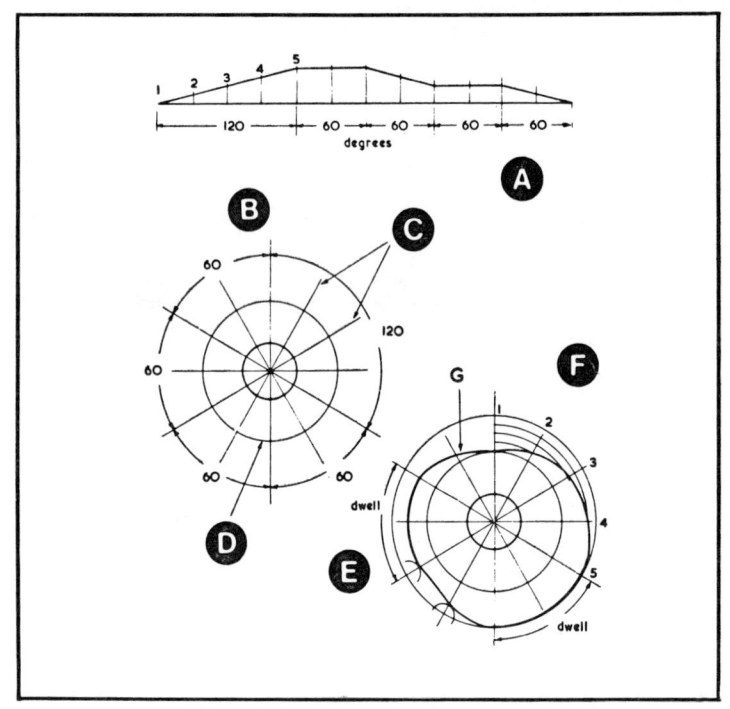

Fig. 17-4. A straight line diagram plots the intended movement of a cam. Then this is transferred to a circular diagram.

for the diameter of the roller. Any differences are probably negligible, but it is good practice to always allow for the roller.

Decide on the size of the cam at the lowest point of travel of the follower (Fig. 17-4D). Mark this on a vertical line. Mark a center at a distance to suit the roller, and draw a full circle through this. You can draw in the dwell positions. Allow for the height they come above the base on the vertical line, including the radius of the roller. Project this around to the appropriate positions (Fig. 17-4E), where you can draw those parts of the cam outlines.

The slopes are actually rounded slightly at their ends to avoid a jerking action. Get the heights of the various intermediate positions on a slope, and mark the matching roller centers on the vertical line. Project around to get points on the cam curve (Fig. 17-4F). One slope is shown numbered. The other slopes are dealt with in the same way. It is the center of the roller which should be marked each time. Then a compass is set to the roller radius used to get a series of curves through which the cam profile will be drawn (Fig. 17-4G). Finally, draw the outline for the whole cam through the contact points.

If a cam displacement diagram is made up of straight lines meeting in angles, there will be a jerking motion at the change of movement (Fig. 17-5A). To avoid that, the change positions are rounded. To get this smooth and correct, the angle of slope is altered slightly to allow for rounding (Fig. 17-5B). Unless there is a particular need for a straight line action, it is better to plot the diagram with a curve. For a simple rise it may be a *harmonic curve,* which changes smoothly. To plot the curve, use a semicircle divided into the same number of divisions as the straight direction of the diagram representing the amount of travel. Then project across to points through which the curve is drawn (Fig. 17-5C).

Sometimes the motion throughout the slope has to be uniform, as when the rod lifts or lowers something steadily. One way of plotting this is to work directly on the cam diagram. Mark equal spaces between high and low on the vertical line. Mark the same number of divisions around the radial amount of rotation of the cam. Project around to the appropriate points and draw the curve (Fig. 17-5D). If there is a roller, allow for that, as in Fig. 17-4.

On a displacement diagram the profile is a parabolic curve, which can be drawn through equally spaced ordinates. The center one is halfway while the others are at graduated proportions vertically—in this case six division heights are spaced at 1:3:5:5:3:1 proportionately vertically (Fig. 17-5E).

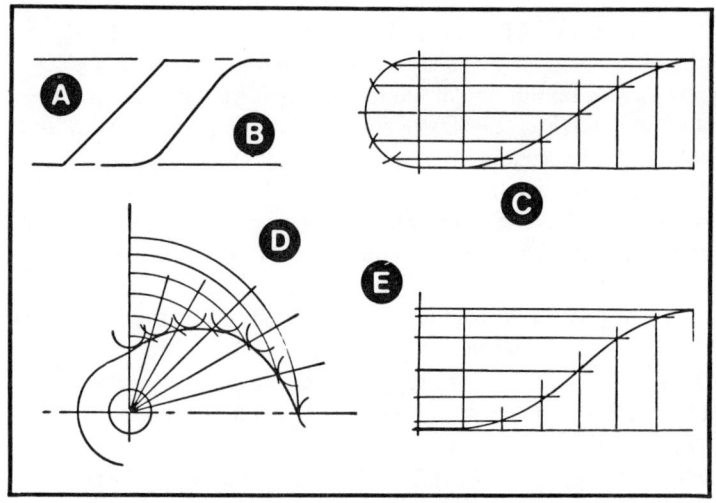

Fig. 17-5. Sudden changes are avoided in the design of a cam, and allowance must be made for the travel of a roller.

There are cylindrical cams with grooves cut around them, so the movement is sideways in relation to the circular motion (Fig. 17-6A). The groove may be likened to an irregular screw thread, but its shape determines movement and pauses in the same way as a plate cam. Its action is easier to understand if one side of the groove is pictured (Fig. 17-6B). The follower is then made to rise and fall as it progresses around an irregular cylinder. Plotting the shape of the cam is very similar to that of a plate cam, but lines are used around the cylinder. If the length of the displacement diagram is the same as the circumference, imagine it wrapped around the cylinder. The other side of the slot in the complete cam is parallel with the first side. It serves to keep the follower in place.

GEARS

Rotary motion from one shaft to another may be transmitted by a belt and pulleys or a chain on sprockets, but most transmission is more positively transmitted by gears. If two smooth wheels come into contact, they may drive by the friction between them. This is sometimes the arrangement in a clutch. If the drive is to be without risk of slipping, there have to be spurs or teeth on the wheels so they interlock (Fig. 17-7A). The circumference of the friction-drive wheel is now the *pitch circle* of the teeth.

The gear teeth actually roll against each other as one wheel drives the other. The design of gear teeth is comparatively complex to get the shapes of meeting teeth correct so they mesh and unmesh properly. The shapes vary according to the gear ratio between the wheels. If one wheel is to drive another at twice the speed of the first, it must have twice the diameter and twice the number of teeth.

There are a number of terms related to gear teeth (Fig. 17-7B) with which a draftsman should be familiar. Usually the draftsman does not have to draw the gear teeth in detail. The curves of the teeth are based on an *involute curve.* If part of a wheel has to be drawn, an acceptable profile can be made with compasses. The pitch of the teeth is measured around the pitch circle, and for drawing purposes the tooth and space at this level are the same (Fig. 17-7C). If a line is drawn as a tangent to the pitch circle through what will be one side of the tooth, another drawn from it at 15° will be through the center to be used for the compass (Fig. 17-7D). Set the compass to 0.125 of the pitch diameter, and mark this distance down the line. A circle through this goes through all the centers to be used by the compass (Fig. 17-7E). Draw the sides of the teeth, with short radial lines and fillets at the root circle (Fig. 17-7F).

Gear cutting details are usually settled in the shop with special gear-cutting equipment, and the draftsman does not have to spend time drawing individual teeth around each gear wheel. It is more common to draw a single gear wheel with everything shown in two views except the teeth (Fig. 17-8A). Instead, the root

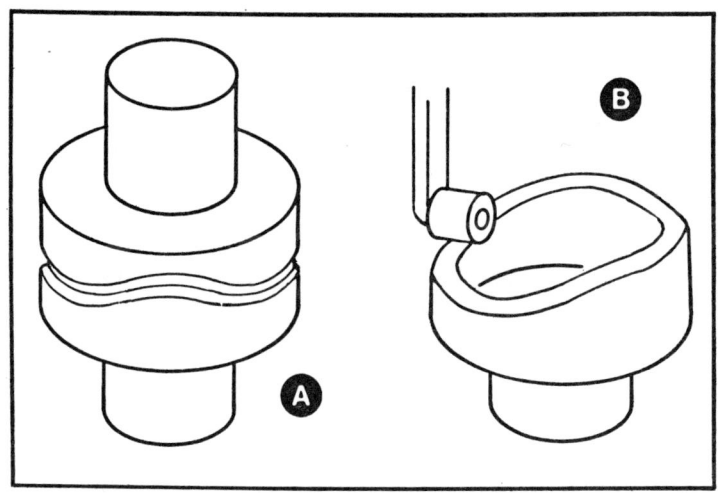

Fig. 17-6. A cam can operate by use of a groove around a cylinder.

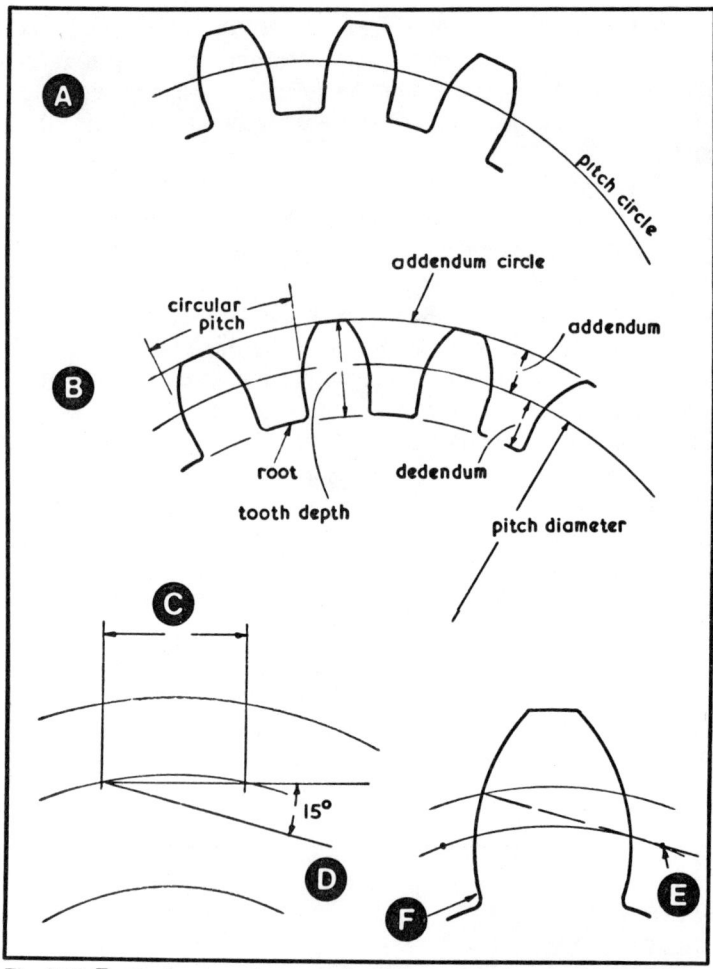

Fig. 17-7. Teeth of a gear wheel are shaped to interlock smoothly (A). There are names for the main parts (B). A good approximation of the tooth shape can be drawn with compasses (C to F).

diameter and pitch diameter are given, but the number of teeth are indicated with any other information tabulated (Fig. 17-8B).

If two gear wheels have to be shown, the drawing is similar, with the pitch circles meeting and circles only drawn for the tops and bottoms of the teeth (Fig. 17-8C).

FINISH MARKS

Many machine parts are cast or otherwise prepared in a way that leaves the surface rough. For some parts of the object that may not matter, but elsewhere the surface has to be finished, either over a large area or just at a part where something else has to bear. There have been many methods of indicating on a drawing that the

number of teeth 40	addend	
tooth depth		

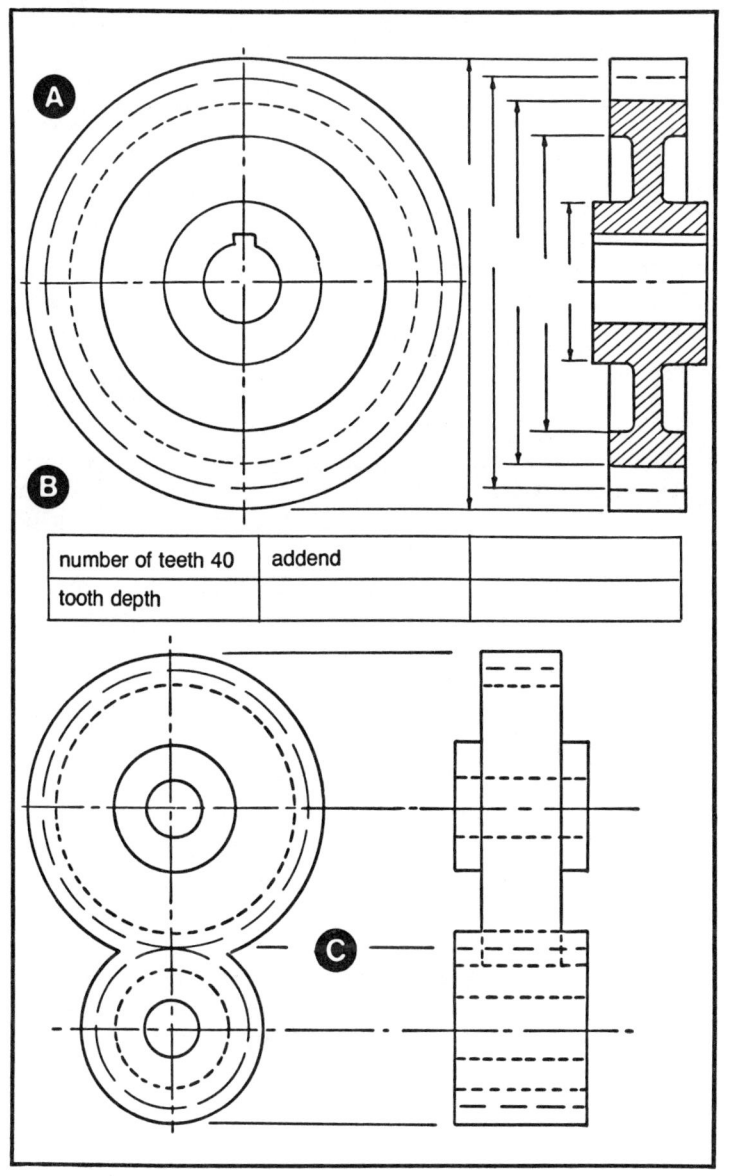

Fig. 17-8. Drawings of gear wheels can be made without showing all the teeth.

257

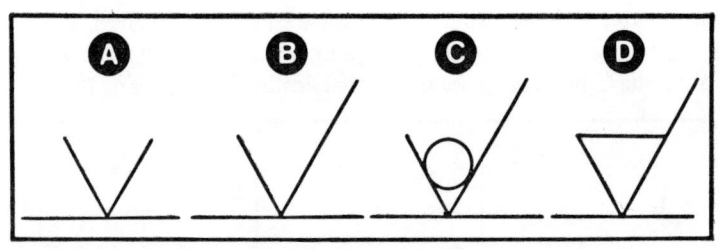

Fig. 17-9. Examples of symbols used to indicate surface finishes on metal.

surface has to be finished in some way, and standards in one shop may differ from those in another. A draftsman should check on what, if any, symbols are used to indicate what is to be done on a surface in a particular office. If a system has to be devised, make sure there is a key on the drawing explaining what each mark means.

A common way of indicating that a surface is to be machined is a V (Fig. 17-9A), usually with equal legs at 60° to each other. Without any further marks, it can be taken to mean that the surface is to be machined all over. There could be a number in the V, so the reader could refer to the key and see what type of finish is meant by that number.

The finish mark recommended to be used with the metric system is a V with one leg not much more than half the length of the other, with the included angle 60° (Fig. 17-9B). If the surface is to be finished without removing metal, a circle is included in the V (Fig. 17-9C). If there is a line across the V, it means that the finish is to be obtained after removing the surface (Fig. 17-9D).

ELECTRICAL ENGINEERING

The field of electronics is probably the most rapidly expanding sphere in the world of industry and engineering. Much of the drafting involved is similar to that for other purposes, particularly when actual layouts are being drawn. Even then there is much greater use made of diagrammatical representations of components. An electrician knows what a particular symbol or diagram means, although it does not represent the appearance the actual component has.

There are many symbols that have become accepted as representing certain electrical components. It is a sort of graphical shorthand. For instance, alternating long thin and short thick lines represents a battery, each pair of lines meaning one cell and the thin lines being positive plates or poles (Fig. 17-10A). A simple diagrammatic switch indicates on and off (Fig. 17-10B), and a lamp shows the two leads going into a filament (Fig. 17-10C).

A circuit is shown as an angular diagram (Fig. 17-10D), although in practice the wires would follow whatever paths were necessary. The circuit diagram shows an electrician how to connect the parts, and he is left to arrange the layout himself. If the wires must be taken along a certain route, there would be another drawing showing this, but not detailing the connections, which would be settled by the circuit diagram.

That is an extremely simple diagram, but complex circuits have to be laid out in a similar way, with symbols or diagrammatic representations linked by lines showing wiring. Much of a draftsman's skill comes in laying out the components in the diagram to allow for the wiring being kept as direct as possible between most components, and to have the minimum of crossings. At one time wires crossing and not touching had a loop in one (Fig. 17-11A). Now they are drawn straight across (Fig. 17-11B), while joins are indicated by a dot at the crossing (Fig. 17-11C).

The symbols used vary according to the type of electrical work. A draftsman concerned with electrical work in a home may use a different set of symbols from the man who is concerned with radio or some advanced piece of electrical gadgetry. Some typical symbols are shown as examples (Fig. 17-12), but there are many more. As can be seen, most of the symbols are fairly easy to draw with instruments. It is possible to get templates through which symbols can be drawn. There is no standard size for symbols. They

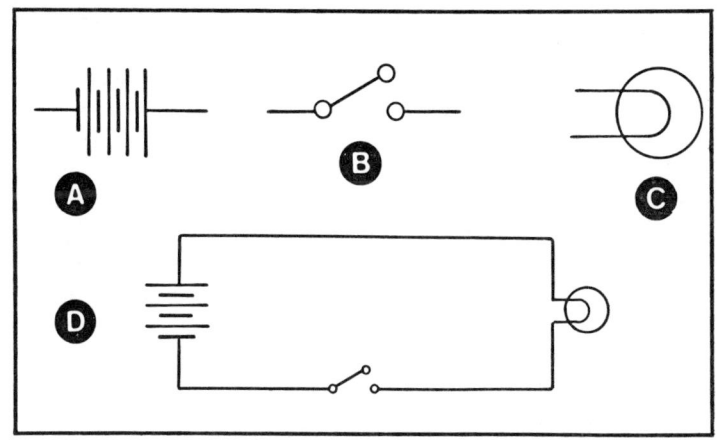

Fig. 17-10. Examples of electrical symbols and a simple circuit.

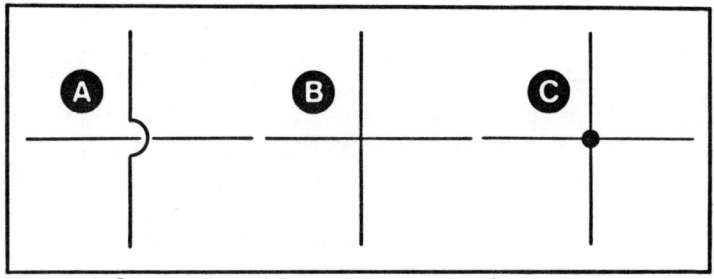

Fig. 17-11. Crossing (A and B) and joining wires (C).

will vary according to the drawing. It may be advisable to keep them small for the sake of neatness, but the details of very tiny ones may be difficult to draw with instruments. They would not be satisfactory freehand.

An example of the skill in laying out complicated wiring in a fairly small space can be seen in the wiring diagram usually provided in an automobile handbook. For neatness in drafting, a layout should be planned so it fills the usual rectangular drawing fairly evenly. It is usually possible to balance the arrangement by the location of tables, title, and any notes required. Allow for some lettering. You may need to expand the meaning of symbols by indicating which light it is, what the capacity of a part is, which of a series it is, or some other piece of information.

A beginning draftsman should practice drawing symbols to various scales. If early drawings are made to a large scale, the way the coils in a transformer or similar thing are built up (Fig. 17-13A), or the angles that are most effective in a resistor (Fig. 17-13B), will be seen. Then they can be drawn smaller, and circuits can be built up (Fig. 17-13C). If you are not originating the work yourself, someone will provide you with a sketch. Otherwise, it is advisable to make a rough sketch yourself. If you find that one wire has to travel all around the diagram, experiment with other arrangements. The layout on the diagram is to show how to connect parts, not how they are actually laid out on the job.

In some diagrams not every wire links shown parts. There may be two wires going away to another component or station, as in telephone work. Make sure that every wire that leads out of the drawing has a note against it saying where it is to go.

AIRPLANE DRAFTING

Compared with most other aspects of engineering, the aerospace industry is new. Its rate of advance is considerably greater compared with general mechanical engineering, which took at least two centuries to evolve. Much of the work connected with airplanes may be regarded as advanced mechanical engineering, but there are many problems. Much of this is due to the need for lightness and compactness, so what has to be contained is a complex mass of things, often with conflicting requirements that have to be compromised. The most successful airplane is the one in which there has been the most satisfactory blending of compromises.

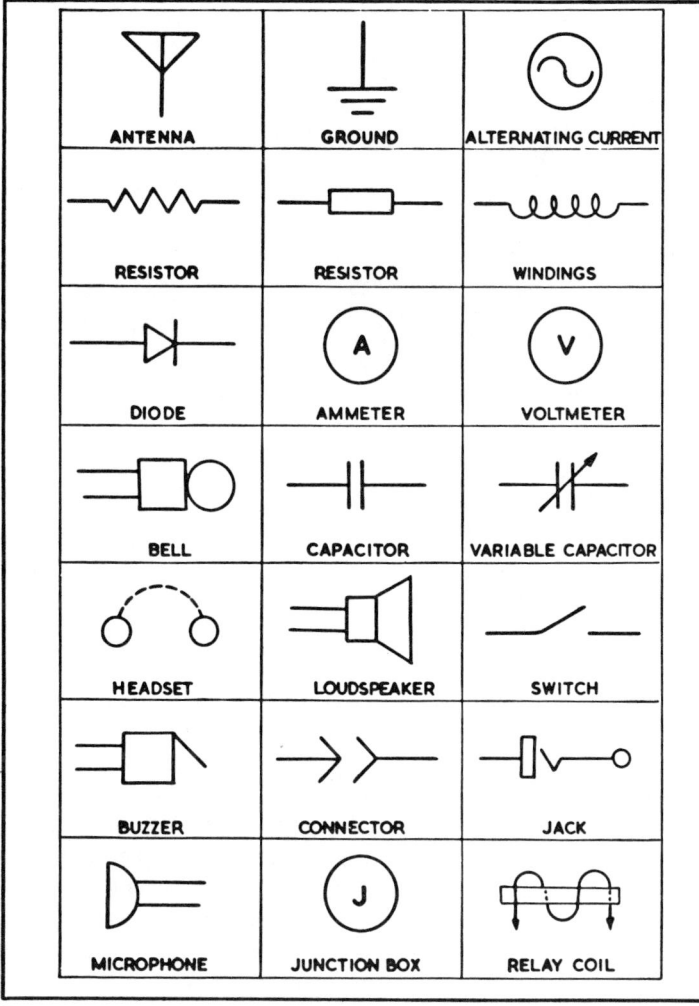

ANTENNA	GROUND	ALTERNATING CURRENT
RESISTOR	RESISTOR	WINDINGS
DIODE	AMMETER	VOLTMETER
BELL	CAPACITOR	VARIABLE CAPACITOR
HEADSET	LOUDSPEAKER	SWITCH
BUZZER	CONNECTOR	JACK
MICROPHONE	JUNCTION BOX	RELAY COIL

Fig. 17-12. A selection of electrical symbols.

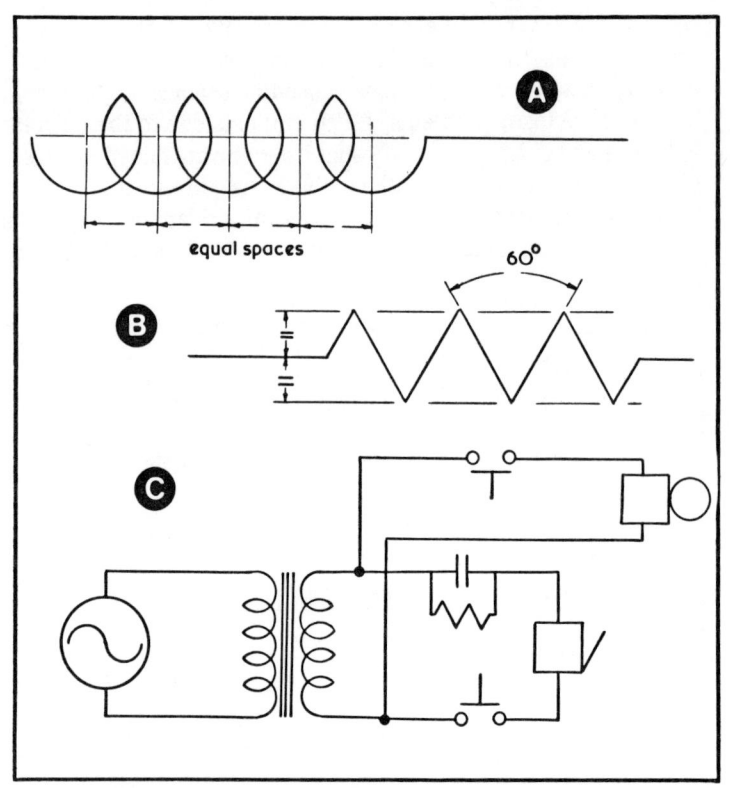

Fig. 17-13. Enlarged details of the drawing of some symbols.

There is a place in airplane drafting for orthographic projection. So many things are involved that any attempt to make an orthographic drawing of all that is in even one place may be too complicated to be of much use.

A side elevation is the best way to show a fuselage exterior. A view from above shows the proportions in that direction, and there may be a front view as in flight or on the ground (Fig. 17-14). These views are useful for general comprehension of the appearance, but they do not usually carry much detail.

Other larger parts or assemblies may be shown by orthographic drawing. Those views would often be accompanied by a more pictorial drawing, either a true isometric or a more artistic rendering. Only in that way can some work be visualized as separated from the other parts around it (Fig. 17-15).

When it comes to the designs of small parts, the drawings are made in the same way as for other mechanical engineering projects. More detail will be shown. In a modern airplane all parts have to be designed with the need for safety combined with lightness. This may not leave much margin for variations that could arise if a detail was left to the man in the shop.

Besides giving every measurement and its tolerances, including even the smallest rounding or fillet, the drawing will usually carry instructions about heat treatment, the way the surfaces are to be finished, and any protective treatment to be applied. There may

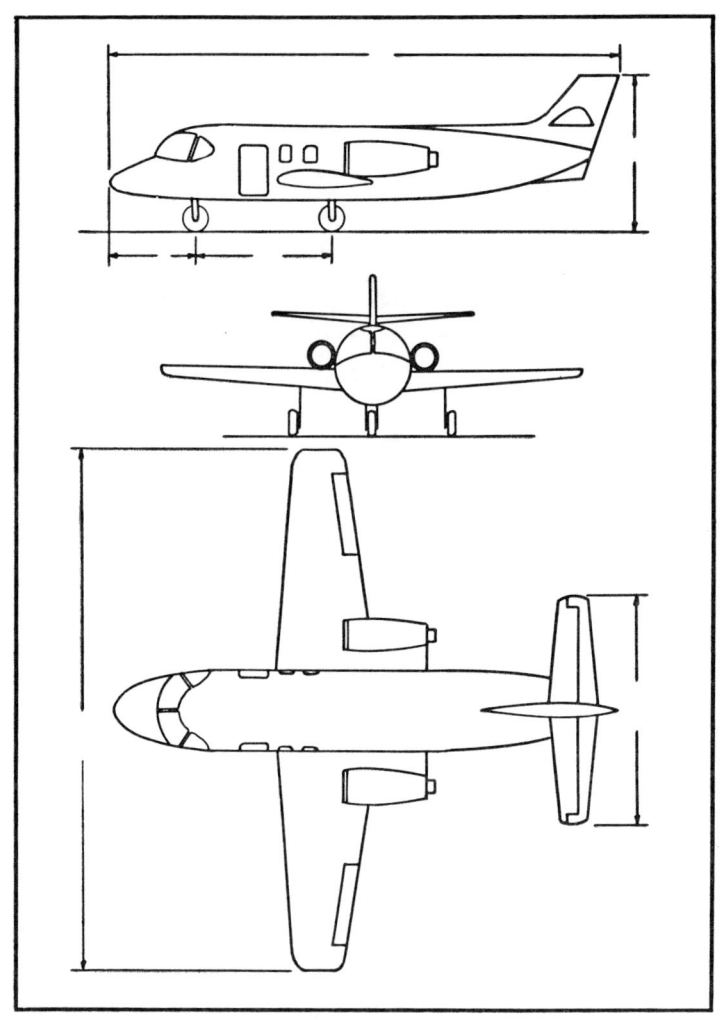

Fig. 17-14. Three views give a general impression and the main sizes of an airplane.

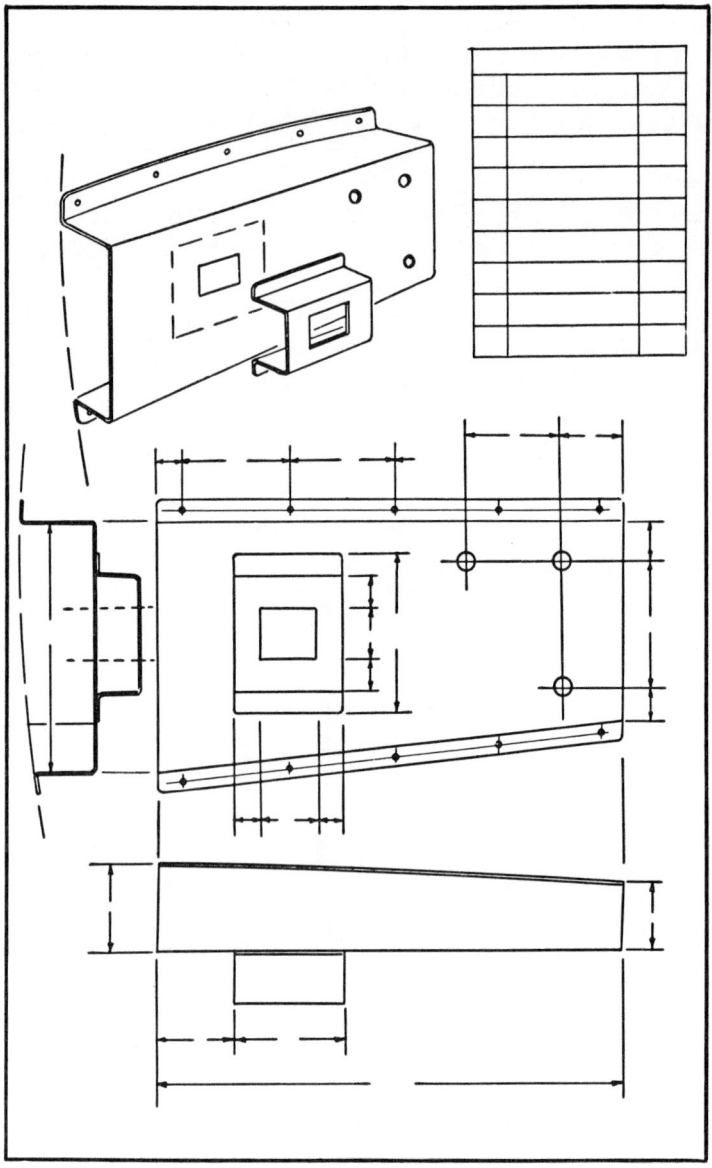

Fig. 17-15. Aircraft detail drawings can be a combination of orthographic and pictorial drawings with tabulated information.

also be references to other drawings. The drawing will certainly have to be checked by someone else concerned with related parts before it goes into the shop.

Chapter 18
Architecture and
Civil Engineering

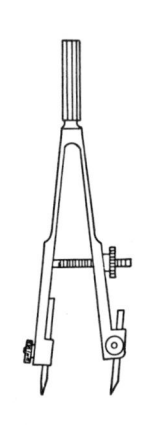

There are a number of differences between drawings of buildings, embankments, and associated activities, and those drawings used for general engineering. The fundamentals are the same. If the user of an architectural drawing saw that it was made by strictly mechanical engineering standards, however, he would feel that it was not what he expected. Conversely, an engineer would be suspicious of a drawing of some mechanical part made in the manner of an architect; yet both might be technically accurate. Structural engineering plays a part in large buildings and many civil engineering projects. Drawings for that work are nearer to those of the mechanical engineering draftsman. As civil engineering is expanded to show large tracts of land, the work of the draftsman becomes mapping. The correct name of the activity is *cartography*.

HOUSE PLAN

In architectural drawing, lines are often allowed to run on, as in the representation of a window (Fig. 18-1A). The wood plan is a more generally used name for the whole collection of drawings of a building, although strictly speaking the work has the same particular meaning as in engineering—a view from above. There is a floor plan for a house, showing the layout (Fig. 18-1B). Although this is to scale, much detail is lacking. The walls may be shown as thick lines, not to a scale thickness, and windows would be indicated diagrammatically. Fixed things, like a bath or a kitchen work top,

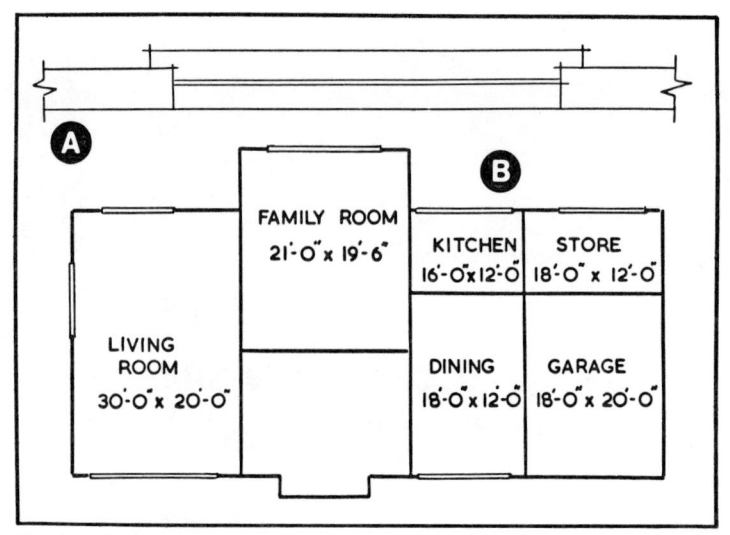

Fig. 18-1. It is customary on some architectural drawings to let lines run on (A). The first drawing shows a floor layout (B).

would be drawn to approximate scale. Internal dimensions would be indicated.

Such a drawing may be in sufficient detail for a builder to lay out the foundations from it, but in the first instance it is a consultation drawing. The customer has to be convinced that this is what is wanted. The builder has to agree that it is practical. A surveyor may have to be satisfied that the house can be built on the land. A local authority may have to give permission for the house to be erected. When the draftsman makes his first plan, he can be fairly certain that the final plan will differ from it, when those concerned have had their opinions analyzed by the architect.

The method of construction has to be shown by a fairly detailed section through a typical wall. There have to be notes explaining some of the parts, as not all readers will be experienced (Fig. 18-2A). A scale has to be chosen that will show details clearly, but there is usually no need to include the whole height. A break in the wall permits a larger scale.

There will be one or more elevations. A simple line drawing (Fig. 18-2B) may be sufficient, but most architects prefer to use shading and will draw in trees or garden features (Fig. 18-2C). It is often the front elevation that is the selling point. This drawing has to be regarded as an advertising exercise as well as a true representation of that view of the house. An architect may then put his own interpretation of the appearance of the house on paper in a perspective drawing, which may come somewhere between true-to-life and something rather fanciful, with more trees, bushes, and people than house detail. This is the architect's concept of the finished house in surroundings that may be imagined rather than expected. As this is the architect visualizing the work, a draftsman does not play a part in it, except that if he is also an artist, he may find employment in producing finished drawings from an architect's sketches.

DIMENSIONS

In architectural drawing there may be an arrowhead on each dimension line, as in engineering (Fig. 18-3A). It is more common to put a figure alongside an unbroken line (Fig. 18-3B). Sometimes a broad V-shaped arrow is used (Fig. 18-3C). A diagonal line across the end of a dimension line may take the place of an arrowhead (Fig. 18-3D). A dot may be used as a terminal mark (Fig. 18-3E). The

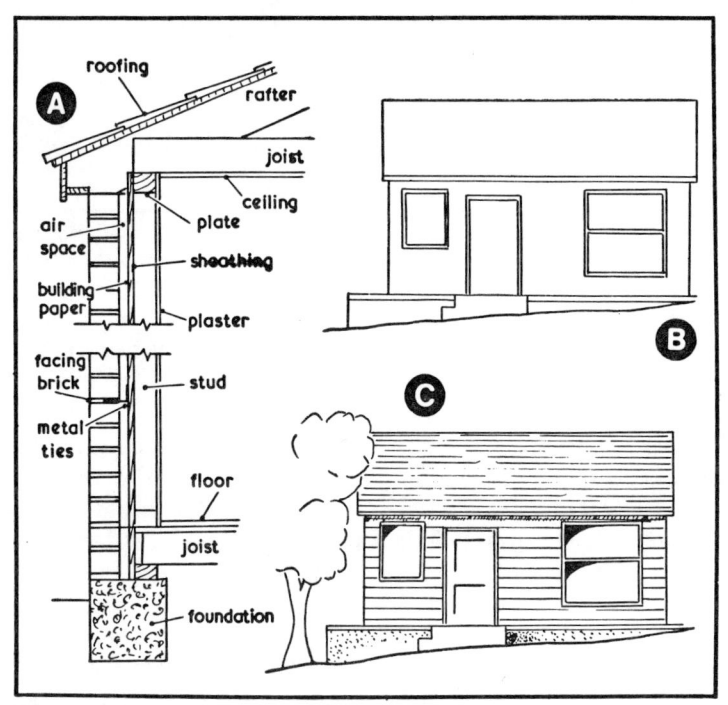

Fig. 18-2. A wall section shows the method of construction (A). A simple elevation (B) shows main features, but this can be worked on further (C) if the drawing is for a customer or others interested.

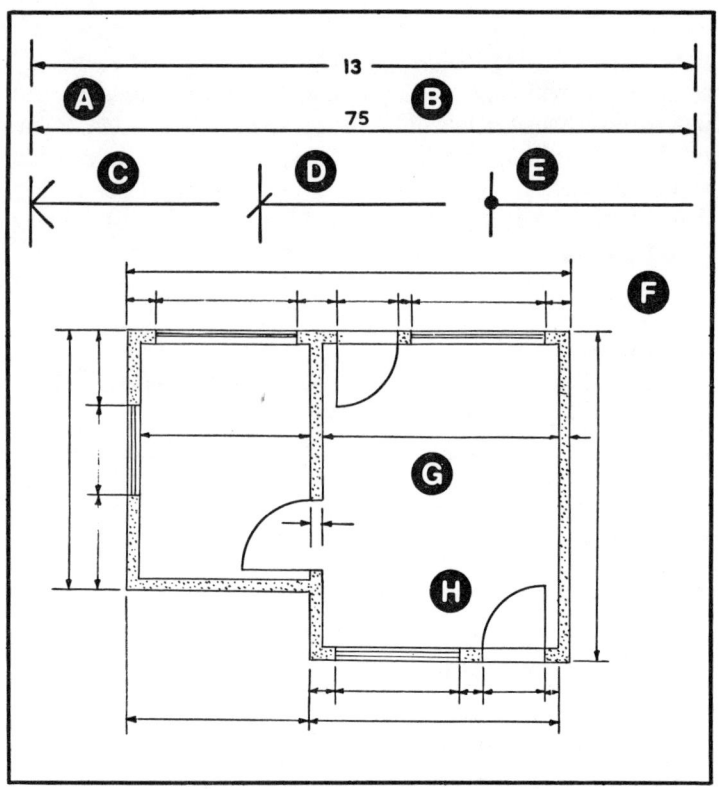

Fig. 18-3. Some architectural draftsmen favor uncommon dimension line terminals. Dimension lines should be complete and clear of main parts of the drawing.

common arrowheads are more usually acceptable. If there is no good reason for doing otherwise, a draftsman should use them.

As with other drawings it is best to keep dimension lines outside the lines representing the object, with overall measurements outside those dealing with shorter distances (Fig. 18-3F). Internal dimensions could be arranged across what would otherwise be a large expanse of blank paper (Fig. 18-3G), rather than project them to the outside. Note that doors are not shown in detail, but the direction they swing is shown (Fig. 18-3H).

The scale of a drawing will have to be settled to suit the size of building and paper. It could be any of the scales used in engineering, but in architecture and civil engineering there is more use of decimal scales such as one-tenth, one-fiftieth, one-hundredth, and much further down the scale for plot and site plans.

Much of a building draftsman's work will be in showing details such as a *cornice* (Fig. 18-4A), the sill construction (Fig. 18-4B), and window details (Fig. 18-4C). There can be much use of isometric or pictorial views, cut away to show several details of construction, in a way that is clearer for a less experienced reader to understand (Fig. 18-4D).

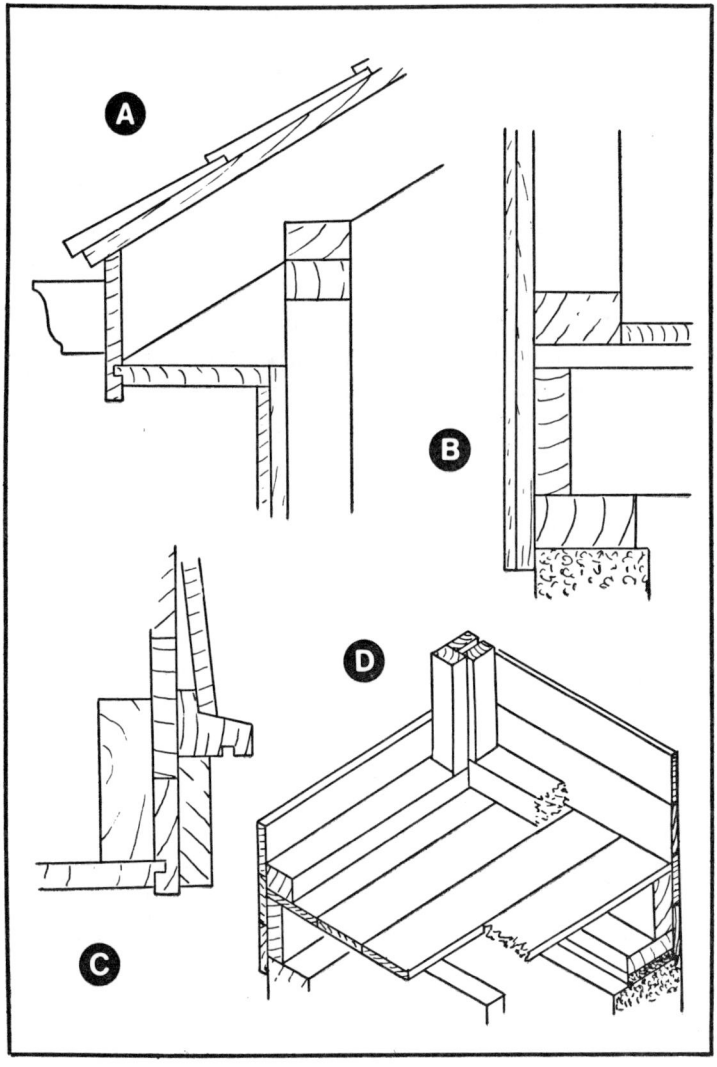

Fig. 18-4. Much of a draftsman's work is in drawing details to guide the building craftsmen.

SITE PLANS

Besides the drawing of a house, there has to be a site plan which shows the position of the building in relation to the rest of the plot it stands on, and any adjoining roads, land, and other features.

This is a plan viewed from above, usually with the building as just a shaded outline (Fig. 18-5A). The limits of the plot of land are shown, with the intended paths and drives and any steps (Fig. 18-5B). If there are roads, they are marked.

If the site is uneven, there could be contour lines. They give the height above sea level—everywhere on a line being at the height marked on it (Fig. 18-5C). Contour lines can usually be obtained from a map or plan already in the hands of the local authority, who may have had to give planning permission.

Also on the plan goes an indication of the direction the plot faces, usually in the form of a compass rose. The name comes from the pattern around a ship's compass, but in this case it can be an arrow in a circle, with N at the tip of the arrow indicating north (Fig. 18-5D). It is also common to put a graphic scale on the drawing, rather than quote a figure that has to be used with a scale rule. A simple graphic scale gives a more immediate means of checking sizes (Fig. 18-5E), either by eye approximately, or more exactly with dividers.

If there are trees on the plot which will remain there after the building has been completed, the drawing would not be complete without them being indicated. A conventional way of showing a tree viewed from above uses a circle with branches radiating from the center (Fig. 18-5F). If there are several trees together forming a piece of woodland, there can be a freehand outline following something like the pattern of branches (Fig. 18-5G).

A further step is the laying out of the site plan of a group of houses. A builder may be developing a large piece of land and erecting several homes of different sorts. There may have to be a swimming pool, recreation facilities, shops, and a church if it is a very large area to be built over. The site plan then shows all the buildings to scale and in their intended positions, with roads, paths, and drives (Fig. 18-6). If the site plan has to sell the idea to others, the drawing could be made more attractive by drawing in trees and bushes (that might or might not eventually be there) and stippling where grass is expected to be.

ANCILLARY SERVICES

Even when the building is not of wood construction, a carpenter plays a large part in house building. A draftsman should

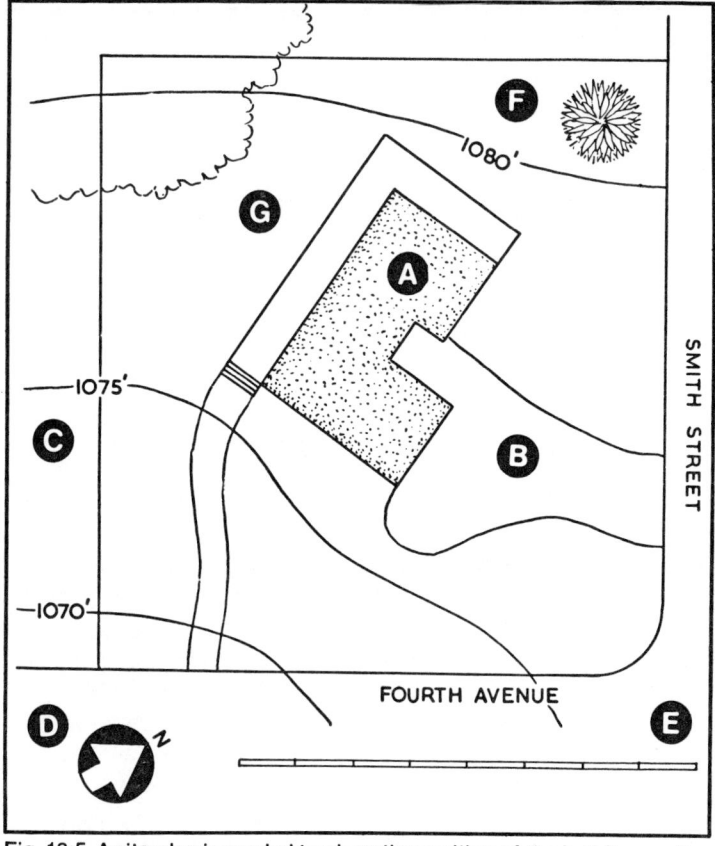

Fig. 18-5. A site plan is needed to show the position of the building on the plot of land.

learn about the basic woodworking techniques, so he appreciates what can and cannot be done with wood. In particular, there are doors and windows which involve sometimes complicated fitting. While some details can be left to the craftsman, the acceptable overall sizes and main features of construction must be known so they can be fitted into other parts of the structure. There are several standardized sizes and spacings that should be observed.

There is a considerable amount of electric wiring about a modern home. How power is brought into the building. The electrical needs in the various rooms have to be considered early in the planning and design stages. Drawing the layout of electrical work is a specialized branch of architectural drafting. There will have to be a floor plan and other views carrying details of the electrical work.

271

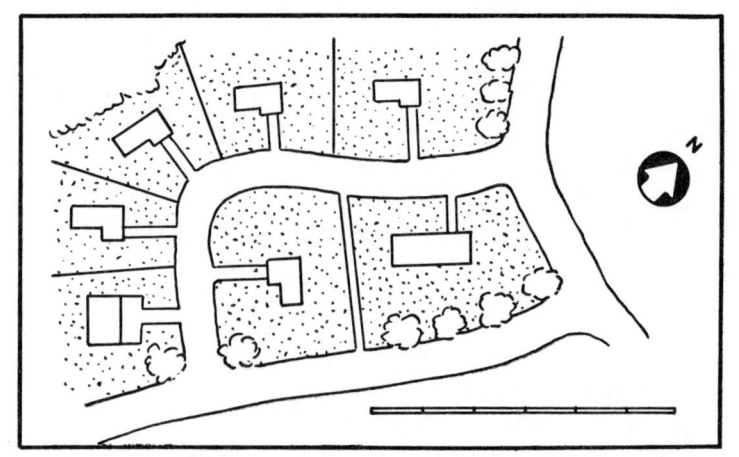

Fig. 18-6. A plan may be needed to show the proposed arrangement of a group of homes.

Symbols are used to show the various outlets and other equipment. Many of these differ from those shown as specimens in the information on electrical engineering. Although there is some standardization of symbols, this is not universal. A draftsman should check what symbols he can use. If there is any doubt, a key to explain the symbols used should be included in the drawing.

Plumbing has to be related to the water supply and the exterior arrangements for sewage disposal. Internal planning in the early stages of design may allow for shorter and more convenient connections outside. Plumbing is more demanding of space in a house, and the layout of pipes should be considered at an early stage. Like the electrical work, there should be a special drawing showing details of the plumbing installation.

In colder climates there may have to be a central heating system, with its boiler, radiators, and pipework laid out on a special drawing in a similar way to the other services. Similarly, the ducting for air conditioning may require special drawings. If there is to be a fireplace, the arrangement of it and its flue and chimney need special consideration, particularly regarding safety. The fireplace is often a decorative feature and will be allowed several drawings.

STRUCTURAL DRAFTING

Where steel is used in a building or for making such things as bridges and towers, the basic parts are standard sections held together by welding, bolting, or riveting. The common sections in many sizes are beams, channels, angles, and tees (Fig. 18-7). Structural assemblies are designed with these sections and flat plates. Joints in many assemblies are made with short cleats of angle section, riveted in both directions.

Much of the detail drafting is concerned with joints, as when a horizontal beam serving as a floor joist has to be connected to a large vertical beam. For a light loading there may be only cleats in the web of the horizontal beam (Fig. 18-8A), but it is better for there to be a seat below (Fig. 18-8B). Apart from additional strength, this is an aid to erectors. Another angle can go above. The angles in the web of the beam may be riveted before the beam leaves the shop. Similarly, the angle forming the seat would be riveted to the vertical beam in the shop. Other riveting would have to be done on site, or bolts may be used. Unless there is a good reason for doing otherwise, the rivets have full round heads, so they are drawn as previously described.

Where plates are used, the designer will have decided on the thickness of plate and the number and size of rivets. A simple strut may have a sheet metal *fishplate* with two rivets each way into the web of a T section upright and an angle section support (Fig. 18-8C).

A common assembly uses two angles with plates between where there are connections. This may make a girder with diagonal braces (Fig. 18-8D).

Structural steel is used to form roof trusses because of its lightness with strength and freedom from warping or rotting.

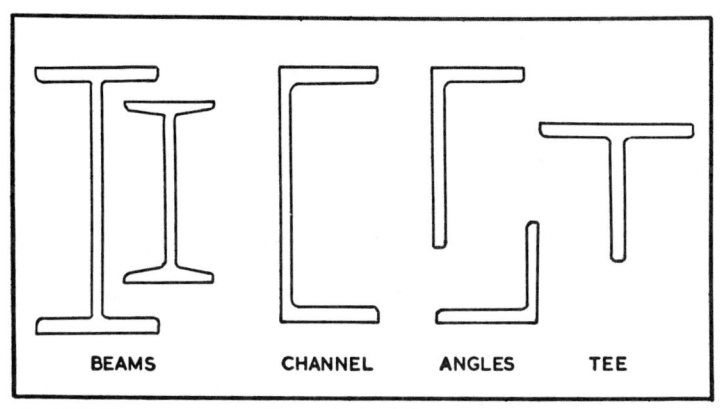

Fig. 18-7. The common sections of rolled steel used in structural engineering.

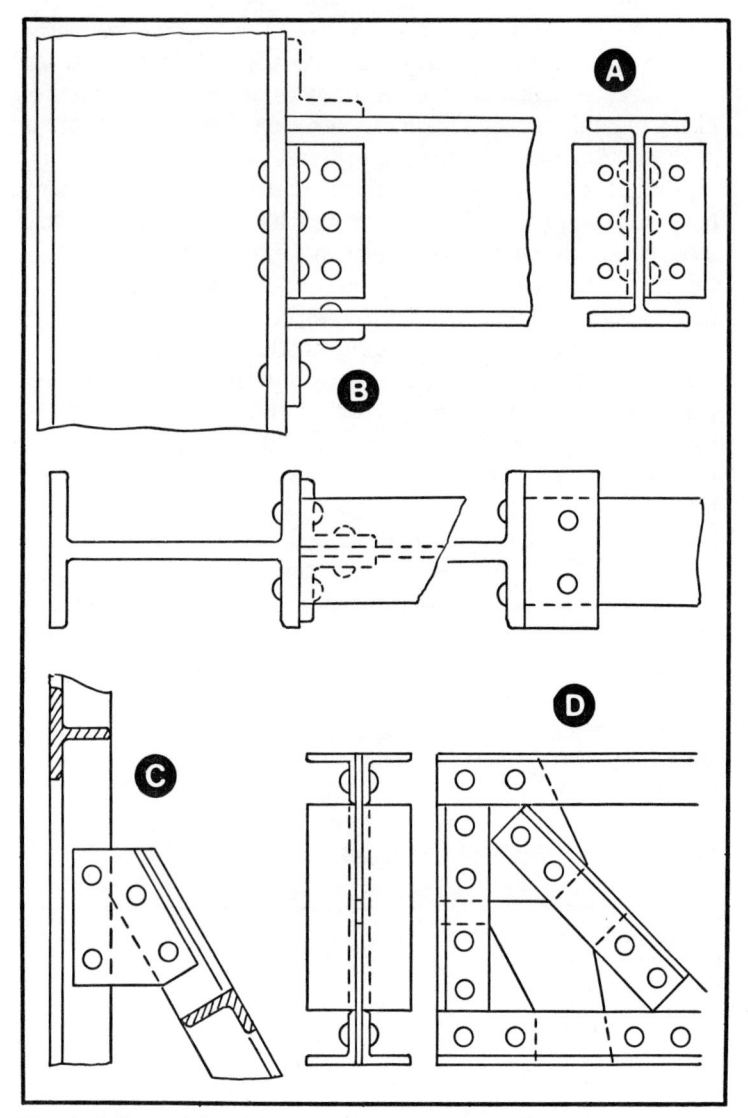

Fig. 18-8. Typical detail drawings in structural engineering.

There are many truss designs (Fig. 18-9), some coming from earlier days of wood architecture.

A typical light truss has a structure of angles linked by riveting to plates. A truss can be assembled completely in the shop, so all parts are riveted. Then holes are allowed for bolting on the purlins to support the roof and for bolting down to the columns or wall plates. It is difficult to achieve absolute precision when fitting a truss swinging from a crane to its supports, and slot holes provide a means of adjustment without the margin of error being large enough to matter. As a truss of this sort is symmetrical, it could be drawn only one side of its center line. The drawing is clearer if the break comes a little further on to the cut side (Fig. 18-10).

The actual sizes of parts will be found in the shop by laying out a half of the part full-size on the floor. This is a special form of drafting called lofting (see Chapter 19). The sizes may be transferred directly to the steel, or wooden templates may be made. If there are many trusses to make, the templates ensure accuracy and speed.

If the drawing is of sufficiently large scale, all of the detail sizes, such as rivet spacings, can be put directly on it. Otherwise, there may be drawings of parts to a larger scale to make things clear. In a shop where much of this work is done, however, those who mark out may be familiar with standard spacings, and these details can be left to them. For instance, rivets may always be 1½d from any edge and 3d apart, where d is the rivet diameter.

CIVIL ENGINEERING

Much of the work of a civil engineering draftsman is concerned with mapping. Up to a certain scale the drawing may be called a plan, while above that it is a map. The dividing line is indistinct. If such things as the widths of roads are drawn to scale, the drawing may be regarded as a plan. If roads are shown by conventional

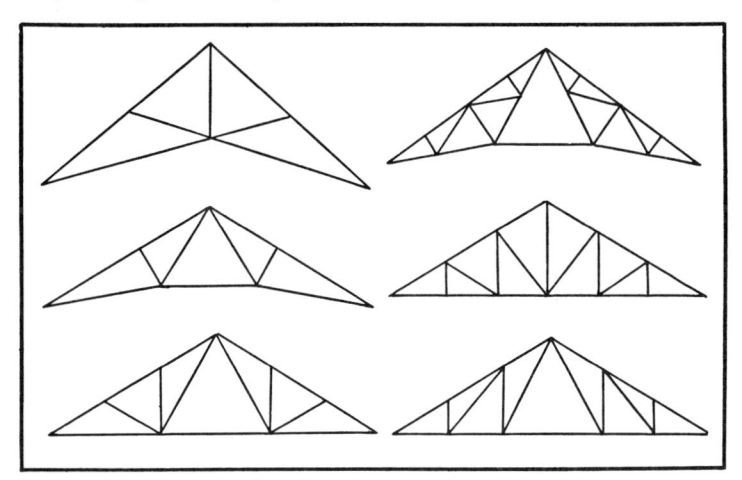

Fig. 18-9. Some forms of steel roof trusses.

parallel lines or a width that bears no relationship to actual measurements, it is a map.

The preliminary work is done by surveyors. The draftsman may have to produce a finished drawing from their sketches. Although a modern surveyor uses many sophisticated instruments, the principle of surveying is based on *triangulation*. The draftsman may have to follow on from the surveyor's field notebook, so he should understand what is involved.

A survey is based on two stations at a known distance apart. Bearings are taken on key points from both stations. With the line between the two stations as base, each pair of bearings produces the other two sides to make a triangle. With three sides known, only one triangle is possible, so where the bearing lines cross is the apex of the triangle and the exact point sighted. If a baseline is drawn to scale to suit the actual distance between the sighting stations, the points on the survey can be repeated and joined to make the plan or map.

As a simple example, suppose a field has to be drawn to scale. The surveyors decide on two points where corners and any other key points can be seen. They mark on the ground these positions at an exact distance apart, preferably a figure that is easy to scale (Fig. 18-11A). With their instrument located exactly over one point, they take a bearing of all the points needed and make a note of them (Fig. 18-11B). They move their instrument to the other end of the baseline and take bearings of the same points from there (Fig. 18-11C).

You draw the baseline to scale, and then repeat the bearings from each end. Where pairs of bearings cross are the points you want (Fig. 18-11D). With all the points located, join them to get the shape and size of the field (Fig. 18-11E).

The bearings may be drawn with the aid of a protractor, but the surveyor will provide figures in minutes as well as degrees. For a more accurate result, you need the angular head of a drafting machine, fitted with a vernier or some other precision means of making fine measurements of angles.

Map drawing involves working neatly with fine detail. Accuracy is very necessary, but it is different from that required in mechanical engineering. Letters and figures may be done with press-on characters. There are a large number of conventional signs, some of which may also be provided in press-on form. Unfortunately, there is no universally adopted system of conventional signs, although most can be understood with a little common

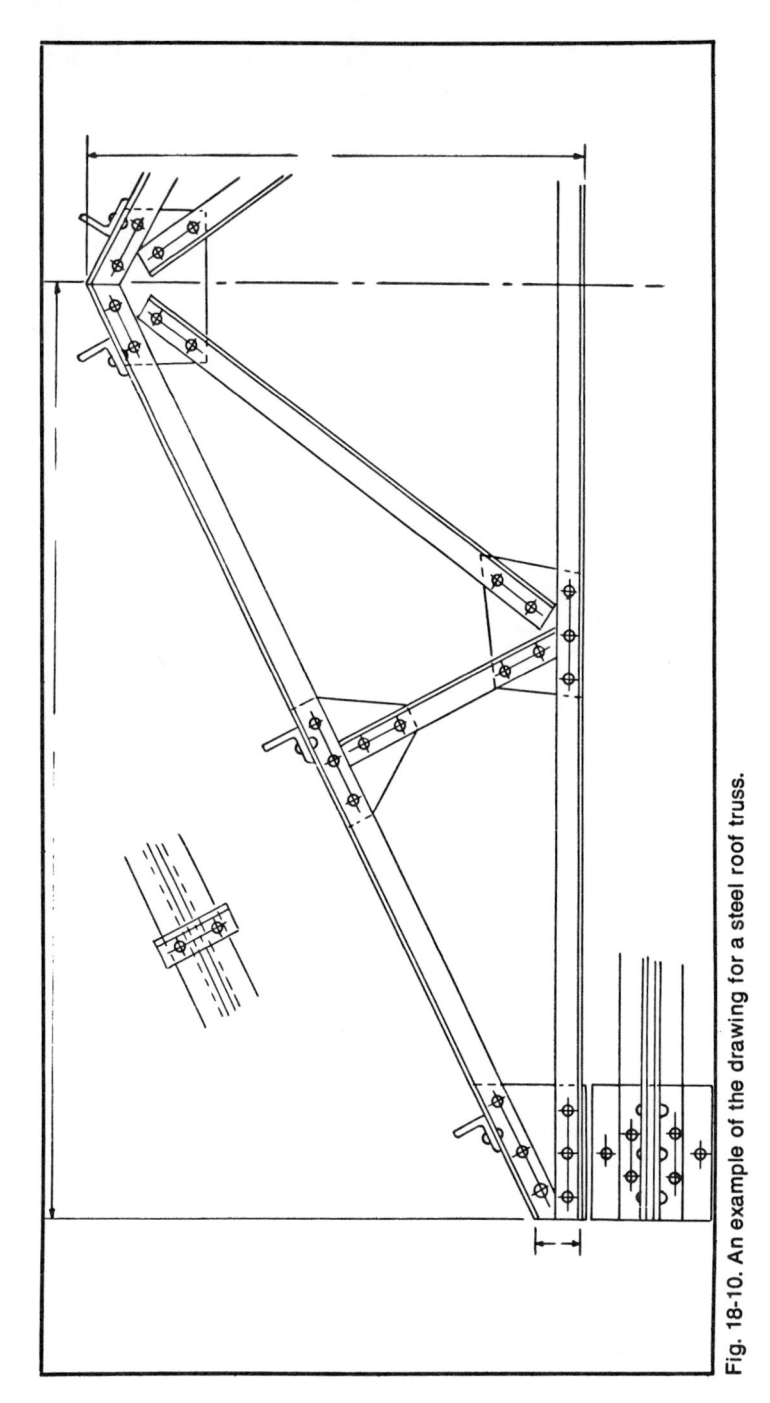

Fig. 18-10. An example of the drawing for a steel roof truss.

sense. If you have your own signs, there should be a key showing their meaning in the margin of the map. If there is no other guide, a government-published map of about the scale you wish to draw will show suitable signs.

Many conventional representations are quite common. Parallel lines filled solidly indicate a road, but variations in width may have different meanings. Open lines may mean a smaller hardtop road or one with a loose surface; that is where the key is needed. Railways are usually single or double lines with short lines across at intervals indicating ties, but not always. Some maps have a very similar way of showing overhead power lines. A church is an

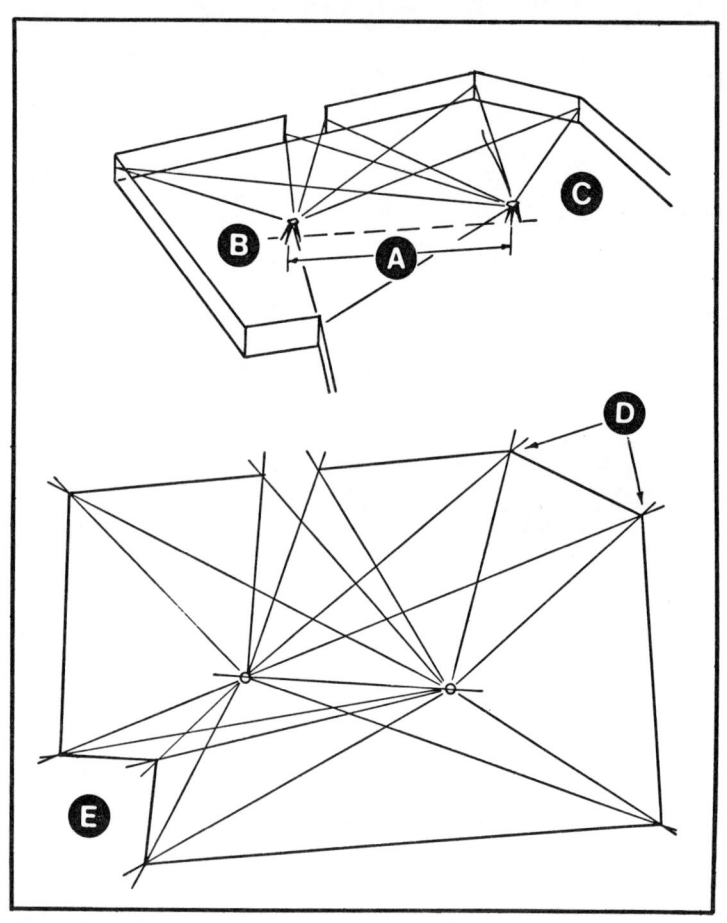

Fig. 18-11. Steps in making a survey (A, B, and C) and producing a drawing from the survey (D and E).

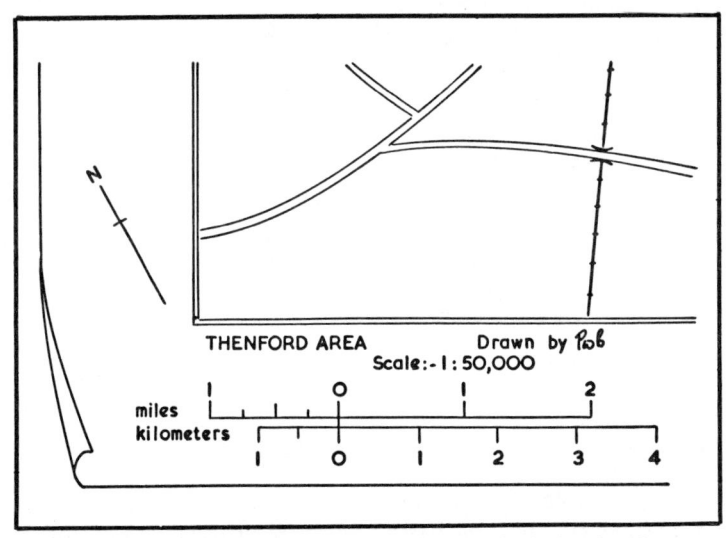

Fig. 18-12. Title, scale, and compass bearing are put in the border of a map.

upright cross. A circle below that indicates that it has a prominent spire. A square would mean a tower.

Any map should be enclosed in a border, and features are usually drawn right up to it. For neatness, there could be a double-line border (Fig. 18-12). You would have more paper outside the border lines than you would for an engineering drawing. Put the title there, together with any other relevant information, including your name and the date. Indicate the scale, usually by a representative fraction and a graphic scale.

The representative fraction is written with a colon instead of a line, such as 1:50,000, which means that the land is 50,000 times the size of the representation on the map. This is 2 centimeters to 1 kilometer or about 1¼ inches to 1 mile. For any figure over 1,000, put a comma before the three zeros. Arrange a graphic scale large enough for distances on the map to be stepped off with dividers. Provide an indication of north, but this need not be an elaborate compass rose.

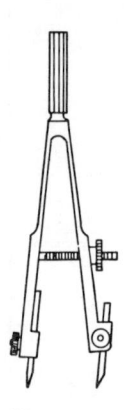

Chapter 19
Lofting

If a structure has many parts, each depending on its relation to other parts, the only safe way to get correct sizes is from a full-size drawing particularly if some or all are curved. If a scale drawing is used, even the thickness of a pencil line can be enough to cause an error when translated to full-size. The problem is particularly acute in shipbuilding and aircraft construction, where compound curves have to be related. Large sheet metalwork may have to be drawn full-size to be certain of getting developments right. In structural engineering, the parts of something like a roof truss (Fig. 18-10) can only be laid out with certainty full-size.

Drawing full-size has to be done on a flat floor. The process is called lofting. The floor may be the actual base of a room or a specially built platform of plywood sheets. Perfect flatness should be the aim, but slight unevenness may not matter. There should be no risk of movement. Ideally the surface would be painted matte black or gray, but for occasional use it is possible to work directly on untreated wood.

The techniques are the same as those of drawing on paper, except for the complications due to size. You cannot put a T square against the edge, so you have to adopt other methods of drawing lines, getting angles, and drawing curves. If there is not already a scale drawing of what you have to set out, prepare one as a guide to what you want to reproduce full-size.

LINES

Straight lines up to about 72 inches may be drawn with a straightedge, which can be any piece of wood planed specially for

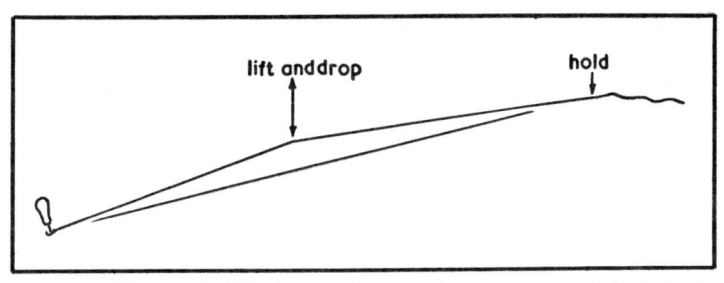

Fig. 19-1. A line of chalk is deposited on the floor when a stretched line is lifted and released.

the particular job. It is unwise to rely on a long piece of wood keeping its shape indefinitely. Changes in its moisture content will cause warping. Longer lines are made with a cord called a *chalk line*. Builders use a rather coarse chalk line. It makes a line which is too thick. A fine, slightly hairy line of enough strength to withstand stretching is needed. Crochet cotton is suitable.

With the line goes a piece of chalk. Have the line on a reel, preferably one with central hollows on each side so it can rotate between finger and thumb. To strike a line, put a small awl or nail through a loop at one end of the line and walk back from it, letting the line run out as you rub it with chalk. Take care not to jerk it, or chalk dust on it will come away. When you have let out enough line, press it to the floor with your thumb as you stretch the line, still without jerking. Lift the line near its center, and it will spring back to deposit a line of chalk on the floor (Fig. 19-1). Make sure the lift is upright. It need only be a few inches. For a very long line an assistant strikes the line, but up to about 15 feet you can reach far enough to lift it with your other hand.

Often the line drawn has to pass through two positions and go beyond them, as when drawing a tangent to two circles. In that case an assistant holds the end that would have been held by the awl, and you both manipulate the line into position before striking it.

RIGHT ANGLES

The first setting out is a baseline far enough to one side to allow all of your drawing to be made on the main area of the floor. In some cases it may be better to strike a center line. For most layouts you need another line at 90° to the baseline; then much of the rest of the work is done by measuring parallel to these lines. Drawing the right angle is the application of one of the methods described in Chapter 7. Most commonly used is the 3:4:5 method.

Decide how far the second line has to go to embrace all that will be drawn, and choose units that will take the arcs as far as that.

Positions can be marked with a pencil, but they will be clearer if they are white. A block of talc or French chalk may be rubbed to a chisel edge with a file (Fig. 19-2A). Positions on a line can then be "pecked" (Fig. 19-2B), with the location at the point of the V. You may use a rule and draw a short line across (Fig. 19-2C).

Mark where the square line is to come on the baseline, and measure four units from it along the line (Fig. 19-2D). A steel tape is the best thing to use. Try to avoid having to build up any length by several shorter measures with a tape or tape rule which is too short, as that can lead to errors.

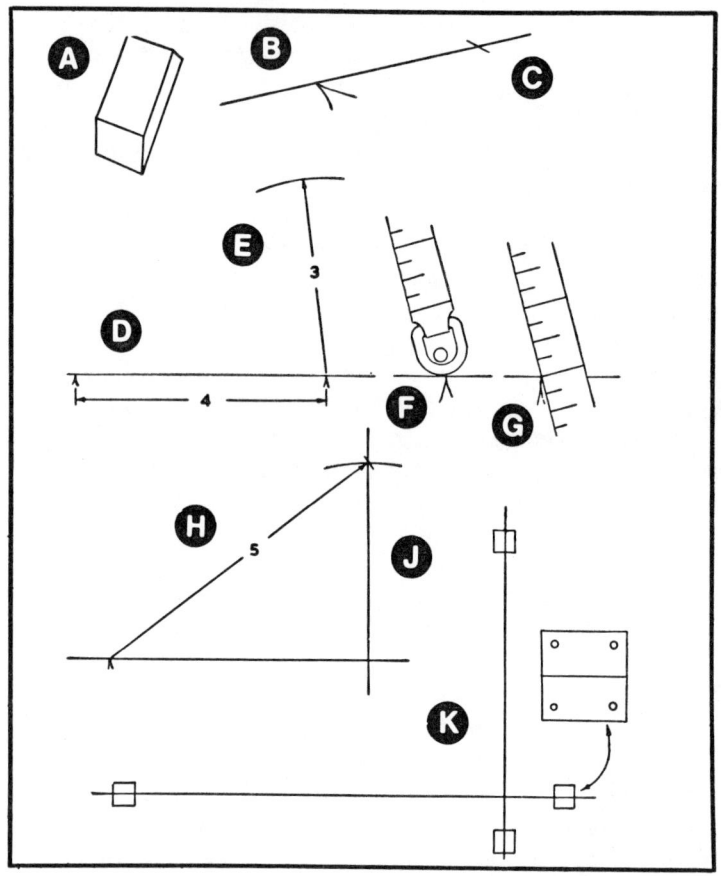

Fig. 19-2. Flat chalk is used for marking (A). Distances can be marked in two ways (B and C). Baselines are set out using the 3:4:5 method (D to K).

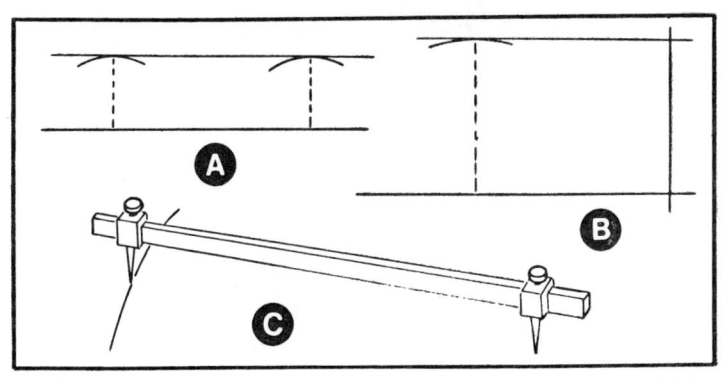

Fig. 19-3. Parallel lines can be marked with arcs. Trammels are useful for this.

Use the tape with the chalk against it to draw an arc three units from the baseline mark that will obviously cross the square line (Fig. 19-2E). If there is a loop at the end of the tape, use an awl with the outside of the loop at the mark (Fig. 19-2F). Otherwise, get an assistant to hold a mark on the tape against the center (Fig. 19-2G). Remember to allow for the change of position along the tape length. Now measure from the second mark on the baseline five units to a point on the arc (Fig. 19-2H), and strike a line through that and the mark on the baseline (Fig. 19-2J). If the floor is to be used regularly for setting out, small metal plates with lines scribed across them could be nailed down as guides for renewing the baseline later (Fig. 19-2K).

Lines parallel with these two baselines may be measured directly if you are sure you are working squarely, as when you measure along a line already known to be square or parallel. Otherwise, it is wiser to draw arcs and strike a line across them (Fig. 19-3A). Sometimes only one arc is needed, when the other position is on a square line (Fig. 19-3B). Trammels are useful tools within the limits of their beams. You can use them like dividers for transferring distances. They can be used as compasses for drawing curves (Fig. 19-3C), but as they scratch the floor you may not want to do that damage. If one trammel head takes a pencil, that will avoid scratches. For clarity, it is advisable to go over the vital part of the pencil curve with chalk as soon as possible.

CURVES

For large curves a compass has to be improvised with a strip of wood, or several overlapping pieces have to be temporarily nailed for a large radius (Fig. 19-4A). The center is an awl pushed through (Fig. 19-4B).

The end of the strip is used with chalk to draw the largest curve, but others needed can be marked with notches along the strip (Fig. 19-4C). Suppose a curved metal angle has to be drawn. You will need curves indicating its heel and toe as well as the line of holes, where they are to come (Fig. 19-4D), so that settles the positions of notches (Fig. 19-4E). If other curves are very much different, it is better to shorten the "compass" or reposition the awl, so you do not have to pull around an excess length of wood. For a very large radius, get one or more assistants to help pull the wood around, so it does not flex.

If the curve is irregular or is part of a circle so large that a strip wood compass is impractical, you have to draw through points plotted in the same way as previously described for developments, cam diagrams, and similar things, except that what you do is much larger. From a scale drawing, it is possible to measure heights on ordinates. Either mark these on the drawing or put them in a table. There are many double curves in setting out the lines of a ship, and the list taken from the scale drawing is called a *table of offsets*. Although these measurements are as accurate as they can be scaled, there will be a few slight errors due to the small drawing size. In setting out full-size, the curve has to be "faired off." From the locations of the positions marked according to the table of offsets, those that are inaccurate can be seen not to come on a fair curve, which will be drawn through the right places.

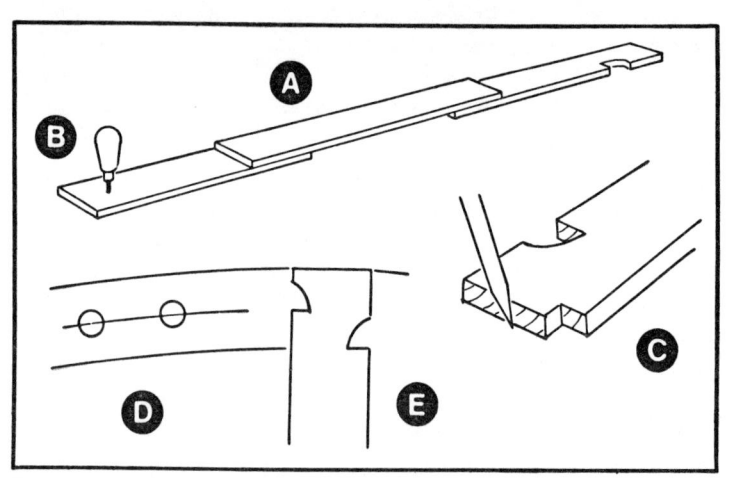

Fig. 19-4. A notched wood strip makes a compass.

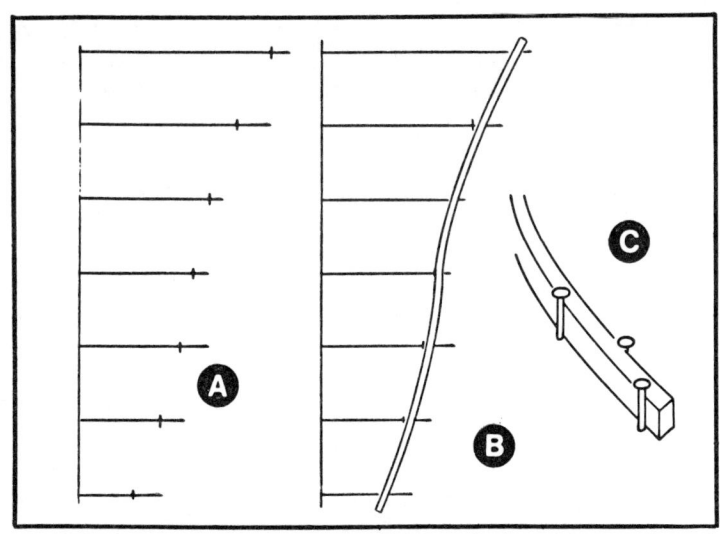

Fig. 19-5. Points on a curve are marked (A), and a batten is bent through them (B). It may be nailed temporarily for drawing against (C).

You plot the positions on the floor (Fig. 19-5A). Then get one or more assistants to help you bend a lath through the points, or as many of them as will match (Fig. 19-5B). Sight along the lath to look for unnatural kinks. You may have to average the curve between marks if few match exactly. Often you can draw around the lath with chalk while it is held. Otherwise, drive nails each side of it to hold it in place before drawing (Fig. 19-5C).

ANGLES

You cannot use a drafting triangle to mark angles on the floor, and the usual protractor is not big enough to be of any use. The common angles of 30°, 45°, and 60° can be marked geometrically. For 30° or 60°, a radius will step off six times around a circumference, but there is no need to draw full curves. From the corner of square lines, use your tape to measure in each direction from the corner positions at least as far as the angled line is to be (Fig. 19-6A). At the same time swing an arc that will obviously contain the angle (Fig. 19-6B). From one of the points, swing a radius across this. A line through this crossing will be at 30° to one line and 60° to the other (Fig. 19-6C). A line at 45° can be found by bisecting a right angle from two points at the same distance from the corner (Fig. 19-6D), or you could draw a line across and divide that by two using measurements (Fig. 19-6E).

285

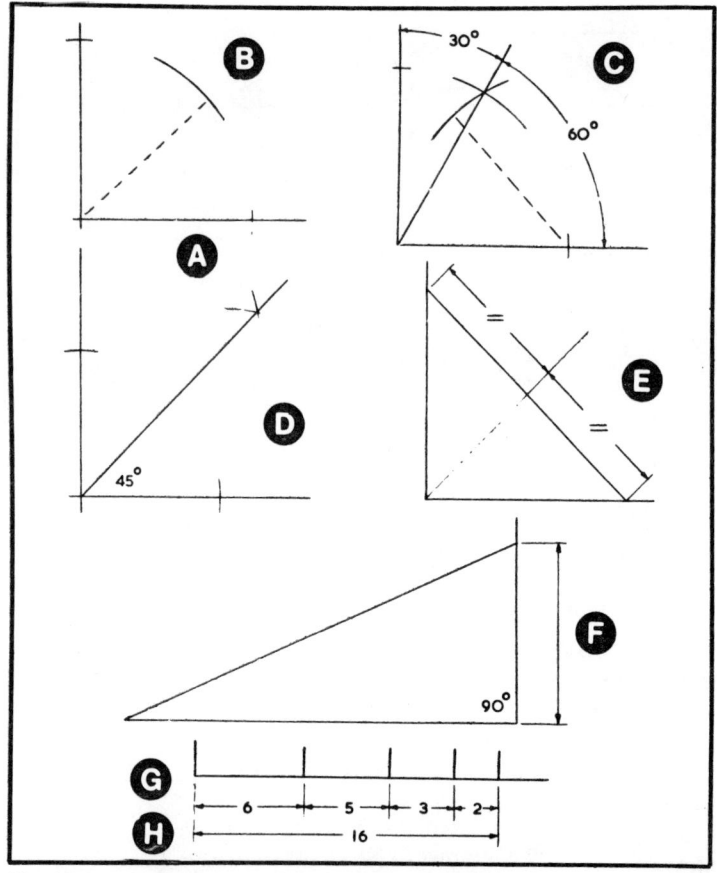

Fig. 19-6. Angles may be marked geometrically (A to E) or by calculation (F). Distances should be measured overall (G and H).

Other angles have to be set out with the use of trigonometrical tables. This method uses the properties of a right triangle. Knowing the distance along the base and the angle required, the tables allow you to calculate the height of the side of the triangle (Fig. 19-6F). As with the other methods, make sure the triangle you set out takes the angular line further than it will be needed. It is wrong to set out anything shorter than required and then rely on extending it. There is less risk of error is the final size is less than your preliminary size. Another cause of error is the stacking up of distances. If you have to mark many positions along a line, do not rely on measuring one from another (Fig. 19-6G). Measure them all from the start (Fig. 19-6H) to avoid a cumulative error.

Chapter 20
Technical Illustrating

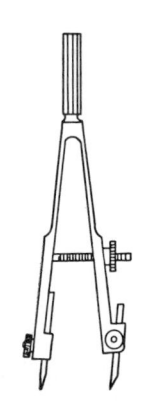

Much illustrating in books and magazines is purely artistic and not the concern of a draftsman. In maintenance manuals, do-it-yourself books, and instructional charts, The illustrations are technical in the sense that they are meant to convey information of mechanical or other practical interest to the reader in a form usually more attractive than in a working drawing. The illustrations can be very similar to working drawings, but they may be made more lifelike with some shading. There could be a hand holding a tool or a person sitting on a tractor.

If a drawing is to have figures or animals on it, the quality of the drawing depends on the artistic ability of the draftsman. Not everyone can manage these drawings. There are other drawings where a pictorial interpretation of something first drawn by orthographic projection may be needed. A parts list may need a picture of an object broken down into its components for identification.

Drawings that are to be reproduced by any of the usual printing processes need to be a good black, so they are normally finished in Indian ink. It is sometimes possible to reproduce pencil, but it is wiser to assume that the drawing should be in ink.

DRAWING SIZES

The first practical consideration is the size of the drawing. A printer can produce the size picture he wants from an original that is bigger or smaller, but the best results come from using an original larger than the final picture. How much bigger depends on the type of drawing, the method of reproduction, and the type of paper to be used for printing. A good quality smooth paper will give good reproductions of very fine detail that would disappear on poorer rough paper. The draftsman gains by drawing bigger. Minor discrepancies will disappear or be lessened by reduction. He has to consider line thickness. As well as reducing overall sizes, the thickness of lines will be reduced in the same proportion. Too fine a line on the original may be lost or only partly reproduced in the final print.

This is a problem where a very large original has to be reduced to page size. What is a perfectly good ink drawing for making same-size prints for shop use may be useless for reducing many times to page size. For most purposes, it is convenient to make the original between 1½ times and two times the final size. The drawings in this book were made 1½ times their final size. The sizes to draw can be obtained by multiplying the final sizes by 1½, but it is simpler to use a diagram (Fig. 20-1A). Draw a rectangle of the final size with a diagonal projecting from one corner. Draw a new border line parallel with the side of the rectangle of the length you want to make the drawing. Project this the other way from the diagonal, and you have the drawing size. At any size up to twice the final one, you can use normal drawing techniques and line thicknesses without fear of trouble in reproduction.

If you already have a drawing very similar to the one you want to draw but the wrong size, it helps in redrawing to use a grid of squares over the drawing, and repeated in the correct proportion as a base for the next drawing. Suppose the existing drawing is three times too large. Squares on it could be 1½-inch, and you prepare to start the new drawing on a similar arrangement of ½-inch squares (Fig. 20-1B). If you note where each line crosses its grid square, you can get it at the same place on a smaller square. You may even alter the relative proportions by having rectangles in the second pattern, while keeping squares on the original. Do not try to carry that alteration too far, however, or you will get problems of distortion.

EXPLODED VIEWS

Care is needed in laying out and planning a disassembled drawing to show components in the most convenient way (Fig. 20-2A). So far as possible, they should be in the correct relative positions and ready to fit together. There is room for some artistic license. The user will not be measuring the drawing. All he wants

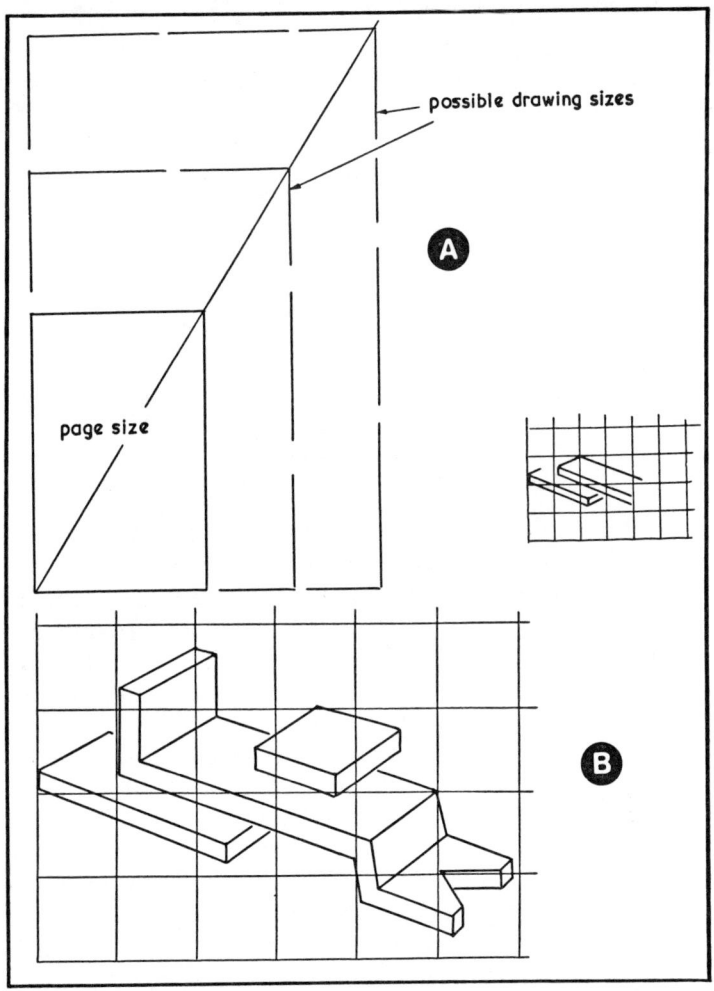

Fig. 20-1. Drawing size can be obtained from page size by using a diagonal (A). A drawing can be made to a different size by using a grid of squares (B).

to do is identify a part. This means that the draftsman does not have to work exactly to scale, although proportions should obviously be about right. It also means that he can draw things so they look right by sketching freehand, followed by lining in with instruments. Things like screw threads are better drawn as they appear than by using any of the drafting conventional representations. The user is expecting to recognize a picture of each part.

Usually each piece has a letter or number, indicated by a leader to a ring (Fig. 20-2B). There will then be a parts list printed

289

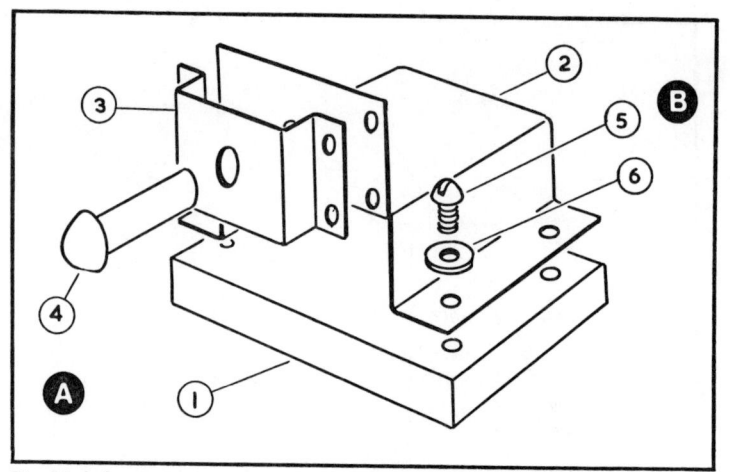

Fig. 20-2. An exploded view with reference serves as a guide to a parts list.

on a facing page or elsewhere, giving a name and parts reference for each identifying letter or number.

Emphasis may be given by shading. In its simplest form the lines indicating outlines are made thicker on the side away from the light (Fig. 20-3A). Other shading may be by ruled lines (Fig. 20-3B), freehand stippling (Fig. 20-3C), or by using a press-on stipple pattern, of regular dots or other marks. There may be solid black shading but too big an expanse of solid black tends to look overpowering and causes some problems in printing.

An ink drawing is made up of black and white. In-between tones are indicated by drawing many fine lines close together or by stippling with spots, but the components of the drawing are still black on white. It is possible to get intermediate shades of gray by air brushing. This is a method of spraying and is more the work of an artist than a draftsman.

PICTORIAL VIEWS

As can be seen by examples in books or magazines, the best way of giving sizes and shapes is by orthographic projection. Details are shown pictorially whenever possible. This means that a draftsman has to use three-dimensional views far more than he would in preparing drawings for use in an industrial shop, particularly if the text and drawings are aimed at an amateur craftsman.

Be prepared to make the drawings without any verticals, if you can get a better view diagonally (Fig. 20-4A). Let lines slope away towards a vanishing point if that makes the picture look more natural. There is no need to rigidly adhere to the conventions of isometric drawing. Let something lead off the edge of a picture or have a ragged break (Fig. 20-4B). If it is metal, use some shading. If it is wood, show the grain (Fig. 20-4C).

Most of these drawings will have references in the text. Your drawing will have a figure number (usually written "Fig. no." or just "Fig.," as in this book). Numbers may go through the book or start afresh with each chapter. There would then be a second number, this one being "Fig. 20 (meaning the chapter)-4." On the drawing comes a reference letter, and this is repeated in the text reference. With anything complicated, you should remember that there are only 26 letters in the alphabet. This total is 24 if you leave out I and O, which might be confused with figures. If there will be more than 24 references, it is better to use numbers. You can draw attention to a number reference on the drawing by ringing it (Fig. 20-4D) or by using a larger letter than elsewhere on the drawing.

GRAPHS

Scientists and others make much use of graphs, but the way they are made is usually unsuitable for reproduction. A draftsman has to make them presentable. The scientist is only concerned with getting results. His graph may not be in the best proportions, and a draftsman can improve its appearance without affecting results.

For instance, on the graph paper the scientist may choose scales that make his plotting very shallow (Fig. 20-5A). If the

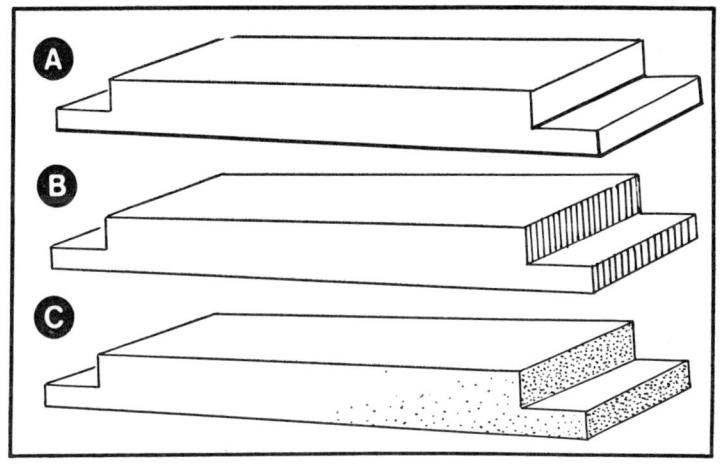

Fig. 20-3. Shading can be by thickening construction lines (A), hatching with lines (B), or by stippling (C).

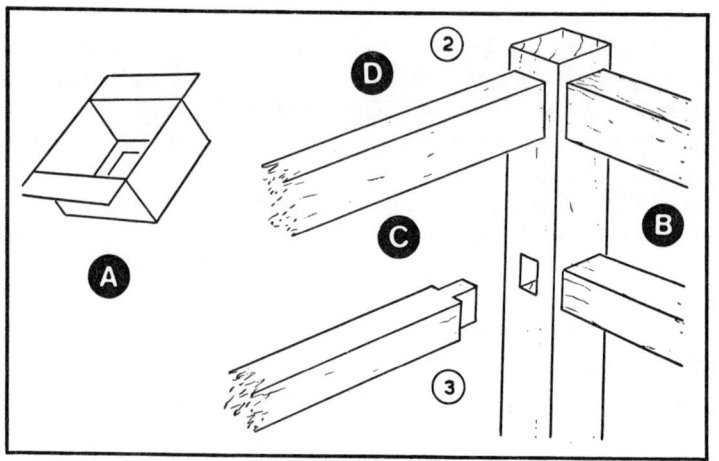

Fig. 20-4. A pictorial view may be tilted (A). Parts can run off (B) or breaks can be used. Reference numbers or letters may be ringed (C and D).

vertical scale is increased, the plotted curve is easier to understand (Fig. 20-5B).

That sort of graph is not very attractive to a layman. It is more common to get the same effect with better visual impact. A sales graph might be a simple line on normal graph paper (Fig. 20-5C), but it would attract more attention if pillars represented each year, with the loss appearing to go below the shelf (Fig. 20-5D). If more precision is needed, there can be columns to the chart heights (Fig. 20-5E) instead of lines.

When something has to be broken down into the parts of a whole, it is sometimes convenient and clearer to use a circle as the whole thing and divide it into segments proportionately (Fig. 20-5F). There may be different shading to distinguish the segments, or you may be able to use colors, as well as printing on the details.

Another form of graph easily understood by a layman has little figures of people or products, each representing a certain number. There could be a line each of men, women, and children, each representing 100 people who had visited an exhibition in a day. There could be lines of cars, trucks, and motorcycles, each representing a certain number using a road during a certain period (Fig. 20-5G).

CHARTS

The skill of a draftsman may be needed to lay out tabulated information. This is straightforward instrument work, but there are problems of proportion, particularly regarding the accommodation of information. Unlike the similar layouts sometimes needed on shop drawings, there can be decorative touches such as double broken borders and different thickness lines (Fig. 20-6A). In some cases the table may not even be upright, or it may appear to be clipped through slots (Fig. 20-6B). The chart may appear to be on a board screwed in place (Fig. 20-6C). These devices are all included

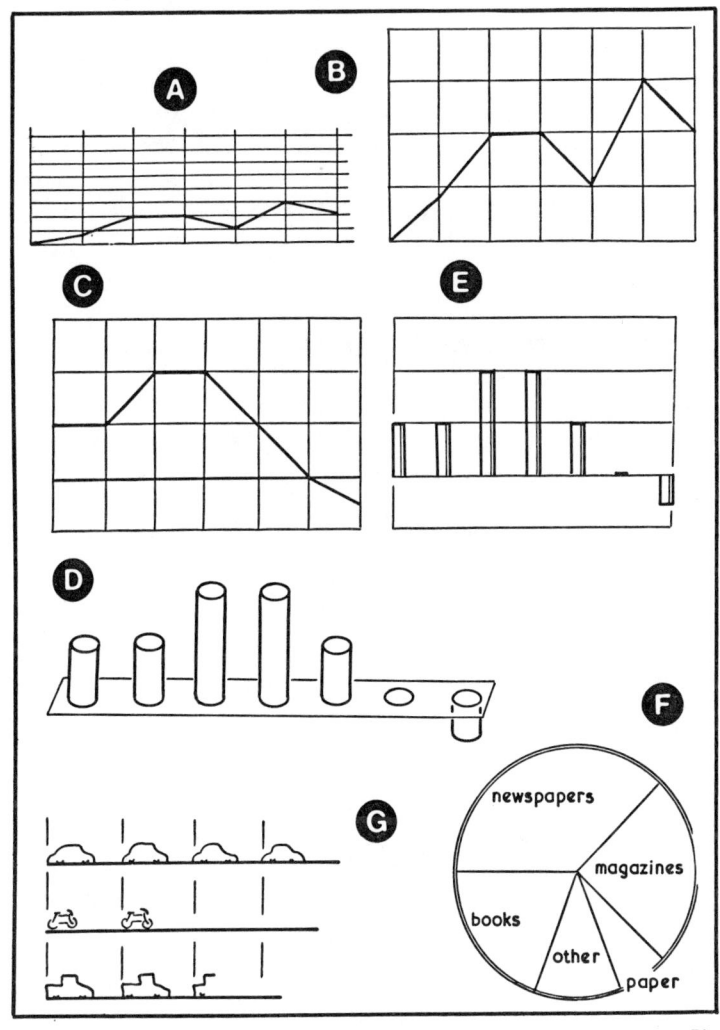

Fig. 20-5. The scale of a badly proportioned craft can be altered (A and B). A line graph (C) can be given visual appeal as a picture (D) or by drawing pillars on the graph lines (E). A round graph shows parts (F). Symbols can indicate numbers of users (G).

to attract attention to what would otherwise not be easily noticed, despite its importance.

Flowcharts may have to be drawn in a way that has some visual appeal. They show a sequence of who does what and how their work is brought together (Fig. 20-7A). In that case they are best formed of lettered labels. If you want to show how parts are brought together to make a whole, it may be better to use pictures (Fig. 20-7B).

If a chart is a one-off for display in an office or shop, there can be color applied, or it may be possible to use colored adhesive tapes. Certain letters and symbols may be obtainable in press-on form to give a better effect, possibly in color, than might be obtained by drawing. If a graph or chart is to be reproduced, watch that proportions are correct, and usually draw it up to twice the size it will eventually be.

STYLE

The designer of a book or magazine may wish to use a similar style throughout. If several draftsmen are involved, it is important that the weight of lines used and the general styles should match. There will be proportions to consider. The book would look unbalanced if something simple was given a large amount of space, while another complicated drawing was squeezed into too small a space.

There must be consistency of lettering throughout, and the layouts of notes and references should be similar. If lettering is all

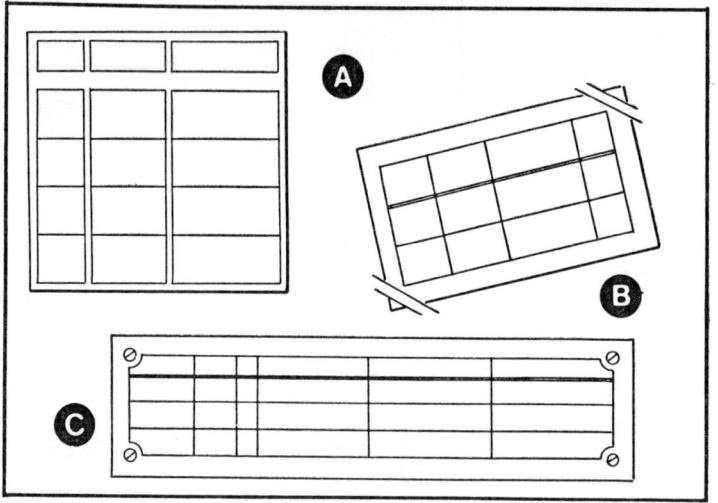

Fig. 20-6. Tables may be more decorative than on working drawings.

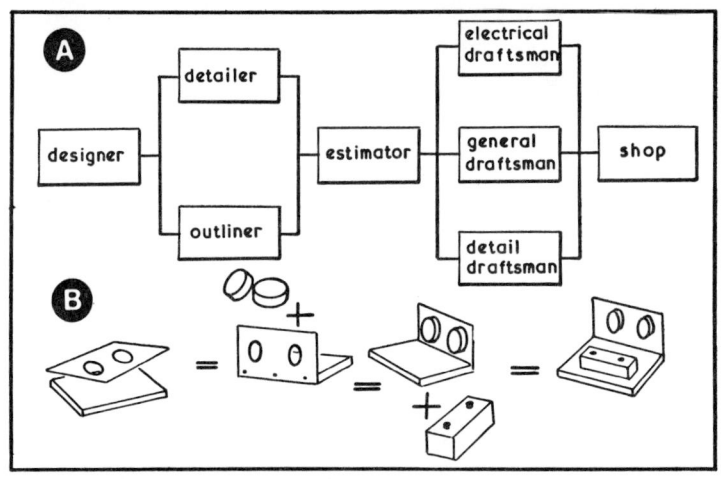

Fig. 20-7. A flowchart may be connected labels (A), or pictures can show stages in assembly (B).

done with a guide, it may not matter who does it. For freehand lettering, though, it may be better for one person to do the lettering on all drawings.

In some publications annotation on the drawings can be typeset by the printer and added to the graphic work. This insures consistency of lettering but adds to costs, so lettering by the draftsman may be preferred.

Not every drawing will be full-page. Sometimes the draftsman may have difficulty in forecasting what size his drawings will be reproduced. Even if he does not know the height, however, he can be fairly certain that most drawings will appear at page width. It helps the publisher and printer if all drawings are made to the same width proportions, even if the heights vary up to page size. All of the drawings in this book were made 6½ inches wide, but the heights varied up to 10 inches for a full page. Some publishers like to put a drawing in a position where the type fits around one side of it, but that arrangement is costly and less popular.

Consider also the shape of a drawing. Ideally it comfortably fills a rectangle, with fairly equal areas of white paper between parts of the drawing. Not every subject lends itself to that treatment, but it is often possible to fill a space with a detail or note, so the general appearance is squared. If there is no other way out, a drawing may have to be L-shaped or some other uneven pattern, but it is better if the majority of drawings can be treated as rectangles when laying out the book.

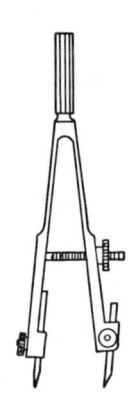

Glossary

acme—Screw thread form with sloping sides and flat top and bottom.

acute angle—Angle less than 90°.

adjustable triangle—Drafting triangle with one corner at 90° and the opposite side adjustable.

aligned dimensions—Arranging the figures and indicating dimensions in line with the dimension lines.

allowance—The minimum clearance between fitting parts.

alloy—Two or more metals mixed together to obtain particular qualities.

ammonia process printing—A method of reproduction in which the image is developed on the sensitive material by ammonia vapor.

anneal—To reduce metal to its minimum hardness, in most cases by heating to redness and allowing to cool slowly. This also removes internal stresses and may be called normalizing.

arc—A curve which is part of the circumference of a circle.

architect's scale—Scale showing relative proportions as required when making drawings of buildings, civil engineering, and similar projects.

assembly drawing—Drawing showing an object made up of many parts in its assembled form.

auxiliary view—Supplementary view of an object, particularly to give the true shape of a slanting surface.

axis—A line about which something rotates. Plural axes.

axonometric projections—The collective name for a cornerwise view, specifically called isometric, dimetric, and trimetric.

babbitt metal—An alloy of antimony, copper, and tin used in bearings.

beam compass—A bar or beam on which compass points can be moved and locked, for drawing larger circles.

bearing—A supporting part, but in particular a support for a revolving shaft.

bevel—A sloping surface.

bevel gears—Gears that mesh with their shafts out of line, usually at 90°.

bird's-eye view—Any view from above.

blind hole—A hole that does not go right through the object.

blueprint—A reproduction of a drawing with the lines white on a blue background.

bolt—A screwed fastener to engage with a nut, but with the threads not taken as far as the bolt head.

bolt circle—The circle on which bolt holes are arranged on the flange of a round part.

bore—The size of a hole. To enlarge a hole to size with a boring bar in a mill or lathe.

boss—A raised circular part on a casting, usually to provide strength or depth for a hole through its center.

bow instruments—Compasses or dividers with spring tops and screw adjustments.

brass—An alloy of copper and zinc.

braze—Join metal parts with a hard solder, usually a form of brass called spelter.

break line—A line indicating that the full length of a part is not shown.

broach—To machine a hole to a shape other than round. The tool for doing this.

bronze—An alloy of copper and tin, with other metals added in some forms. Then the name may be prefixed with the name of that metal.

brownline print—One form of reproduction with brown lines on a white background.

buff—To polish on a rotating fabric wheel using an abrasive compound.

burnish—To smooth a surface by rubbing with hard material.

burr—A rough or jagged edge turned over after punching or other machining.

bush or bushing—A hollow sleeve forming a bearing or a guide for a rod or round tool.

buttress—A screw thread form angled to withstand a greater thrust one way.

cabinet drawing—An oblique drawing with the diagonal parts drawn to half scale.

caliper, calliper—An adjustable measuring device with two legs, used for checking internal or external thicknesses or diameters.

cam—A part mounted on a revolving shaft with an outline that changes the motion of a part bearing on it to a reciprocating or to and fro action.

carriage bolt—A bolt with a shallow round head and a short square neck below to prevent the bolt turning in wood.

cartography—The making of maps.

case harden—A treatment of iron to give it a thin surface of steel.

casting—To pour molten metal into a mold of the intended shape and leave it to harden.

center drill—A drill to make a small countersunk hole to match the center of a lathe. Also called a "Slocumb" drill.

center line—A line about which the drawing, or part of it, is symmetrical. Drawn with long and short dashes.

chain dimensions—Arranging dimension lines point to point in a series.

chamfer—A bevel on an edge or end.

circular pitch—The distance from a point on a gear tooth to a similar point on the next tooth, measured around the pitch circle.

civil engineering—That branch of engineering concerned with land, roads, and structures such as bridges, embankments, and similar projects.

compass—The drawing instrument used for drawing circles, with a pencil or pen on one leg rotating about a point on the other leg.

concentric circles—Circles of different sizes drawn about the same center.

cone—A body tapering from a broad base to a point. It may be assumed to be round unless otherwise described. A square cone may be called a pyramid.

construction lines—The visible outlines of a drawing. Lines drawn to lay out the views on a drawing, on which final lines are made.

contours—Lines on a map passing through points which are all the same height.
core—In casting metal, a form made of sand to produce holes in the casting. It can be broken up and removed after the metal has set.
counterbore—To enlarge the end of a hole so a larger part of a screw or rod can be set in.
countersink—Bevel the end of a hole so the tapered head of a screw or other round part can fit in.
cross-hatching—Shading with a series of close lines or other marks to indicate that the surface has been cut, usually to show the type of material, according to the system of marks used.
crown—In engineering, a rounded top, particularly of a pulley, so the belt rises to the highest point.
cube—A body with flat sides, parallel and with the corners square, having the same dimensions along all edges.
cutaway section—Part of the drawing of an object that has been cut away to clarify details in that plane.
cutting plane—The imaginary plane that cuts through an object to produce a section to be drawn.
cycloidal curve—The curve followed by a point on a wheel rolling along a straight line.
cylinder—A solid or hollow body with circular ends and parallel sides.

datum—Point or level from which measurements are taken. Plural data.
degree—The unit for measuring angles, with 360 degrees to a full circle, and indicated by the symbol "°."
detail paper—A drawing paper which has sufficient transparency for prints to be made, or heavier paper which is opaque and may be green or buff in color.
development—The shape obtained by opening out the surface of an object, such as the flat outline of a piece to fold into a box.
diameter—The distance across a circle through the center.
diazo—A method of reproducing drawings, giving a dark line on a white background.
die—A tool used in a machine to cut or press out a shape. A tool used for cutting a thread on a rod.
die casting—A casting of metal or plastic, made by pouring the molten material into a metal mold, usually under pressure.

dimension lines—equilateral triangle

dimension lines—Lines terminating in arrowheads, used to indicate sizes or locations.

dimetric—Drawing similar to an isometric view, but with the lower angles due to a lower view point.

dividers—Instruments like compasses, but with two points, used for checking and marking distances on a drawing.

dowel—A wood or metal cylindrical pin used for joining parts together.

draft—The tapered sides of a foundry pattern, which allows it and the casting to be withdrawn from the sand. Depth, particularly of the underwater part of a hull.

drafting machine—Device mounted on a drawing board to combine the functions of T square and triangles as well as protractor.

drafting media—Any material on which a drawing is made.

drawing board—A wooden board on which paper can be mounted for drawing. Usually portable and intended to be used with a T square.

drill—A revolving tool for making a hole. For metal, it is usually described as a twist drill.

drop bow compass—A compass for pencil or pen intended for drawing very small circles. The part carrying the pen or pencil slides on the part with the center point.

drop forging—A method of forming hot metal between dies under pressure.

dual dimensioning—Giving sizes according to two systems such as inches and millimeters.

dusting brush—Brush kept for removing erasing particles and other dust from a drawing.

dyeline printing—General name for methods of reproducing drawings by exposing to light followed by developing.

elevation—A drawing of a side view of an object.

ellipse—An enclosed curve made about two foci or obtained by cutting diagonally across a cylinder. The appearance of a circle viewed from the side.

engineer's scale—A scale rule with its edges divided proportionately to full size.

equilateral triangle—A triangle which has sides of equal length and all corner angles the same.

erasing machine—A powered device rotating rubber or other material for erasing.
erasing shield—A metal or plastic shield with holes in it. It is used to protect surrounding areas when erasing.
exploded view—A drawing showing the separate parts that make up an assembly, apart but in the correct relation to be brought together.
extension lines—Thin lines leading to a drawing of an object, either to link two views or to serve as reference points for dimension lines.

face—A flat surface. The process of producing a flat surface on a lathe or other machine tool.
FAO—An indication that the object is to be finished all over.
fasteners—Collective name for nuts, bolts, screws, etc.
file—A tool with many cutting edges used on metal or plastic. The action of using the tool. Storing drawings or other papers.
fillet—An inside rounded corner between two surfaces.
finish marks—Symbols and other indications of what treatment the various surfaces of the object shall be given.
fit—The degree of tightness or looseness of parts that come together.
fixture—A device for gripping or locating a part, particularly when it is being machined.
flange—An extension in the form of a rim, usually as a thickening at the end of a pipe or tube.
focus—A point about which a curve is drawn. Plural foci.
foreshortening—The apparent reduction in length, due to the angle at which the line is viewed.
forge—To form hot metal into shape by hammering. The fireplace in which the metal is heated.
French curve—A plastic or wood shape used for drawing around to get irregular curves.
front elevation—The main view from the front on a drawing.
front view—Alternative name for the front elevation.
functional drawing—A drawing made to convey the required information with the minimum number of views and lines.

galvanize—A method of coating iron with zinc to prevent rust.
gasket—A thin piece of material placed between two thicker parts to produce a tight fit when squeezed together.

301

gauge, gage—A device for checking the size of an object. The action of measuring with a gauge. A dialed instrument used for measuring various things.

gear—A toothed wheel used to engage with other gears or similarly toothed parts to transmit motion.

gib key—A key with a raised head. It is used to lock a wheel to a shaft.

Gothic lettering—Plain upright lettering.

graphic chart—Chart laid out to show the relationship between two or more functions or actions.

graphic scale—A scale drawn on a map or similar drawing so it can be used for reference.

grind—Use an abrasive wheel to remove metal.

heat treat—Alter the state of a metal by the use of heat. In particular to harden, temper, or anneal steel.

helix—The shape obtained by winding a wedge shape around a cylinder. The path of a screw thread. Sometimes wrongly called a spiral.

hexagon—A six-sided figure. The shape of many bolt heads and nuts.

horizontal plane—A plane surface parallel with the horizon.

horizontal projectors—Projectors from the top surface of an object.

hyperbola—The shape produced on the section by making a vertical cut through a cone to one side of the apex.

Indian ink—A dense black ink used for drawing.

involute—The path followed by a point on the circumference of a circle when the layer on which the point is located is unwound.

irregular curve—Alternative name for a French curve.

ISO—Abbreviation for the International Standards Organization which has agreed on metric standards.

ISO metric—Meaning that the standards used are those of the agreed standard, but care is needed not to say the term as one word, when it could be confused with isometric.

isometric drawing—Drawing made as a view towards one corner with the other sides drawn at 60° to the vertical or horizontal axis.

isometric projection—A similar view to an isometric drawing, but with all sizes reduced to allow for foreshortening, to give an overall size nearer that which would actually be viewed.

isosceles triangle—A triangle with two sides of equal length and therefore two equal angles.

jig—A device for holding either the working tool or the piece being worked on. Usually serving as a guide for repetitive work.

kerf—The cut made by a saw. The width of that cut.

key—A piece used to lock a wheel hub to a shaft by fitting into grooves.

keyway—The groove or slot into which a key fits.

knuckle—A rounded edge. A form of rounded screw thread.

knurl—Dents, grooves, or raised parts to provide a grip, particularly around a knob or handle.

lead—The amount that a nut advances on a bolt. It is the same as the pitch for a single-start thread, but is two or three times the pitch for double or treble thread.

leader—Line with an arrowhead used to indicate a part on an object, often relating a note or reference letter to the part.

left-hand thread—A screw thread in which the external thread advances into the internal thread when it is turned counterclockwise. All general purpose threads are right-handed.

Lewis key—A key for linking a wheel hub to a shaft which is entered into angled slots, so it is better able to transmit the thrust in one direction of rotation.

light table—A drawing table that can be illuminated from below so as to make tracing or the use of overlays easier to do accurately.

limit—In engineering drawing, an indication of the largest and smallest sizes which are acceptable instead of the nominal size given.

linen—The cloth used for making tracing cloth.

location dimension—A dimension that gives distances between parts of an object, as distinct from those which indicate sizes.

lofting—Making a drawing full-size. In particular, laying out something which is so large that it has to be drawn on the floor, as in structural engineering and shipbuilding.

lower case letters—Small letter as distinct from capital (upper case) letters. The names come from printing, where the type is kept in upper and lower cases.

lug—An extending ear from a larger part, where it may serve as a handle, the location of a hole or dowel, or a link with another part.

machine screw—A threaded rod with a head, which may be in many forms. A machine screw differs from a bolt in being threaded for the full length, while a bolt is only threaded a short distance from the end.

malleable casting—A toughened casting which will withstand hammering. Some castings tend to be brittle and may crack under hammering.

mechanical drawing—Drawing with instruments instead of freehand, as done by an artist.

metric scale—A scale using metric measurements, either full-size or proportionally.

microfilm—Photographic film used to copy drawings to a small scale for storage and reproduction purposes.

micrometer caliper—Device for measuring to fine limits, using an accurately cut fine screw thread and calibrations on an enclosing thimble.

mill—Machine a surface in a machine using a rotating toothed cutter.

minute—One-sixtieth part of a degree used in measuring angles.

miter—The angle between two meeting parts. This is 45° in a square corner.

multi-view drawing—Alternative name for orthographic projection.

neck—A groove cut around a cylindrical part, often at a change of size or where a thread is cut close under a screw head.

nonaligned dimension—Dimension arranged without regard to the arrangement of other dimensions.

nonisometric lines—Lines on an isometric drawing which are not parallel with any of the main axes.

oblique drawing—A pictorial view made by drawing lines obliquely to indicate thickness from a front view similar to that of an orthographic drawing.

oblique projection—Drawing with projectors at angles other than 90° to surfaces of the object.

obtuse angle—An angle larger than 90°.

octagon—A shape with eight straight sides.

offset section—One of a series of sections of an object.

ordinates—Projected lines on which to plot positions, usually of positions on a curve.

orthographic projection—The common form of mechanical drawing, with views drawn as seen squarely and in the correct relation to each other.

overlay—A transparent sheet placed over a drawing, usually carrying another drawing which has to be related to the first.

pantograph—An instrument with crossing arms that can be adjusted to enlarge or reduce a drawing in a desired proportion. The name may be also be given to a similar arrangement of crossing arms used to transfer motion in other devices.

parabola—The curve obtained by making a cut diagonally across a cone.

parallel projection—The use of parallel lines at right angles to the surface of the object. Alternative name for orthographic projection.

parallel straightedge—Part of a drafting machine in which the straightedge is controlled at the ends by cords which allow it to move up and down a drawing board with a parallel motion.

partial section—A section showing only part of an object.

peen, pein—To stretch or spread a head or edge with the ball peen of an engineer's hammer.

pentagon—A five-sided figure.

perpendicular—A line drawn at right angles to another line.

perspective—A view as naturally observed, with lines going away from the viewer tapering as they go towards a vanishing point.

phantom lines—Second or alternative position of a moving part.

phantom section—A hidden section.

Phillips head—A screw head with a star-shaped socket instead of a slot of a screwdriver.

photo drawing—A drawing made from a photograph or a photograph on which more details have been drawn.

photostat—A photographic reproduction of a drawing made by a special process, with dark lines on a white background and the same size or bigger or smaller.

phototracing—A photographic copy of a drawing on tracing film.

pickle—An acid bath for cleaning metal.

pictorial drawing—Any three-dimensional drawing, particularly one intended to give an impression of the object as it actually appears.

pitch—The distance from the top of a screw thread to its top in the next revolution. The same as lead in a single thread, but not if it is double or treble, when it is half or one-third.

plan—A view of an object from above. General name for a drawing, particularly one showing the layout of houses or other buildings or a large-scale map of a locality.

plane—A flat surface.

plane of projection—Imaginary surface at right angles to the projection lines.

planimeter—Device that measures areas by tracing outlines.

plate—Thin sheet material. To coat a metal object with another metal.

polish—Smooth with very fine abrasive to achieve a high luster.

polyester—A man-made plastic material used for making tracing film.

polygon—A many-sided figure.

primary center line—The main center line of a drawing, where there are others for secondary parts.

prism—Object with parallel ends and parallel sides often, but not necessarily, of triangular section.

profile—Side view of an object.

proportional dividers—Dividers with points at both ends and an adjustable pivot so it can be arranged to set the size between the points at one end in the desired proportion to the points at the other end.

protractor—Instrument for measuring angles, usually a full circle or a semicircle calibrated around the edge in degrees.

punch—To pierce thin metal by driving a tool through it. The tool for that purpose, as well as others for marking centers, driving nails and pins, etc.

pyramid—A square cone.

quadrant—A 90° part of a circle, embraced by two diameters and the part of the circumference.

rack and pinion—An arrangement with a toothed gear wheel engaging with a straight rack having matching teeth.

radial point projection—Method of projecting one view from another by using projection lines carried around a corner with compasses.

radius—The distance from the center of the circumference of a circle.

ream—To bring a hole to the exact size by finishing with a fluted rotating cutter.

rectangle—A shape with parallel sides and 90° corners, but not a square.

reference—Additional information or an indication that this can be obtained on another drawing or part of that drawing.

revolution—Moving an object with a circular motion through a selected number of degrees.

revolved section—A section included in a drawing, but at right angles to the part to which it refers, usually to indicate the shape across the piece at that point.

revolved view—A view incorporated in a drawing with orthographic views, but showing the object as viewed from an angled direction.

rhomboid—A shape with opposite sides parallel and the same length, but different in each direction, having its corners not at right angles. May be regarded as a rectangle pushed out of shape.

rhombus—A shape with opposite sides parallel and all four the same length, but with the corners not at right angles. May be regarded as a square pushed out of shape.

right angle—An angle which is square, or at 90°.

right angle projection—Alternative name for orthographic projection.

right-hand thread—A screw thread where the external thread enters the internal thread when it is turned clockwise. The normal direction for general purpose screwed parts.

Roman letters—An upright form of decorative lettering, rarely used in general drawing, but sometimes in titles of architectural drawings.

root—The bottom of a screw thread.

round—In engineering the curve between two outside surfaces, instead of a chamfer or leaving the edge sharp.

ruling pen—A pen with blades that can be adjusted in their distance apart and used with Indian ink for drawing lines.

scale—A straightedge marked for measuring, either full-size or in a proportion. Scale is also the relationship between the size of the drawing and the size of the object it represents.

scalene triangle—A triangle with all sides and angles different.

schematic drawing—A drawing showing parts of a system, using symbols and a stylized layout.

screw—A threaded rod or a matching hole, with ridges cut around and following a helical path.

section—A view showing the effect of an imaginary cut on an object, particularly to show some hidden detail or shape.

section lining—hatching with close parallel lines or in some other way to indicate a cut surface.

septagon—A seven-sided figure.

setscrew—A headless screw, with a slot or socket for driving.

shaft—Round rod on which pulleys, gears, and other parts are mounted to rotate and transmit power.

shear—Cut material between two blades.

shim—Thin metal used between parts as packing to adjust spacing.

ship's curves—A form of large French curves particularly suited to drawing the lines of a ship.

SI—The initials of the organization responsible for the details of the simplified metric system coming into worldwide use.

side elevation—The view of an object from the side, usually at right angles to the front elevation.

side view—The view from the side, usually at right angles to the front view. Also called side elevation or profile view.

sketch—A preliminary drawing, partially or entirely freehand.

slide rule—A rule with scales sliding alongside each other, each arranged in a logarithmic way so they can be used for calculations.

socket head—A head, usually on a screw, with a socket to take a wrench for turning it. May be square, hexagonal, or a special shape.

spline—A groove like a long keyway to engage with other parts.

spot face—To finish a circular face slightly below the surrounding parts, so a bolt or other part through a hole there will have a flat seating to pull against.

square—Besides meaning a shape with equal sides and corners at right angles, the word is used to indicate that a part or line is at 90° to another.

stacked dimensions—Dimension lines arranged uniformly outside each other parallel with the face to which they refer.

staggered dimensions—Having a series of dimensions arranged so none of them are along the same straight line.

stat—Abbreviation for photostat.

stet—Printer's term meaning let it stand—ignore an alteration.

straightedge—A straight instrument used for drawing lines. If it is calibrated along the edge for measuring, it is also a scale. In draftsmanship it is unusual to call it a rule or ruler.

stud—A rod screwed through its length or only at the ends. In effect it is a headless bolt that will take a nut at either end.

symmetry—Having opposite sides that are mirror images of the other. On a drawing, the fact that parts are symmetrical is indicated by a center line.

tangent—A line that touches, but does not cut, the circumference of a circle.

tap—A tool for cutting a thread in a hole. The action of using it. Alternative name for a faucet.

taper pin—Round bar with a uniform taper and used to secure some assemblies.

technical drawing—Drawing with instruments to provide measurements and details of an object, for practical and not decorative purposes.

technical fountain pen—A pen for technical drawing, with a supply of ink carried in its body and a fine tube feeding it to the drawing surface.

technical illustrating—Adaption of drafting to the illustration of technical books and articles to provide pictures that can be understood by lay people or semiskilled workers, and that can usually be expected to improve the appearance of the page.

temper—Adjustment of steel by heat treatment to a semihard condition to suit the purpose to which the part will be put. Particularly applicable to tools.

template, templet—A shape that can be drawn around or through to repeat frequently used or standard shapes.

thread—The helical ridges around a rod or hole that make a screw.

through hole—A hole that goes right through an object.

title block—A standard form of general title and other details in the corner of a drawing.

tolerance—How much a size may deviate from the measurement given. The difference between the limits. The allowance made for fitted parts.

top view—View from above. Also called a plan view.

tracing cloth or film—The transparent medium used for drawings which are to be reproduced. The cloth is treated to provide a drawing surface. The film, usually polyester plastic, has largely superseded cloth.

tracing paper—Semitransparent paper which is not as durable as tracing cloth or film.

trammel—An instrument for drawing ellipses, consisting of a cross frame and a type of beam compass that moves in it, with a pen or pencil at its end. The name may also be used for the large beam compass without the frame.

trapezium—A four-sided figure with none of the sides parallel or the same length.

trapezoid—A four-sided figure with top and bottom parallel, but the sides sloping in.

triangle—A three-sided figure. Also the name for the three-sided instruments used by a draftsman. They may also be called set squares.

triangular scale—A scale with a grooved triangular section that permits more scale sizes to be calibrated along the edges.

trimetric—A view similar to isometric except the viewpoint is assumed to be slightly off-center.

truncation—Cutting off, such as a plane cutting off the top of a cylinder or a cone.

T square—The tool used as a straightedge bearing against the side of a drawing board for drawing lines and as a base for other instruments.

turning—The process of making round objects by rotating them against a tool in a lathe.

unidirectional dimensioning—Having all the figures indicating dimensions the same way up, usually to be read when the bottom of the drawing is horizontal.

unified—One standard used for screw threads.

upper case letters—Capital letters, to distinguish them from small letters, which are lower case. These are printer's terms indicating which cases the type is kept in.

vellum—Name sometimes given to drawing paper.

vernier—An arrangement of a small auxiliary scale alongside a main one to permit very fine measurements with the caliper or other measuring tool.

vertex—Point where two lines meet to form an angle.
visible line—Lines drawn solid to indicate that they can be seen in that view. Hidden lines are dotted (short dashes).

washer—A ring of metal or other material that may come between a bolt head or a nut and the bearing surface.
weld—Join pieces of metal by heating them to a fusing temperature and hammering them together. Join metal parts with a flame by fusing metal into the seam.
Woodruff key—A key for locking a pulley hub to a shaft, made so that its lower part fits into a curved slot in the shaft, but there is a flat top to fit into the hub.
worm's eye view—A drawing showing how the object appears from a low viewpoint.

X and Y axes—When there have to be reference letters, it is common to choose letters from the end of the alphabet to avoid confusion with the more commonly used earlier letters.

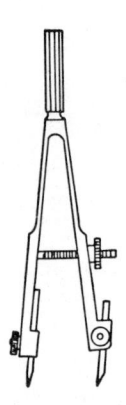

Index

A
American Welding Society	195
Ancillary services	270
Angles	121, 285
Angular instrument, using	71
Arcs, drawing	69
using	68
Arrowheads	125
Art gum	18
Awls	97

B
Bend allowances	162
Bisecting	79
Blind hole	184
Blue printing	12
Bolts	185
Borders	66
Breadth	116

C
Cam, displacement	253
plate	251
Cams	250
Cartographer	10
Cartography	265
Castellated nut	187
Cavalier	221
Chains	248
Chamfer	122
Charts	292
Circles	119
drawing	69
Circles drawn obliquely	218
Communications	8
Compass, ink	67
pencil	67
Compasses	21
drop bow	28
friction joint	25, 70
using	66
Cones, other truncated	169
round	169
Conic developments	166
Conic projections, truncated	148
Cornice	269
Cross-hatching	128
Curved developments	164
Curves	34, 283
helical	178
helix	178
using	73
involute	255
Cutting plane	126

D
da Vinci, Leonardo	7
Designing	236
Detail paper	12, 198
Developments	156
Diameter, major	179
minor	179
pitch	179
Diazo process	209
Dimensions	113, 248, 267
locating	116
small	118
Dividers	22
friction joint	70
using	68
Drafting	1
airplane	260
geometry	77
history	5

machines	12
structural	272
Draftsman	1
Draftsmanship	1
Drawing, chair	240
engineering	244, 247
laying out	231
number of	232
oblique	213
three-dimensional	211
Drawing boards	11
Drawing instruments	11
Drawing sizes	287

E

Elevation, front	2
side	2
Ellipses	36, 95
Engineering, aerospace	248
aircraft	248
architecture	265
civil	265, 275
electrical	248, 258
hydraulic	248
Erasers	17
Erasing	207
Errors of parallax	48
Exploded drawings	9

F

Fasteners	178
Fine scale	50
Finish marks	256
Fishplate	273
Fractions	108
French curves	37
Furlongs	248

G

Gears	254
Graphic scales	48
Graphs	291
Greeks	7
Guidelines	105

H

Handling instruments	63
Hexagons	86
Hexagons, drawing	87
House plan	265
Hydraulic engineering	7
Hyperbola	150
Hypotenuse	83

I

Indian ink	38, 198
Industrial Revolution	6
Ink, drawing with	202
Ink instruments	38
Inking	197
Internal cuts	174
Isometric, axes	223
curves	225

J

Junctions	174

K

Keys, flat	193
Lewis	195
saddle	193
Woodruff	195

L

Lead	179
Length	116
Lettering	101
down-stroke	104
Gothic	102
inclined	103
italic	103
one-stroke	104
Roman	102
single-stroke	104
slanting	103
upright	102
Letters, lower case	102
upper case	102
Lines	280
break	58
center	56
dimension	56, 114
hidden	54
main	64
phantom	56
reference	56
section	58
Lofting	280

M

Material lists	232
Material schedules	232
Medullary rays	130
Mice	36
Micrometer	52

N

Neutral axis	162
Nib	41
Nuts	185

O

Octagons	86, 147
drawing	88
Oilstone	66

P

Pantograph	14
Parabola	150
Parts, curved	142
multiple	130
Pencils	16
Pens	39
ruling	199
technical fountain	201
Pentagons, drawing	88
Pitch circle	254
Plinth	240
Pneumatic work	10
Polygons	84
regular	85
Projections, conic	147
isometric	4
oblique	4
orthographic	2
pictorial	4
third-angle	3
Proportions	46, 105
Protractors	30

R

Reproduction	208
Right angles	77, 281
Rivets	189
cone head	190
flat head	191
pan head	190
raised or oval head	190
round or button head	190
truss or wagon box	191
Romans	7
Rule	20
Ruling pen	39

S

Scales	6, 20, 44
Screw threads	178
Section lining	128
Sectioned materials	130
Sections	126
revolved	134
thin	137
Set squares	19
Shapes	172
Site plans	270
Spiral	178
Splines	36
Spring bows	27
Straightedges	20
Style	102, 294

T

T squares	12
Tab washer	187
Table of offsets	284
Tangents	90
Technical illustrating	287
Templates, using	73
Thickness	116
Thread,	180
British Standard Whitworth	180
buttress	180
representing	181
Tolerances	122
Tracing materials	197
Trammel	25
Triangles	19, 82
equilateral	83
isosceles	83
right	83
Triangulation	276
Tubes	157
Turnbuckle	181

U

Unidirectional	228

V

Vellum	198
Vernier gauge	52
Views, auxiliary	139
axonometric	221
choice of	229
exploded	288
isometric	221
pictorial	212, 290
revolved	153
side	2
trimetric	221

W

Welding	195